Linear Methods
of Applied Analysis

Linear Methods
of Applied Analysis

ALLAN M. KRALL
Pennsylvania State University

 1973

Addison-Wesley Publishing Company
Advanced Book Program
Reading, Massachusetts

London • Amsterdam • Don Mills, Ontario • Sydney • Tokyo

Library of Congress Cataloging in Publication Data

Krall, Allan M
   Linear methods of applied analysis.

   Includes bibliographies.
   1. Mathematical analysis. 2. Differential equa-
tions. I. Title.
QA300.K64   515'.35   73-13753
ISBN 0-201-03902-8
ISBN 0-201-03903-6 (pbk.)

ABCDEFGHIJ-MA-79876543

Reproduced by Addison-Wesley Publishing Company, Inc., Advanced Book Program, Reading, Massachusetts, from camera-ready copy prepared by the author.

American Mathematical Society (MOS) Subject Classification Scheme (1970):
33–01, 33A65; 34–01, 34A10, 34A30, 34B05; 35–01, 35A22, 35D05, 35E05, 35J05, 35K05, 35L05; 42–01, 42A60, 42A68; 47–01, 47B05, 47B10, 47B15

Manufactured in the United States of America

This book is dedicated to my father

PROFESSOR H.L. KRALL

The encouragement he has given throughout the
years can never be adequately described.

Table of Contents

INTRODUCTION

The purpose of this book is to present to students in mathematics, engineering, and the physical sciences at the upper undergraduate and beginning graduate levels an introduction to part of what is fashionably called today applied mathematics. As anyone who attempts to define the term soon learns, applied mathematics is a large rather nebulous region sitting between mathematics and various areas of application. In fact a few years ago a group of applied mathematicians met in Colorado and tried to define it. Their results were less than satisfactory.

At most universities, however, there is a course covering the material presented here. It is usually given the title of "methods of mathematical physics", "boundary value problems of mathematical physics", or some variation involving all or some of these words. The course may be taught by a physicist, a mathematician, or an engineer with slightly different emphases. In all cases, the material consists of some portion of mathematics which has applications to the physical sciences: the basic facts concerning ordinary and partial differential equations and the background necessary for proper understanding. These are the subjects of this book.

The point of view taken is that of the mathematician, not that of the physicist or engineer. That is, our interest is primarily, but not exclusively, with the mathematics. Applications as such occupy a secondary role as motivation and as a check to see if what we do is reasonable (The term is deliberately vaguely used.) We shall use theorems to state our results and proofs to illustrate the techniques involved, as a mathematician would.

Finally, we recommend that those who wish to see a rigorous development of the applications consult courses in physics and engineering. In fact, such courses are much more difficult than the mere mathematics would indicate, and it is this (among other reasons) which makes physics and engineering legitimate fields. We shall find that the mathematics alone is sufficient to occupy most of our time.

The first portion of the book is concerned with certain necessary inequalities, linear spaces and operators: the background necessary for studying that which follows. Next, by using the contraction mapping theorem we show the existence and uniqueness of solutions to systems of ordinary differential equations with special emphasis on equations of second order.

Next follows the Stone-Weierstrass theorem, which is of fundamental importance, but is not fully appreciated by most appliers of mathematics and then an introduction to Hilbert space, the primary setting for most of the remainder of the book. We next study both the regular and singular Sturm-Liouville problems, with the examples given by the classical orthogonal polynomials and the Fourier integral.

The second portion of the book uses a distribution setting to discuss the classical partial differential equations of second order occurring in mathematical physics: the Laplace equation, the heat equation, and the wave equation. Special emphasis is placed upon series solutions and the Green's functions for Dirichlet, Neumann and mixed boundary conditions. The results of the earlier portions of the book are used throughout these discussions.

The book has the form, therefore, of something akin to a ladder. The

base consists of linear spaces, and at the top sit the partial differential
equations.

The book is designed to serve a number of different courses. Specifically
it can be used as

1. A supplement to an elementary course on vector spaces and matrices
    by using sections I 1-4, II 1-4, VII 1-6, VIII 1-6, IX 1-4.

2. A text for a course on Hilbert spaces and linear operators by using
    sections I 1-4, II 1-4, V 4,5, VII 1-7, VIII 1-9, IX 1-5, X 2,3, XV 3

3. A first semester course on ordinary differential equations by using
    sections I 1-4, II 1-4, III 1-4, IV 1-5,7, V 1-5, VII 1-6,
    VIII 1-4, IX 1-5.

4. A second semester course on ordinary differential equations by using
    sections X 1-5, XI 3, XII 1-6, XIV 1-4, (Appendix I 1-4 if time
    permits).

5. A one semester course on partial differential equations by using
    sections XIII 1-4, XIV 1-3,5, XV 1-5, XVI 1-5, XVII 1-4.

6. A one year course on ordinary and partial differential equations
    and related areas. The topics found most appropriate are
    found in I 1-4, II 1-4, III 1-4, IV 1-5,7, VI (reading assignment),
    VII 1-7, VIII 1-6 (7-9 reading assignment), IX 1-5, X (reading
    assignment), XI 2,3, XIII 2-4, XIV 1-5, XV 1-5, XVI 1-5 if time
    permits, XVII 1-4.

A word is in order concerning the origin of this book. While the author was a graduate student at the University of Virginia, he had the exceptional privilege of attending a class in ordinary and partial differential equations given by Professor Marvin Rosenblum. Professor Rosenblum's class was later used most successfully as a model for a similar course taught by the author at The Pennsylvania State University, and the notes resulting from the author's attempt at Penn State are the basis for this book. The author would like to express his grateful appreciation for the inspiration given by Professor Rosenblum some years ago.

In addition the author would like to express his thanks to Professor H.L. Krall and Dr. Richard C. Brown who read the manuscript and made many excellent suggestions.

The typist, Miss Brenda Snyder, deserves special credit for an excellent job of deciphering the original handwritten copy. Her careful job during the many hours she spent in tedious labor is much appreciated.

                                                              AMK

I. SOME INEQUALITIES

We are all familiar with the triangle inequality in its geometric form as well as its expression in terms of analytic geometry. This fundamental rule has a number of useful extensions. In addition there are some other very useful inequalities, which are closely related to the triangle inequality. We state them with a suitable derivation or proof immediately following.

I.1. Young's Inequality.

I.1.1. Theorem. (Young's Inequality). Let x and y be strictly increasing continuous functions, defined on $[0,\infty)$, which satisfy $x(0) = 0$, $y(0) = 0$, $x(y(t)) = y(x(t)) = t$ for all t in $[0,\infty)$. If $a > 0$, $b > 0$, then

$$a \cdot b \leq \int_0^a x(t)dt + \int_0^b y(t)dt.$$

Pictorially the theorem says the following: Consider the graph of x:

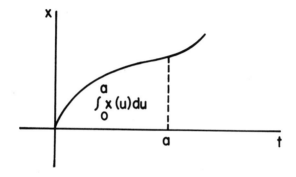

If x(t) = u is thought to be the independent variable, then y(x(t)) =

y(u) = t is the dependent variable, and we have the following graph, drawn

in reverse:

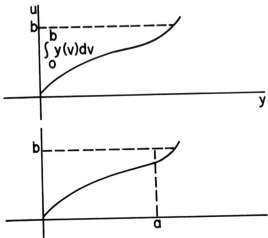

or

The theorem states that the integrals represent a larger area than the

rectangle ab.

Proof. Either b ≧ x(a), which implies a ≦ y(b), or a ≧ y(b), which

implies x(a) ≧ b. We may, therefore, without loss of generality assume

the first. Then

$$\int_0^a x(u)\,du + \int_0^b y(v)\,dv$$

$$= \int_0^a x(u)\,du + \int_0^{x(a)} y(v)\,dv + \int_{x(a)}^b y(v)\,dv.$$

The first two integrals combine to equal a·x(a). The third is greater than

[b−x(a)]y(x(a)), or [b−x(a)]·a.

Thus

$$\int_0^a x(u)\,du + \int_0^b y(v)\,dv \geq a\cdot b.$$

In evaluating the first two integrals, we have relied heavily upon the graph. To produce a completely analytic proof requires additional assumptions such as found in the book Integration, by E.J. McShane, or a somewhat longer argument as in the article "An Analytic Proof of Young's Inequality" by J.B. Diaz and F.T. Metcalf (The American Mathematical Monthly vol. 77(1970), 603-609). The reader should consult these references to see the proofs presented there.

I.1.2. Examples. The following two examples show that functions x and y really do exist. The second is of particular interest, since it is essential for what follows afterward.

1. Let $x(t) = \log(t+1)$ and $y(t) = e^t-1$. It is easy to verify that they satisfy all the hypotheses of I.1. Young's inequality then states that

$$a\cdot b \leq \int_0^a \log(t+1)\,dt + \int_0^b (e^t-1)\,dt,$$

$$a\cdot b \leq (a+1)\log(a+1)-a+e^b-1-b.$$

If $a+1 = u$ and $b+1 = v$, we have as a corollary

$$u \cdot v \leq u \log u + e^{v-1},$$

when $u \geq 1$ and $v \geq 1$.

2. Let $x(t) = t^{p-1}$ and $y(t) = t^{q-1}$, where $p > 1$, $q > 1$, and $1/p + 1/q = 1$. Again it takes only a minor calculation to show that x and y satisfy the hypotheses of I.1. Young's inequality then states that

$$a \cdot b \leq \int_0^a t^{p-1} dt + \int_0^b t^{q-1} dt,$$

or

$$a \cdot b \leq \frac{a^p}{p} + \frac{b^q}{q}.$$

This formula also holds for complex numbers in the form

$$|ab| \leq \frac{|a|^p}{p} + \frac{|b|^q}{q}.$$

I.2. Hölder's Inequality. We can use the previous example to derive in two cases what is commonly known as Hölder's inequality. The right sides are generalizations or variations of the standard Euclidean distance formula.

I.2.1. Theorem (Hölder's Inequality, the Discrete Case). Let $1 < p < \infty$, $1 < q < \infty$, and $1/p + 1/q = 1$. Let a_i, b_i, $i = 1,\ldots,n$, be complex numbers. Then

$$\left| \sum_{i=1}^n a_i \bar{b}_i \right| \leq \left[\sum_{i=1}^n |a_i|^p \right]^{1/p} \left[\sum_{i=1}^n |b_i|^q \right]^{1/q}.$$

I.2.2. Theorem (Hölder's Inequality, the Continuous Case). Let $1 < p < \infty$, $1 < q < \infty$, and $1/p + 1/q = 1$. Let f, g be complex valued, measurable functions on the interval $[a,b]$. Then

$$\left| \int_a^b f(t)\overline{g}(t)dt \right| \le \left[\int_a^b |f(t)|^p dt \right]^{1/p} \left[\int_a^b |g(t)|^q dt \right]^{1/q}.$$

Because the proofs of these two theorems are essentially the same, we shall only present a proof of the discrete case.

Choose λ and μ to be real valued such that $\lambda \left[\sum_{i=1}^n |a_i|^p \right]^{1/p} = 1$ and $\mu \left[\sum_{i=1}^n |b_i|^q \right]^{1/q} = 1$. In example I.1.2, part 2, we set $a = \lambda a_i$ and $b = \mu \overline{b}_i$ and sum over i. The result from I.1.2, part 2, is

$$\lambda\mu \sum_{i=1}^n a_i \overline{b}_i \le \frac{\lambda^p \sum_{i=1}^n |a_i|^p}{p} + \frac{\mu^q \sum_{i=1}^n |b_i|^q}{q} = 1.$$

Substitution for λ and μ completes the proof.

I.2.3. Theorem (Schwarz's Inequality, the Discrete Case). Let a_i, b_i, $i = 1,\ldots,n$, be complex numbers. Then

$$\left| \sum_{i=1}^n a_i \overline{b}_i \right| \le \left[\sum_{i=1}^n |a_i|^2 \right]^{1/2} \left[\sum_{i=1}^n |b_i|^2 \right]^{1/2}.$$

I.2.4. Theorem (Schwarz's Inequality, the Continuous Case). Let f, g be complex valued, measurable functions on the interval $[a,b]$. Then

$$\left| \int_a^b f(t)\overline{g}(t)dt \right| \le \left[\int_a^b |f(t)|^2 dt \right]^{1/2} \left[\int_a^b |g(t)|^2 dt \right]^{1/2}.$$

These are, of course, special cases of Hölder's inequality (p = q = 2).
In the discrete case, when n = 2 or 3, division by the terms on the right
side yields

$$|\cos \phi| \leq 1,$$

where ϕ is the (complex) angle between the vectors (a_1,a_2) and (b_1,b_2)
in two dimensions, or between (a_1,a_2,a_3) and (b_1,b_2,b_3) in three dimensions. If $\sum_{i=1}^{n} a_i b_i = 0,$ when n = 2 or n = 3, then $\cos \phi = 0,$ and
the vectors are orthogonal. We carry this terminology to the other cases
also to say that if the left side of Hölder's or Schwarz's inequality is 0,
the elements involved are orthogonal.

I.3. <u>Minkowski's Inequality</u>. If the preceeding was in some way a measure
of the (complex) angle between elements, Minkowski's inequality corresponds
to the triangle inequality, and states, roughly speaking, that the sum of the
distances of two sides of a triangle is greater than the length of the third
side.

I.3.1. <u>Theorem (Minkowski's Inequality, the Discrete Case)</u>. Let $p \geq 1$,
and let $a_i,$ $b_i,$ i = 1,...,n, be complex numbers. Then

$$\left[\sum_{i=1}^{n} |a_i+b_i|^p\right]^{1/p} \leq \left[\sum_{i=1}^{n} |a_i|^p\right]^{1/p} + \left[\sum_{i=1}^{n} |b_i|^p\right]^{1/p}.$$

I.3.2. Theorem (Minkowski's Inequality, the Continuous Case). Let $p \geq 1$,
and let f, g be complex valued measurable functions on the interval [a,b].
Then

$$[\int_a^b |f(t)+g(t)|^p dt]^{1/p} \leq [\int_a^b |f(t)|^p dt]^{1/p} + [\int_a^b |g(t)|^p dt]^{1/p}.$$

Again the proofs of these two theorems are so similar we discuss only the
discrete case.

When $p = 1$, the result follows immediately from the triangle inequality
for complex numbers. When $p > 1$, first assume that $\sum_{i=1}^{n} |a_i+b_i|^p \neq 0$. Then

$$\sum_{i=1}^{n} |a_i+b_i|^p = \sum_{i=1}^{n} |a_i+b_i|^{p-1} |a_i+b_i| ,$$

$$\leq \sum_{i=1}^{n} |a_i+b_i|^{p-1} |a_i| + \sum_{i=1}^{n} |a_i+b_i|^{p-1} |b_i| .$$

We now use Hölder's inequality, and assume that q satisfies $1/p + 1/q = 1$.
Then

$$\sum_{i=1}^{n} |a_i+b_i|^p \leq [\sum_{i=1}^{n} |a_i+b_i|^{(p-1)q}]^{1/q} [\sum_{i=1}^{n} |a_i|^p]^{1/p}$$

$$+ [\sum_{i=1}^{n} |a_i+b_i|^{(p-1)q}]^{1/q} [\sum_{i=1}^{n} |b_i|^p]^{1/p}.$$

Finally, we divide both sides by $[\sum_{i=1}^{n} |a_i+b_i|^{(p-1)q}]^{1/q}$. Noting that
$(p-1)q = p$ and $1 - 1/q = 1/p$, we have

$$[\sum_{i=1}^{n} |a_i+b_i|^p]^{1/p} \leq [\sum_{i=1}^{n} |a_i|^p]^{1/p} + [\sum_{i=1}^{n} |b_i|^p]^{1/p}.$$

I.4. <u>A Relation between Different Norms</u>. The expression $[\sum_{i=1}^{n} |a_i|^p]^{1/p}$

represents length in a sense and is called in mathematical language a <u>norm</u>.

(The same is also true in the continuous case.) It is easy to see that,

when $p = 2$ and $n = 2$ or $n = 3$, Euclidean length is represented. On the

other hand, in a city with north-south and east-west running streets the

distance from one place to another is represented by $p = 1$, $n = 2$. The

limit as p approaches ∞ is best illustrated in the continuous case,

since

$$\lim_{p \to \infty} [\int_a^b |f(t)|^p dt]^{1/p} = \underset{t \in [a,b]}{\text{ess sup}} |f(t)|,$$

where ess sup stands for the maximum excluding sets of measure zero (zero

length). If the functions represented by f are continuous, then ess sup

is the ordinary maximum, and convergence under this norm is uniform. Those

who have had advanced calculus will recall that uniform limits of continuous

functions are also continuous, so continuity is preserved. The values of

p between 1 and 2 and between 2 and ∞ may be thought of as intermediary.

In the discrete case there is a rather unique relation between norms,

for different values of p. It follows as Corollary I.4.5.

I.4.1. <u>Lemma</u>. If $u \geq 0$, $v \geq 0$, and $0 < r < 1$, then $(u+v)^r \leq u^r+v^r$.

Proof. If $u = 0$ or $v = 0$, the result is trivial. If $u \neq 0$, let $t = v/u$. The original inequality is thus equivalent to $(1+t)^r \leq 1+t^r$. Let

$$f(t) = (1+t)^r - 1 - t^r \ .$$

Then

$$f'(t) = r[1/(1+t)^{1-r} - 1/t^{1-r}] < 0 \ .$$

Thus $f(t)$ is decreasing. Since $f(0) = 0$, f is negative when $t > 0$, and the result follows.

I.4.2. **Lemma.** Let A_i, $i = 1,\ldots,n$, be non-negative numbers, and let $0 < r < 1$. Then

$$[\sum_{i=1}^{n} A_i]^r \leq \sum_{i=1}^{n} A_i^r .$$

Proof. This follows from successive applications of Lemma I.4.1.

$$[\sum_{i=1}^{n} A_i]^r = [\sum_{i=1}^{n-1} A_i + A_n]^r \leq [\sum_{i=1}^{n-1} A_i]^r + A_n^r \ ,$$

$$\ldots \leq \sum_{i=1}^{n} A_i^r .$$

I.4.3. **Theorem.** Let $0 < p < q$, and let a_i, $i = 1,\ldots,n$, be complex numbers. Then

$$[\sum_{i=1}^{n} |a_i|^q]^{1/q} \le [\sum_{i=1}^{n} |a_i|^p]^{1/p}.$$

Proof. Let $r = p/q$ and let $A_i = |a_i|^q$ in Lemma I.4.2. Then

$$[\sum_{i=1}^{n} |a_i|^q]^{p/q} \le [\sum_{i=1}^{n} |a_i|^p].$$

Taking the p-th root of both sides completes the proof.

The reader is invited to examine what the statement would be in the continuous case to determine whether or not it is always true. In particular let the interval $[a,b] = [0,1]$, and compare it with Schwarz's inequality, where one of the functions in Schwarz's inequality identically equals 1.

I.4.4. <u>Theorem</u>. <u>Let</u> $1 < p < q$, <u>and let</u> a_i, $i = 1,\ldots,n$, <u>be complex</u> <u>numbers</u>. <u>Then</u>

$$[\sum_{i=1}^{n} |a_i|^p]^{1/p} \le [\sum_{i=1}^{n} |a_i|^q]^{1/q} \cdot n^{1/p-1/q}.$$

Proof. We apply Hölder's inequality to find

$$\sum_{i=1}^{n} |a_i|^p = \sum_{i=1}^{n} |a_i|^p \cdot 1 ,$$

$$\le [\sum_{i=1}^{n} (|a_i|^p)^{q/p}]^{p/q} [\sum_{i=1}^{n} 1^{\frac{q}{q-p}}]^{\frac{q-p}{q}} ,$$

$$= ([\sum_{i=1}^{n} |a_i|^q]^{1/q})^p \, n^{\frac{q-p}{q}} .$$

Taking the p-th root,

$$\left[\sum_{i=1}^{n} |a_i|^p \right]^{1/p} \leq \left[\sum_{i=1}^{n} |a_i|^q \right]^{1/q} n^{1/p-1/q}.$$

I.4.5. Corollary. Let $1 < p < q$, and let a_i, $i = 1,\ldots,n$, be complex numbers. Then the norms $\left[\sum_{i=1}^{n} |a_i|^p \right]^{1/p}$ and $\left[\sum_{i=1}^{n} |a_i|^q \right]^{1/q}$ are equivalent. That is,

$$\left[\sum_{i=1}^{n} |a_i|^q \right]^{1/q} \leq \left[\sum_{i=1}^{n} |a_i|^p \right]^{1/p} \leq \left[\sum_{i=1}^{n} |a_i|^q \right]^{1/q} \cdot n^{1/p-1/q}.$$

Thus when considering finite dimensional vector spaces, the particular choice of norm is immaterial. Convergence with respect to one implies convergence with respect to any other. Therefore we denote all these norms $\left[\sum_{i=1}^{n} |a_i|^p \right]^{1/p}$ by the single expression $\|a\|_{R^n}$ $(a = (a_1,\ldots,a_n))$.

While Theorem I.4.3 fails to have a continuous counterpart, Theorem I.4.4 does when the measure of the interval is finite.

I.5. An Abstract Integral. In this section we introduce a function which is an abstraction of the definite integral for (Riemann or Lebesgue) integrable functions. Because of the abstract approach, however, it also has other applications.

We let X be a set of points. X may be an interval, a discrete set of points, or both. It may be a collection of intervals. At this point, the type of set is immaterial.

I.5.1. Definition. We denote by F a collection of functions, defined on X, with the following properties.

    1. If f and g are in F, then

$$0 \leqq f < \infty,$$

$$0 \leqq g < \infty.$$

    2. If f and g are in F, then $\alpha f + \beta g$ is in F for all $\alpha \geqq 0, \beta \geqq 0$.

    3. If f and g are in F, then $f \cdot g$ and f^{α}, $0 \leqq \alpha < \infty$, are in F.

I.5.2. Definition. We denote by I a function, defined on F, with the following properties. For each f and g in F,

    1. $0 \leqq I(f) < \infty$,

    2. $I(\alpha f) = \alpha I(f)$ when $0 \leqq \alpha < \infty$,

    3. If $f \leqq g$, then $I(f) \leqq I(g)$,

    4. $I(f+g) \leqq I(f) + I(g)$.

I.5.3. Definition. We let $\|f\|_{p} = [I(f^{p})]^{1/p}$, $1 \leqq p < \infty$,

$$\|f\|_{\infty} = \sup_{X} f.$$

I.5.4. Theorem. Let $1 < p, q < \infty$ and $1/p + 1/q = 1$ or let $p = 1, q = \infty$. If f, g are in F, then

$$I(f \cdot g) \leq \|f\|_p \|g\|_q.$$

This is Hölder's inequality in this setting.

I.5.5. Theorem. Let $p \geq 1$. If f, g are in F, then

$$\|f+g\|_p \leq \|f\|_p + \|g\|_p.$$

This is Minkowski's inequality in this setting.

I.5.6. Theorem. Let $0 < p < 1$. If f, g are in F, then

$$\|f+g\|_p^p \leq \|f\|_p^p + \|g\|_p^p.$$

The proofs of these theorems are similar to those previously given and are left as exercises. In particular the last, where $0 < p < 1$, is of interest since the p-th power is present on both sides.

I.5.7. Examples of the Abstract Integral. We have already presented two examples of the abstract integral during the first part of the chapter. To express the first in the notation of I.5, let $X = \{1, 2, \ldots, n\}$, the first n integers, let F be the collection of non-negative functions on X, and let

$$I(f) = \sum_{i=1}^{n} f(i).$$

If $f(i)$ is denoted by a_i, the connection is obvious.

To express the second, let $X = [a,b]$. F is then the non-negative functions defined on X. Finally let

$$I(f) = \int_{a}^{b} f(t)dt.$$

A third example, not illustrated earlier, may be expressed in the following manner. Let X be any set, let F be the non-negative functions defined on X, and let

$$I(f) = \sup_{t \in X} f(t).$$

A glance shows that $I(f)$ satisfies conditions 1-4 of I.5.2. Thus $I(f)$ generates a new type of norm: the sup norm or uniform norm.

I. Exercises

1. Give some other examples of Young's inequality. Do your examples have
    physical or mathematical significance?

2. Give a proof of Hölder's inequality in the continuous case.

3. What happens to Hölder's inequality when p approaches 1 and q
    approaches ∞ in the discrete case? In the continuous case? State
    and prove a theorem concerning each case.

4. Give a proof of Minkowski's inequality in the continuous case.

5. Find a specific counterexample to Theorem I.4.3 in the continuous case.
    Prove Theorem I.4.4 in the continuous case.

6. Give a direct proof of Theorems I.5.4, I.5.5 and I.5.6.

7. Investigate the third example of I.5.7 in detail. Show that the various
    inequalities are valid in this example.

References

1. N. Dunford and J.T. Schwartz, "Linear Operators, vol. I," Interscience,
        New York, 1958.

2. P.R. Halmos, "Finite Dimensional Vector Spaces," Van Nostrand, Princeton
        1958.

3. E. Hewitt and K. Stromberg, "Real and Abstract Analysis," Springer-
        Verlag, New York, 1965.

4. E.J. McShane, "Integration," Princeton University Press, Princeton,
        N.J. 1944.

5. F. Riesz and B. Sz-Nagy, "Functional Analysis," Frederick Ungar, New
        York, 1955.

6. M.H. Stone, "Linear Transformations in Hilbert Space," American Mathe-
        matical Society, New York, 1966.

7. A.E. Taylor, "Introduction to Functional Analysis," John Wiley and Sons,
        New York, 1958.

II. LINEAR SPACES AND LINEAR OPERATORS

The present chapter is concerned with the basic setting for a great deal of modern mathematical analysis and applied mathematics: the linear or vector space. In a linear space addition, subtraction, magnification and contraction of elements are all possible, and sometimes even multiplication between elements is possible. In what follows we will use the standard terminology employed by mathematicians, although not in any subtle way. In this way the standard mathematical language should become familiar, and will be part of the reader's vocabulary.

II.1. Linear Spaces.

II.1.1. Definition. A complex (real) linear space, denoted by X, is a set of elements with the operations of addition and scalar multiplication defined. If x and y are in X and α is a complex (real) number, then x+y and αx are in X. Further, the following axioms are satisfied.

1. Addition is associative and commutative: If x, y and z are in X, then

$$(x+y)+z = x+(y+z)$$

and

$$x+y = y+x.$$

2. There exists an element θ in X such that

$$x+\theta = x$$

for all x in X.

3. For each x in X there is a unique element $-x$ such that

$$x+(-x) = \theta.$$

4. If α and β are complex (real) numbers, and x and y are in X, then

$$\alpha(x+y) = \alpha x + \alpha y,$$

$$\alpha(\beta x) = (\alpha\beta)x,$$

$$1 \cdot x = x,$$

$$0 \cdot x = \theta,$$

$$(\alpha+\beta)x = \alpha x + \beta x..$$

Subtraction is then defined by

$$x-y = x+(-y).$$

It is an easy computation to show that $(-1)x = -x$, $\alpha\theta = \theta$, $x+y = x+z$ implies $y = z$, and when $\alpha \neq 0$, $\alpha x = \alpha y$ implies $x = y$.

II.1.2. Definition. A nonempty subset, M, of a linear space X is called a linear manifold, or a subspace of X, if when x and y are in M, and

α is any complex (real) number, then αx and $x+y$ are in M.

We note that if M is a linear manifold of X, then M is also a linear space. The element θ is a subspace of every linear space.

We call the subspace M of X a proper subspace when M $\subsetneq$ X.

In layman's terms, a linear space is a set of elements where the usual operations of addition, subtraction and scalar multiplication (multiplication by complex (real) numbers) are permissible with the results still within the set.

II.1.3. Examples. 1. The simplest examples of linear spaces are the finite dimensional vector spaces with components of the form $(x_1, x_2, \ldots, x_n)$. The most familiar of these are the two and three dimensional Euclidean spaces. Although the reader may not have considered them as such previously, the real[*] and complex number systems are also linear spaces. We recall that multiplication is defined in one, two and three dimensional spaces by ordinary multiplication or by a cross product.

2. A more complicated example is that of an infinite dimensional vector space with components of the form $(x_1, x_2, \ldots, x_n, \ldots)$, or perhaps continuing in both directions with components $(\ldots, x_{-n}, \ldots, x_{-2}, x_{-1}, x_0, x_1, x_2, \ldots, x_n, \ldots)$.

---

[*] The real number system, as a linear space, can only be multiplied by real scalars.

3. The simplest types of subspaces in the finite dimensional case are those sets of components with 0 at certain coordinate points. For instance the set of all vectors of the form $(x_1,0,x_3)$ is a two dimensional subspace of a three dimensional linear space.

4. In the infinite dimensional case an interesting and very useful subspace is found by considering the set of all vectors with only a finite number of nonzero components. That is, vectors of the form $(x_1,x_2,\ldots,x_n,\ldots,0,0,0,\ldots,)$.

5. Finally a linear space of a different sort is found by considering the set of functions which are continuous over a finite interval. A subspace of this space would be, for instance, those continuous functions which vanish at some point, or vanish over some fixed interval within the original interval. The space of continuous functions is infinite dimensional.

We have been using the term _dimension_ intuitively in the discussion of these examples. We now make its meaning precise.

II.1.4. _Definition._ _Let S be any nonempty subset of a linear space X, and let M be the set of all finite linear combinations of elements of S. That is, x is in M if and only if $x = \sum_{i=1}^{n} \alpha_i x_i$, where α_i, $i = 1,\ldots,n$, are complex (real) numbers and x_i, $i = 1,\ldots,n$, are in S. Then M is a linear manifold in X. M is said to be spanned by S._

M is the smallest linear manifold containing S.

II.1.5. <u>Definition</u>. <u>A finite set</u> of <u>elements in a linear space X, $\{x_i\}_{i=1}^n$</u>,
<u>is said to be linearly dependent if there exists a set of complex (real)</u>
<u>numbers $\{\alpha_i\}_{i=1}^n$</u>, <u>not all 0, such that</u>

$$\sum_{i=1}^n \alpha_i x_i = \theta.$$

   If a set of elements is not linearly dependent, then it is linearly
independent.

   An infinite number of elements is linearly independent if every finite
subset is linearly independent.

   It is easy to show that a finite set of elements $\{x_i\}_{i=1}^n$ is linearly
independent if and only if

$$\sum_{i=1}^n \alpha_i x_i = \theta$$

implies $\alpha_i = 0$, $i = 1,\ldots,n$. We also note that a linearly independent set
cannot contain θ.

II.1.6. <u>Theorem</u>. <u>Suppose $\{x_i\}_{i=1}^n$ is a set of elements in a linear space</u>
<u>X with $x_i \neq \theta$, $i = 1,\ldots,n$. The set is linearly dependent if and only if</u>
<u>one of the elements x_k is in the linear manifold spanned by $\{x_i\}_{i=1}^{k-1}$.</u>

Proof. Suppose the set $\{x_i\}_{i=1}^n$ is linearly dependent. Then there exists
a <u>smallest</u> k such that $\{x_i\}_{i=1}^k$ is dependent $(k \leq n)$. Thus there exist
complex (real) numbers $\{\alpha_i\}_{i=1}^k$, not all zero such that

$$\sum_{i=1}^{k} \alpha_i x_i = \theta.$$

Clearly $\alpha_k \neq 0$, since $\{x_i\}_{i=1}^{k-1}$ is an independent set. Therefore, we have

$$x_k = \sum_{i=1}^{k-1} (-\alpha_i/\alpha_k) x_i.$$

The converse is trivial.

Therefore, we are able to eliminate those elements which are not needed. Once this is done, the following definition is natural.

II.1.7. __Definition.__ __Let X be a linear space, and suppose that there exists__ __a positive integer n such that X contains a set of n linearly indepen-__ __dent elements, but every set of n+1 elements is linearly dependent. Then__ __X is finite dimensional, and n is the dimension of X. If a linear space__ __is not finite dimensional, it is infinite dimensional.__

If every set of n+1 elements in a linear space X is linearly depen-dent, then for any arbitrary element x, together with a fixed set of n linearly independent elements, is linearly dependent. This implies that there exist complex (real) numbers $\{\alpha_i\}_{i=1}^{n}$, not all zero such that

$$x = \sum_{i=1}^{n} \alpha_i x_i.$$

II.1.8. __Definition.__ __A finite set S of a linear space X is said to be__ __a basis for X if S in linearly independent, and the linear manifold__ __spanned by S is X.__

We note that the coefficients $\{\alpha_i\}_{i=1}^{n}$ which are used to express an

arbitrary element   x   in   X   are dependent upon   x   and are unique.   They are
said to be the coordinates of   x   with respect to the basis   S.

II.1.9. <u>Examples</u>. 1. n-dimensional complex (real) vector spaces consisting
of elements of the form $(x_1, x_2, \ldots, x_n)$ have dimension   n.   The set of
elements $(1,0,0,\ldots,0), (0,1,0,\ldots,0), \ldots, (0,0,0,\ldots,1)$ form a basis,
since if x_i stands for the element which has all zeros except in the i-th
position where it has a 1, the element $(\alpha_1, \alpha_2, \ldots, \alpha_n) = \sum_{i=1}^{n} \alpha_i x_i$.

2. The space of continuous functions, defined on the interval [-1,1],
denoted by C[-1,1], has infinite dimension. We shall show later by means of
the Stone-Weierstrass theorem that $\{1, t, t^2, \ldots, t^n, \ldots\}$ form a basis for
C[-1,1].

Sometimes it is convenient to use different bases for representing
elements in a linear space. Of course, then each element will have a different set of coordinates for each basis and will look quite different in each
representation. The space changes its appearance, while it is, in fact, the
same space.

In addition, sometimes different linear spaces can be shown to be
essentially the same. That is, except for the notation used in expressing
them, they are similar in every way. Such spaces are said to be isomorphic.

II.1.10. <u>Definition</u>. <u>Two complex (real) linear spaces   X   and   Y   are</u>
<u>linear space isomorphic to each other if there exists a 1 to 1 correspondence</u>
<u>between them, denoted by ⇔,   such that</u>

1. x ⇔ y implies αx ⇔ αy for all complex (real) numbers α and all

    x in X, y in Y.

2. x_1 ⇔ y_1 and x_2 ⇔ y_2 implies that $x_1 + x_2$ ⇔ $y_1 + y_2$ for all x_1,

    x_2 in X and y_1, y_2 in Y.

II.1.11. Theorem. The correspondence

$$x \Leftrightarrow (\alpha_1, \alpha_2, \ldots, \alpha_n)$$

between elements x of an n-dimensional complex (real) linear space X and
their coordinates with respect to a fixed basis is an isomorphism between X
and the n-dimensional vector space $K^n(R^n)$, where K^n is the complex n-dimensional Euclidean space (R^n is the real n-dimensional Euclidean space).

In other words, when dealing with finite dimensional spaces, we may as
well consider n-dimensional Euclidean space. Any other such space is
isomorphic to it.

II.2. Linear Operators. The basic object of our attention throughout the
book will be that of the linear operator or transformation. Linear operators
are first encountered in elementary vector calculus in the form of matrices.
Indeed every linear operator on a finite dimensional vector space is uniquely
defined by its matrix representation with respect to any fixed basis. There
are other linear operators which are not so easily represented. Differentiation is such an operator. Integration is another. A boundary form is still
another.

II.2.1. <u>Definition.</u> Let X and Y be complex (real) linear spaces, and
let D be a subset of X. Suppose A is a function with domain D in X
and range in Y. Then A is an operator or transformation from X to Y.

If Y is the space of complex (real) numbers, then A is a functional.

An operator B is an extension of an operator A, if the domain of
A is contained in the domain of B, and Ax = Bx for all x in the domain
of A. We write in this case A ⊂ B.

An operator A with domain D in X and range in Y is linear, if
D is a linear manifold in X, if

$$A(\alpha x) = \alpha(Ax)$$

for all x in D and all complex (real) numbers α, and if

$$A(x_1+x_2) = Ax_1 + Ax_2$$

for all x_1, x_2 in D.

If A is linear and Y is the space of complex (real) numbers, then
A is a complex (real) linear functional.

We next characterize linear functionals and linear operators, defined
on sets D in X, when D is an n-dimensional space.

II.2.2. <u>Theorem.</u> Let A be a linear functional with domain D in X,
and let $\{x_i\}_{i=1}^n$ be a basis for D. Further let $Ax_i = \xi_i$, i = 1,...,n.

Then for each $x = \sum\limits_{i=1}^{n} \alpha_i x_i$ in D,

$$Ax = \sum_{i=1}^{n} \alpha_i \xi_i.$$

This follows immediately from the linearity of A.

II.2.3.   Theorem.   Let A be a linear operator with domain D in X and range also in X, and let X be n-dimensional. Further let $\{x_i\}_{i=1}^{n}$ be a basis for X. Then

$$Ax_j = \sum_{i=1}^{n} a_{ij} x_i$$

$j = 1,\ldots,n$, uniquely defines the n^2 quantities $\{a_{ij}\}_{i,j=1}^{n}$ (a_{ij}) is the matrix of A with respect to the basis $\{x_i\}_{i=1}^{n}$.

If $Ax = y$ and $x = \sum\limits_{i=1}^{n} \alpha_i x_i$, $y = \sum\limits_{i=1}^{n} \beta_i x_i$, then

$$\beta_i = \sum_{j=1}^{n} a_{ij} \alpha_j,$$

$i = 1,\ldots,n$, or as matrices $(a_{ij})(\alpha_j) = (\beta_i)$, or

$$\begin{pmatrix} a_{11} & \cdots & a_{1n} \\ \vdots & & \vdots \\ a_{n1} & \cdots & a_{nn} \end{pmatrix} \begin{pmatrix} \alpha_1 \\ \vdots \\ \alpha_n \end{pmatrix} = \begin{pmatrix} \beta_1 \\ \vdots \\ \beta_n \end{pmatrix}.$$

Again this follows immediately from the linearity of A.

II.2.4. Definition. Let A and B be operators with domains D_A and D_B in X and ranges in Y. Then

   1. By A+B we mean the operator with domain $D_{A+B} = D_A \cap D_B$, defined by

$$(A+B)x = Ax + Bx$$

       for all x in D_{A+B}.

   2. Let α be a complex (real) number. Then by αA we mean the operator with domain $D_{\alpha A} = D_A$, defined by

$$(\alpha A)x = \alpha(Ax)$$

       for all x in $D_{\alpha A}$.

   3. We let 0 be the zero operator. That is, the domain of 0 is X and

$$0x = \theta',$$

       where θ' is the zero element in Y.

II.2.5. Theorem. Within the appropriate domains the following hold

   1. (A+B)+C = A+(B+C),

   2. A+B = B+A

3. $\alpha(A+B) = \alpha A + \alpha B$,

4. $1 \cdot A = A$,

5. $0 \cdot A \subset 0$,

where A, B, C are operators from X to Y, and α is a complex (real)
number.

The proofs are left to the reader.

II.2.6. <u>Definition</u>. Let X, Y, Z be linear spaces. Suppose B is an
operator with domain D_B in X and range in Y. Suppose A is an operator
with domain D_A in Y and range in Z. By AB we mean the operator with
domain $D_{AB} = \{x \mid x \in D_B, \ Bx \in D_A\}$ and range in Z, defined by

$$ABx = A(Bx)$$

for all x in D_{AB}.

II.2.7. <u>Definition</u>. By the identity operator I in X we mean the
operator with domain $D_I = X$, defined by

$$Ix = x$$

for all x in $D_I = X$.

II.2.8. Theorem. Within the appropriate domains, the following hold

   1. (A+B)C = AC+BC,

   2. A(B+C) = AB+AC,

   3. IA = AI = A.

In the preceeding, the operator A was a transformation from a set D_A in X   into Y. We may ask: Given the operator A and an element y in its range in Y, is it possible to recover the element x in X which yields y under transformation by A? That is, given Ax = y, is it possible to find x? If this is so, A is said to possess an inverse.

II.2.9. Definition. The operator A possesses an inverse, denoted by A^{-1}, if for each y in the range of A, R_A, there exists a unique element x in D_A such that Ax = y. In that case we define A^{-1} by letting $A^{-1}y = x$.

II.2.10. Theorem. The operator A possesses an inverse if and only if for all x_1, x_2 in D_A, $Ax_1 = Ax_2$ implies $x_1 = x_2$.

Proof. If A possesses an inverse, and $Ax_1 = Ax_2$, then the element $y = Ax_1 = Ax_2$ corresponds to a unique element. So $x_1 = x_2$. Conversely, if for some y in R_A there exist x_1, x_2 such that $Ax_1 = y$ and $Ax_2 = y$, then $Ax_1 = Ax_2$, and $x_1 = x_2$. A^{-1} then exists by definition.

II.2.11. Theorem. Suppose A^{-1} exists. Then

1. $D_A = R_{A^{-1}}$ (the range of A^{-1}),

    $D_{A^{-1}} = R_A$ (the range of A),

2. $A^{-1}Ax = x$ for all x in D_A,

3. $AA^{-1}y = y$ for all y in $D_{A^{-1}}$.

4. $(A^{-1})^{-1} = A$.

Proof. 1. The second part follows from the definition. To show the first part, we note first that $R_{A^{-1}} \subset D_A$, since each y in $D_{A^{-1}}$ is representable by y = Ax for some x in D_A, and $A^{-1}y = x$ by definition. If x is in D_A but not in $R_{A^{-1}}$, then y = Ax is in R_A. Thus y is in $D_{A^{-1}}$, and $A^{-1}y = x$, giving a contradiction.

2. This follows from the definition.

3. If y is in $D_{A^{-1}}$, then y = Ax for some x in D_A. Then $A^{-1}y = A^{-1}Ax = x$. Thus $AA^{-1}y = Ax = y$.

4. If x is in D_A, then $A^{-1}[(A^{-1})^{-1}-A]x = [A^{-1}(A^{-1})^{-1}-A^{-1}A]x = \theta$. $[(A^{-1})^{-1}-A]x = \theta$ for all x in D_A, and $(A^{-1})^{-1} = A$.

II.2.12. Theorem. Let A transform $D_A = X$ into X. Then A^{-1} exists and $D_{A^{-1}} = X$ if and only if there exist operators B and C such that

        AB = CA = I.

Proof. Suppose A^{-1} exists and $D_{A^{-1}} = X$. Then $A^{-1}A = AA^{-1} = I$.

Conversely, assume $AB = CA = I$. Then $D_B = D_A = X$. Since $AB = I$, the range of A is also X. Now suppose $Ax = Ay$. Then $CAx = CAy$, or $x = y$. Thus A^{-1} exists on the range of A, which is X. To show $A^{-1} = B = C$, we note $AB = I$ implies $A^{-1}AB = A^{-1}I = A^{-1}$. But $A^{-1}AB = IB = B$. So $A^{-1} = B$. Similarly, $CA = I$ implies $CAA^{-1} = IA^{-1} = A^{-1}$. But $CAA^{-1} = CI = C$, and $A^{-1} = C$.

Although this theorem appears to be quite elegant, its use is quite limited.

II.2.13. <u>Definition.</u> <u>If an operator A satisfies $D_A = D_{A^{-1}} = X$, then A is regular.</u>

II.2.14. <u>Examples.</u> The simplest examples of regular operators are those which are represented by nonsingular matrices on finite dimensional vector spaces. Another type is defined on $C[-1,1]$ by multiplication by any nonzero continuous function. Unfortunately a large and very interesting class of operators, those defined by differentiation, are not regular, even if they possess an inverse. Their domains must be restricted to sufficiently differentiable functions, and therefore are not the entire space.

We conclude this section with two theorems which apply to differential operators as well.

II.2.15. <u>Theorem.</u> <u>Let A be a linear operator from X to Y. Then</u>

the range of A is a linear manifold in Y, and $A\theta = \theta'$.

II.2.16. Theorem. The inverse of a linear operator A from X into Y exists if and only if $Ax = \theta'$ implies $x = \theta$. When A^{-1} exists, it is a linear operator from Y to X.

We leave the proofs as exercises.

II.3. Norms and Banach Spaces. We have already encountered norms briefly in Chapter I. They are, roughly speaking, functions which measure distances between elements of a linear space. They must, of course, have certain characteristics. It is these which we now make precise.

II.3.1. Definition. Let X be a linear space. Let x and y be in X, and let α be a complex (real) number. A norm on X, denoted by $\|\cdot\|$, is a function whose domain is X, whose range is the nonnegative real numbers, and which satisfies

1. $\|x+y\| \leq \|x\| + \|y\|$,

2. $\|\alpha x\| = |\alpha| \|x\|$,

3. $\|x\| \geq 0$, and $\|x\| = 0$ if and only if $x = \theta$.

X (together with its norm $\|\cdot\|$) is called a normed linear space.

$\|x\|$ is the length of the element x, and $\|x-y\|$ is the distance from x to y.

II.3.2. Examples. We have already seen several examples in Chapter I. Let us recall them briefly.

1. If S is a set of points, let X be the linear space of functions defined on S. We then let F = {|f|, f ∈ X}, and suppose there exists an abstract integral I, defined on F, such that I, F satisfy the condition of I.5. If p ≥ 1, then

$$\|f\|_p = I(|f|^p)^{1/p}$$

satisfies the definition II.3.1.

2. For instance, if S = {1,2,...,n},   f(i) = x_i, and

$$\|x\|_{R^n} = [\sum_{i=1}^{n} |x_i|^p]^{1/p},$$

then X is a normed linear space.

3. If S = [a,b], and X consists of all complex valued functions such that $|f|^p$ is integrable, then

$$\|f\|_p = [\int_a^b |f|^p dt]^{1/p}$$

defines a norm and X, under this norm, is a normed linear space.

4. If X consists of the continuous functions defined on the interval [a,b], then

$$\|f\|_\infty = \sup_{t \in [a,b]} |f(t)|$$

defines a norm on X. In this case X is called C[a,b]. The norm is the sup norm or uniform norm.

If we are in possession of some device for measuring distances, we can consider easily the concept of convergence in a manner suggested by the properties of convergence of the real number system.

II.3.3. Definition. A sequence $\{x_n\}_{n=0}^{\infty}$ in a normed linear space X is called a Cauchy sequence if and only if

$$\lim_{m,n \to \infty} \|x_m - x_n\| = 0,$$

where m and n may vary independently.

A normed linear space X is complete if and only if every Cauchy sequence converges to an element x in X. That is, $\lim_{n \to \infty} \|x_n - x\| = 0$.

A complete normed linear space is called a BANACH space.

The reader might now ask what is so special about Cauchy sequences, when it is convergence which counts?

II.3.4. Theorem. In a Banach space X, a sequence is a Cauchy sequence if and only if it is a convergent sequence.

Proof. If $\{x_n\}_{n=0}^{\infty}$ is a Cauchy sequence, then by definition it is convergent. If $\{x_n\}_{n=0}^{\infty}$ is convergent to x in X, then if n and m are sufficiently large, $\|x_n - x\|$ and $\|x_m - x\|$ may be made arbitrarily small. Since

$$\|x_n - x_m\| \leq \|x - x_n\| + \|x - x_m\|,$$

$\{x_n\}_{n=0}^{\infty}$ is a Cauchy sequence.

Therefore a Banach space is one in which the operations of addition, subtraction, scalar multiplication AND taking limits are all possible. In other words, everything works!

II.3.5. <u>Examples</u>. 1. The space of complex numbers with absolute value as it norm is a complex.Banach space. The space of real numbers with absolute value as its norm is a real Banach space.

2. The n-dimensional linear vector spaces, consisting of elements of the form $x = (x_1, x_2, \ldots, x_n)$ with norm defined by

$$\|x\|_{R^n} = [\sum_{i=1}^{n} |x_i|^p]^{1/p}, \quad p \geq 1,$$

are Banach spaces. The previous examples consist of the cases $n = 1$.

3. $C[a,b]$ is a Banach space under the uniform norm, $\|f\| = \sup_{t \in [a,b]} |f(t)|$, since uniformly convergent continuous functions have a continuous function as their limit.

4. $C[0,1]$, with norm

$$\|f\|_p = [\int_0^1 |f(t)|^p dt]^{1/p}, \quad p \geq 1,$$

forms a normed linear space, but that space is NOT complete.

To show not, let f_n be defined by

$$f_n(t) = \begin{cases} 0 & , & 0 \leq t \leq \frac{1}{2} - \frac{1}{n}, \\ n(t - \frac{1}{2} + \frac{1}{n}) & , & \frac{1}{2} - \frac{1}{n} \leq t \leq \frac{1}{2}, \\ 1 & , & \frac{1}{2} \leq t \leq 1. \end{cases}$$

Then

$$\|f_n - f_m\|_p < [\int_{\frac{1}{2}-\frac{1}{n}}^{\frac{1}{2}} 2^p dt]^{1/p} = 2\cdot(1/n^{1/p}),$$

which approaches 0, as n approaches ∞. Thus $\{f_n\}_1^\infty$ is a Cauchy sequence. But its limit,

$$f(t) = \begin{cases} 0, & 0 \le t < \frac{1}{2}, \\ 1, & \frac{1}{2} \le t \le 1, \end{cases}$$

is discontinuous and is not in C[0,1].

This sort of occurrence, the failure of a normed linear space to be complete, is quite common, especially if the elements have such properties such as being several times differentiable, vanishing at a particular point, or being integrable when raised to a power greater or equal to one. This makes things a little awkward, particularly when taking limits, but this awkwardness is only temporary.

II.3.6. Theorem. Let X be a normed linear space. There exists a Banach space Y such that Y contains X isomorphically, preserving the operations of addition, subtraction and scalar multiplication.

Proof. If X is complete then Y = X. If X is not complete, let Y be the space of all Cauchy sequences of X. $Y = \{y | y = \{x_i\}_{i=0}^\infty, \{x_i\}_{i=0}^\infty$ is a Cauchy sequence in X.} Addition in Y is defined as follows. If

$y_1 = \{x_i^1\}_{i=0}^{\infty}$ and $y_2 = \{x_i^2\}_{i=0}^{\infty}$, then $y_1 + y_2 = \{x_i^1 + x_i^2\}_{i=0}^{\infty}$. Scalar multiplication is defined by $\alpha y = \{\alpha x_i\}_{i=0}^{\infty}$. Two elements y_1 and y_2 are equal in Y if $\lim_{i \to \infty} \|x_i^1 - x_i^2\|_X = 0$, where the subscript X denotes the norm in X. Finally the norm in Y is defined by

$$\|y\|_Y = \lim_{i \to \infty} \|x_i\|.$$

X is imbedded in Y by the correspondence $y_x = \{x\}_{i=0}^{\infty}$. That is, each x_i in the Cauchy sequence is equal to x. We note $\|y_n\|_Y = \|x\|_X$.

Finally if $\{y^j\}_{j=0}^{\infty} = \{\{x_i^j\}_{i=0}^{\infty}\}_{j=0}^{\infty}$ is a Cauchy sequence in Y, then $\{y^j\}_{j=0}^{\infty}$ converges to $y^0 = \{x_i^1\}_{j=0}^{\infty}$. To show this we note the following: Since $\{y^j\}_{j=0}^{\infty}$ is a Cauchy sequence in Y, if $\varepsilon > 0$ is arbitrary, then there exists an N_1 such that if $j,k > N_1$, then $\|y^j - y^k\|_Y < \varepsilon$. This means that for i sufficiently large, say $i > N_2$, that $\|x_i^j - x_i^k\|_X < \varepsilon$, when $i > N_2$, $j,k > N_1$. But the i-th element in the sequence y^0 is x_i^i. Thus $\|x_i^j - x_i^i\|_X < \varepsilon$, if $i,j > \max(N_1, N_2)$. (i.e., let $k = i$). Thus $\lim_{j \to \infty} \|y^j - y^0\|_Y = 0$, and $\{y^j\}_{j=0}^{\infty}$ converges to y^0 in Y. Y is complete.

Thus the failure of a normed linear space X to be complete is only a minor inconvenience. By the device just used X can be imbedded in a Banach space with little effort.

For the most part as we progress we shall assume that X is complete.

We conclude this section with a convergence theorem, which is most conveniently set in a Banach space.

II.3.7.   Theorem.   Let   X   be a Banach space and $\{x_i\}_{i=0}^{\infty}$ be a sequence of

elements in   X   satisfying $\sum_{i=0}^{\infty} \|x_i\| < \infty$.   Then the sequence

$$y_n = \sum_{i=0}^{n} x_i, \quad n = 0,1,\ldots$$

converges to an element

$$y_0 = \sum_{i=0}^{\infty} x_i$$

in   X.

Proof.   Since $\sum_{i=0}^{\infty} \|x_i\| < \infty$, $s_n = \sum_{i=0}^{n} \|x_i\|$ is a Cauchy sequence.   Now if
$n \leq m$,

$$\|y_n - y_m\| = \|\sum_{i=n+1}^{m} x_i\| \leq \sum_{i=m+1}^{n} \|x_i\|.$$

Thus $\{y_n\}_{n=0}^{\infty}$ is a Cauchy sequence in   X.   Since   X   is complete, $\{y_n\}_{n=0}^{\infty}$
converges to an element in   X,   which is obviously y_0.

II.4.   Operator Convergence.   If   A   is a linear operator from a normed linear

space   X   into another normed linear space   Y   (X   may be the same as   Y.),

it is easy to use the norms provided to define a norm for   A.   Since the set

of operators is already obviously a linear space, the space of operators

becomes a normed linear space itself.

II.4.1. <u>Definition.</u> Let X be a normed linear space. The set of elements x in X which satisfy, for a fixed element x_0 in X, $\|x-x_0\| < \delta$ is called the open sphere of radius δ with center x_0.

II.4.2. <u>Definition.</u> Let X be a normed linear space. Let A be an operator with domain D_A in X and range in a normed linear space Y. Then A is continuous at an element x_0 in X if

$$\lim_{x \to x_0} \|Ax - Ax_0\|_Y = 0,$$

where x, x_0 are in D_A. That is, when given any $\varepsilon > 0$, there exists a $\delta > 0$ such that if $\|x-x_0\|_X < \delta$, then $\|Ax-Ax_0\|_Y < \varepsilon$, we say A is continuous at x_0.

If A is continuous at all elements x in D_A, then A is continuous.

II.4.3. <u>Definition.</u> Let X be a normed linear space. Let A be an operator with domain D_A in X and range in a normed linear space Y. Suppose further that

$$M = \sup_{\substack{x \in D_A \\ x \neq \theta}} \|Ax\|_Y / \|x\|_X < \infty.$$

Then A is bounded by M. We say that $M = \|A\|$, where the symbol $\|\cdot\|$ denotes the operator norm. We will sometimes write $\|\cdot\|_{op}$, just as we have written $\|\cdot\|_X$ and $\|\cdot\|_Y$ to distinguish between norms in X and Y.

There is still a third concept concerning linear operators, which we will require shortly, that of a Lipschitz condition.

II.4.4. <u>Definition</u>. <u>Let X be a normed linear space. Let A be an</u>
<u>operator with domain D_A in X and range in a normed linear space Y. Then</u>
<u>A satisfies a Lipschitz condition with Lipschitz constant K if</u>

$$\|Ax-Ay\|_Y \leq K\|x-y\|_X$$

<u>for all x, y in D_A.</u>

II.4.5. <u>Theorem</u>. <u>Let X and Y be normed linear spaces, and let A be</u>
<u>a linear operator with domain $D_A = X$ and range all of Y. Then the following</u>
<u>statements are equivalent.</u>

1. <u>A is continuous at x_0 in X.</u>

2. <u>A is continuous.</u>

3. <u>A is bounded.</u>

4. <u>A satisfies a Lipschitz condition.</u>

Proof. We shall show 1 implies 2, 2 implies 3, 3 implies 4, and 4 implies 1.

Suppose A is continuous at x_0 and that y approaches y_0. Now

$$Ay-Ay_0 = A(y-y_0+x_0)-Ax_0.$$

Since $(y-y_0+x_0)$ approaches x_0 and A is continuous at x_0, $A(y-y_0+x_0)$ approaches Ax_0. Thus the right side above approaches θ', the zero element in Y, and Ay approaches Ay_0.

Suppose next that A is continuous but not bounded. Then for each integer n there exists an element x_n such that

$$\|Ax_n\|_Y > n\|x_n\|_X .$$

Letting $y_n = x_n/n\|x_n\|_X$, we see that $\|y_n\|_X = 1/n$ and $\|Ay_n\|_Y > 1$. As n becomes large, y_n approaches θ, and, since A is continuous, Ay_n approaches $A\theta = \theta'$. But $\|\theta'\|_Y = 0$. Since $\|Ay_n\|_Y > 1$, it cannot approach 0, and we have a contradiction.

If A is bounded, then

$$\|Ax\|_Y \leq \|A\|\|x\|_X$$

for all x in X. We replace x by $x-y$ to find

$$\|Ax-Ay\|_Y \leq \|A\|\|x-y\|_X .$$

A satisfies a Lipschitz condition with Lipschitz constant $\|A\|$.

Finally, if A satisfies a Lipschitz condition, it is obviously continuous.

The operator norm has a number of interesting properties. We state

some of them for linear operators which transform the space X into itself,
although most can be easily extended.

II.4.6. Definition. We denote by $L(X,X)$ the linear space of bounded
linear operators which transform the normed linear space X into itself.

II.4.7. Theorem. Let X be a normed linear space, let A and B be in
$L(X,X)$, and let α be a complex (real) number. Then the operator norm
$\|\cdot\|$ has the following properties.

   1. $\|A+B\| \leqq \|A\| + \|B\|$,

   2. $\|\alpha A\| = |\alpha|\|A\|$,

   3. $\|A\| = 0$ if and only if $A = 0$, the zero operator.

$L(X,X)$ is a normed linear space.

Proof. Only the first property is not obvious. It follows from the fact

$$\sup_{\substack{x \in X \\ x \neq \theta}} \frac{\|(A+B)x\|}{\|x\|} \leqq \sup_{\substack{x \in X \\ x \neq \theta}} \frac{\|Ax\|}{\|x\|} + \sup_{\substack{x \in X \\ x \neq \theta}} \frac{\|Bx\|}{\|x\|} \; .$$

II.4.8. Theorem. Let X be a normed linear space. Let A and B be in
$L(X,X)$. Then

   1. $\|AB\| \leqq \|A\|\|B\|$,

   2. $\|A^n\| \leqq \|A\|^n$, $n = 0,1,\ldots$.

Proof. The first is true, since

$$\|ABx\| \leq \|A\|\,\|Bx\|,$$

$$\leq \|A\|\,\|B\|\,\|x\|.$$

We divide by $\|x\|$ and maximize. The second is a corollary of the first.

The additional property $\|AB\| \leq \|A\|\,\|B\|$ shows that $L(X,X)$ is more than just a normed linear space. The name given to spaces with this addition is that of an algebra. $L(X,X)$ is a <u>normed linear algebra</u>.

Throughout this section we have only assumed that X is normed linear space, but not necessarily complete. With the additional assumption of completeness we can say even more about $L(X,X)$.

II.4.9. <u>Theorem</u>. <u>Let X be a Banach space. Then $L(X,X)$ is also a Banach space.</u>

Proof. The only part to show is the completeness of $L(X,X)$. Suppose $\{A_n\}_{n=1}^{\infty}$ is a Cauchy sequence in $L(X,X)$. Then if m, n are sufficiently large, $\|A_n - A_m\|$ can be made arbitrarily small. If x is an arbitrary element in X, then

$$\|A_n x - A_m x\| \leq \|A_n - A_m\|\,\|x\|,$$

so $\{A_n x\}_{n=1}^{\infty}$ is a Cauchy sequence in X. Since X is complete, $\{A_n x\}_{n=1}^{\infty}$ converges to an element we call $A_0 x$. Clearly $A_0 x$ is defined for all x in X. It is also easy to see A_0 is linear, since each of the sequence

$\{A_n\}_{n=1}^{\infty}$ is linear. To show A_0 is bounded, we observe that for any   x   in X

$$\|A_0 x\| \leq \|(A_0 - A_m) x\| + \|A_m x\|.$$

By definition, if   m   is sufficiently large, $\|(A_0 - A_m) x\|$ can be made
arbitrarily small, say less than $\varepsilon\|x\|$,   where ε is arbitrarily small, but
fixed. Since $\|A_m x\| \leq \|A_m\|\|x\|$,

$$\|A_0 x\| < [\|A_m\| + \varepsilon]\|x\|,$$

and A_0 is in $L(X,X)$.

We say when $L(X,X)$ is a Banach space with additional properties of
Theorem II.4.8, that $L(X,X)$ is a <u>Banach algebra</u>.

II.4.10.   <u>Examples</u>. 1. The simplest example of a Banach algebra $L(X,X)$
is represented by the set of all   n × n   matrices. The norm in $L(X,X)$ will
depend, of course, upon which norm is chosen for   X.

2. If   X   consists of infinite dimensional vectors $(x_1, x_2, \ldots, x_n, \ldots)$.
then each element of $L(X,X)$ can be represented an infinite matrix. In
order to assure its boundedness, however, some additional assumptions are
needed concerning the components of the matrix. We invite the reader to
explore this problem in greater depth.

II. Exercises

1. Show that in a linear space $(-1)x = -x$, $\alpha\theta = \theta$, $x+y = x+z$ implies $y = z$, and $\alpha x = \alpha y$ implies $x = y$ when $\alpha \neq 0$.

2. What do subspaces of two and three dimensional Euclidean spaces look like geometrically?

3. Show that in three dimensional Euclidean space, the set of elements $(1,0,0)$, $(0,1,0)$, $(0,0,1)$ is linearly independent and is a basis. What occurs in higher dimensions?

4. Show that with respect to a fixed basis in a finite dimensional linear space, the coefficients necessary to express an arbitrary element are unique.

5. Show that K^n (the complex linear space with dimension n) is isomorphic to R^{2n} (the real linear space with dimension 2n).

6. Give detailed proofs of Theorems II.2.2, II.2.3, II.2.5 and II.2.8. What are the various domains in each case?

7. Let the linear operator A transform the n-dimensional space X into itself. Show that A^{-1} exists if and only if $\det(a_{ij}) \neq 0$, where (a_{ij}) is the matrix representation for A with respect to a fixed basis.

8. On the space $C[-1,1]$, let the linear operator A be defined by $Ax = m(t)x$, where $m(t)$ is continuous and nonzero on $[-1,1]$. Show that A^{-1} exists.

9. Give detailed proofs of Theorems II.2.15 and II.2.16.

10. Let A and B be linear operators on the linear space X such that
    $D_A = D_B = X$ and BA = I. Prove or disprove that AB = I.

11. Let A and B be linear operators on the linear space X such that
    $D_A = D_B = X$ and B is regular. Show that A is regular if and only
    if AB is regular. Show then that $(AB)^{-1} = B^{-1}A^{-1}$. Show that if
    AB is regular and AB = BA, then A and B are regular.

12. In Theorem II.3.6, what is Y when X is the linear space of rational
    numbers?

13. In Theorem II.3.6, what is Y when X is C[0,1] with the norm

$$\|f\|_p = [\int_0^1 |f|^p dt]^{1/p}, \quad p \geq 1?$$

14. Give several (at least 5) examples of Banach algebras.

References

1. S. Banach, "Operations Lineaires, "Monografje Matematyczne, Warsaw, 1932.

2. N. Dunford and J.T. Schwartz, "Linear Operators, vol. 1," Interscience,
       New York, 1958.

3. P.R. Halmos, "Finite Dimensional Vector Spaces," Van Nostrand, Princeton,
       1958.

4. E.L. Ince, "Ordinary Differential Equations," Dover Publications, New
       York, 1956.

5. M.A. Naimark, "Normed Rings," P. Noordhoff, Groningen, The Netherlands,
       1964.

6. F. Riesz and B. Sz-Nagy, "Functional Analysis," Frederick Ungar, New
       York, 1955.

7. M.H. Stone, "Linear Transformations in Hilbert Space," American
       Mathematical Society, New York, 1966.

8. A.E. Taylor, "Introduction to Functional Analysis," John Wiley and Sons,
       New York, 1958.

III.   EXISTENCE AND UNIQUENESS THEOREMS

In this chapter our ultimate goal is to show the existence and uniqueness of solutions to certain ordinary differential equations. To do so we use the setting of the previous chapter, a Banach space, and a device known as a contraction mapping. The results are then applied to an integral equation. These are in turn applied to certain ordinary differential equation. Finally, the results are extended and refined.

III.1.   The Contraction Mapping Theorem. A contraction mapping is a transformation which reduces the distance between elements. It has, in addition, another extremely important property. It possesses a unique fixed point, that is, an element which is transformed into itself. It is this fixed point we wish to find. It will ultimately turn out to be the solution we seek when considering ordinary differential equations.

III.1.1.   Theorem (The Contraction Mapping Theorem). Let $0 \leqq M < 1$. Let X be a Banach space, and let A be an operator in X with domain D_A having the following properties.

1. x_0 is in D_A.

2. $\{x; \|x-x_0\| \leqq \|Ax_0\|/(1-M)\} \subset D_A$.

Then the equation

$$x = x_0 + Ax$$

has a unique solution x, which is in D_A.

Proof. If $M = 0$, Ax is constant, and $x = x_0 + Ax$ is uniquely defined.

If $M > 0$, we define the sequence $\{x_n\}_{n=0}^{\infty}$ by

$$x_1 = x_0 + Ax_0,$$

$$x_2 = x_0 + Ax_1,$$

$$\ldots$$

$$x_{n+1} = x_0 + Ax_n,$$

$$\ldots \ .$$

We shall perform an induction on the statements

a. $\|x_{n+1} - x_0\| \le \sum_{j=0}^{n} M^j \|Ax_0\|$

(so $\{x_n\}_{n=1}^{\infty}$ are all in D_A),

b. $\|x_{n+1} - x_n\| \le M^n \|Ax_0\|$.

Now $\|x_1 - x_0\| = \|Ax_0\|$. Thus a is true for $n = 0$. b is obviously true
for $n = 0$. Assume a and b are true for n, and consider the statements
for $n+1$.

$$\|x_{n+2}-x_0\| = \|x_{n+2}-x_{n+1}+x_{n+1}-x_0\|,$$

$$\leq \|x_{n+2}-x_{n+1}\| + \|x_{n+1}-x_0\|,$$

$$\leq \| Ax_{n+1}-Ax_n\| + \sum_{j=0}^{n} M^j\|Ax_0\|,$$

$$\leq M \| x_{n+1}-x_n\| + \sum_{j=0}^{n} M^j\|Ax_0\|,$$

$$\leq \sum_{j=0}^{n+1} M^j\|Ax_0\|.$$

$$\|x_{n+2}-x_{n+1}\| \leq \|Ax_{n+1}-Ax_n\|,$$

$$\leq M\|x_{n+1}-x_n\|,$$

$$\leq M^{n+1}\|Ax_0\|.$$

The statements are thus true for n+1, and, therefore, for all n.

Next consider the series

$$s = \sum_{n=0}^{\infty} (x_{n+1}-x_n).$$

$$\|s\| \leq \sum_{n=0}^{\infty} \|x_{n+1}-x_n\|,$$

$$\leq \sum_{n=0}^{\infty} M^n\|Ax_0\|,$$

$$\leq \|Ax_0\|/(1-M).$$

We see by II.3.7 that s exists and is in X. If we let $s = x-x_0$, then

$$x = \sum_{n=0}^{\infty} (x_{n+1}-x_n)+x_0$$

is in D_A. Since the series telescopes,

$$x = \lim_{n\to\infty} x_n.$$

Further, since

$$\|Ax_n-Ax\| \leq M\|x_n-x\|,$$

$$Ax = \lim_{n\to\infty} Ax_n.$$

Thus we have

$$x = \lim_{n\to\infty} x_{n+1} = \lim_{n\to\infty} (x_0 + Ax_n)$$

$$= x_0 + \lim_{n\to\infty} Ax_n = x_0 + Ax.$$

x   is a solution.

If y is another solution, then

$$\|x-y\| = \|Ax-Ay\|,$$

$$\leq M\|x-y\|,$$

$$< \|x-y\|,$$

a contradiction.

III.1.2. **An Example.** If $X = R^1$, the real number system, then

$$x = x_0 + Ax,$$

$$\|Ax - Ay\| \leq M\|x - y\|, \quad 0 \leq M < 1,$$

corresponds to the following picture.

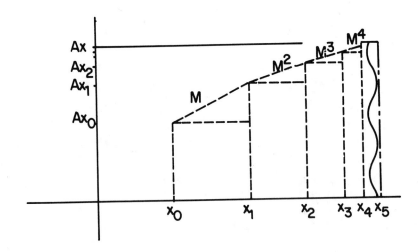

The slope with each approximation decreases at least by a factor of M.

The next theorem, a corollary of the contraction mapping theorem, the Neumann expansion, is encountered in the solution of integral equations.

III.1.3. **Theorem (Neumann Series).** Let X be a Banach space, and let A be linear operator in $L(X,X)$ such that $\|A\| < 1$. Then the equation

$$x = x_0 + Ax$$

has a unique solution

$$x = \sum_{n=0}^{\infty} A^n x_0$$

for each x_0 in X,

We leave the proof to the reader. Before we see some examples we prove the following variation of Theorem III.1.3.

III.1.4. Theorem. Let X be a Banach space, let A be a linear operator in $L(X,X)$, and let λ be a complex (real) number such that $|\lambda| > \|A\|$. Then

1. $(\lambda I - A)$ is regular (see II.2.13).

2. $\|(\lambda I - A)^{-1}\| \leq (|\lambda| - \|A\|)^{-1}$.

3. $(\lambda I - A)^{-1} y = \sum_{n=0}^{\infty} (A^n / \lambda^{n+1}) y$ for all y in X.

Proof. We solve the equation

$$(\lambda I - A)x = y$$

for x to find

$$x = (1/\lambda)y + (1/\lambda)Ax,$$

which is of the form of the previous theorem. Since $\| (1/\lambda)A\| = \|A\|/|\lambda| < 1,$
there is a unique solution

$$x = \sum_{n=0}^{\infty} (A^n/\lambda^{n+1})y.$$

This shows

$$(\lambda I-A)^{-1} = \sum_{n=0}^{\infty} (A^n/\lambda^{n+1}).$$

Since

$$\| \sum_{n=0}^{\infty} (A^n/\lambda^{n+1})\| \le \sum_{n=0}^{\infty} \|A\|^n/|\lambda|^{n+1} = 1/[\,|\lambda|-\|A\|\,],$$

$(\lambda I-A)^{-1}$ is defined for all y in X and is bounded. Thus we have proved
1, 2 and 3 in reverse order.

III.1.5. <u>Examples</u>. 1. Let $X = R^1$, the Banach space of real numbers. If
A is a linear operator with domain $D_A = R^1$, range in R^1, then
$Ax = A(x \cdot 1) = x(A1) = cx$, where the constant $c = A1$, and $\|A\| = |c|$. If
we solve $(\lambda I-A)x = y$, we find $x = [1/(\lambda-c)]y$. This, of course, is valid
for all $\lambda \ne c$. Theorem III.1.4 only shows the equation is solvable if
$|\lambda| > |c|$. In that case,

$$y/(\lambda-c) = \sum_{n=0}^{\infty} (c^n/\lambda^{n+1})y.$$

2. A Volterra Equation. This example is so important it deserves to

have its own section number. It is quite vital to what lies ahead. Consider

the (nonlinear) Volterra equation

$$(V) \qquad x(t) = x(\tau) + \int_{\tau}^{t} f(s,x(s))ds,$$

where t, τ lie in an interval I, f is continuous on a rectangle R with

center $(\tau, x(\tau))$ and satisfies

$$|f(s,x)-f(s,y)| < K|x-y|$$

for some constant K and all (s,x), (s,y) in R.

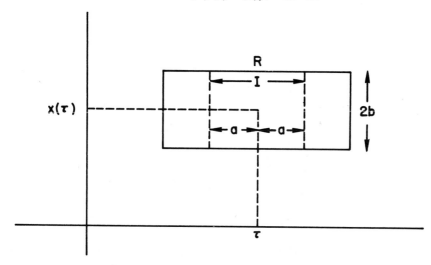

We look for conditions sufficient to guarantee the convergence of the

sequence $\{x_n\}_{n=0}^{\infty}$, where

$$x_0(t) = x(\tau) = \xi,$$

$$x_1(t) = \xi + \int_\tau^t f(s,x_0(s))ds,$$

$$\cdots$$

$$x_{n+1}(t) = \xi + \int_\tau^t f(s,x_n(s))ds,$$

$$\cdots \cdot$$

The procedure we shall follow will be to choose a small enough so that when s, t are in I, the sequence converges uniformly (that is, in the Banach space C[I]).

Let

$$Ax(t) = \int_\tau^t f(s,x(s))ds,$$

where $D_A = \{x; \sup_{t\in I}|x(t)-\xi| \leq b\}.$

If x, y are in D_A, then

$$\|Ax-Ay\| = \sup_{t\in I} \left|\int_\tau^t [f(s,x(s))-f(s,y(s))]ds\right|,$$

$$\leq \sup_{t\in I} \int_\tau^t |f(s,x(s))-f(s,y(s))|ds,$$

$$\leq \sup_{t\in I} \int_\tau^t K|x(s)-y(s)|ds,$$

$$\leq K\|x-y\| \sup_{t\in I} |t-\tau|,$$

$$\leq Ka\|x-y\|.$$

In order to satisfy the hypotheses of Theorem III.1.1, we require $Ka = \rho < 1$
and that $\|Ax_0\|/(1-\rho) < b$. Since

$$\|Ax_0\|/(1-\rho) = \sup_{t \in I} \left| \int_\tau^t f(s,\xi)ds \right| /(1-\rho)$$

$$< Ma/(1-\rho),$$

where $|f(s,t)| < M$ when (s,t) is in R, our requirement will be met if

$$Ma/(1-\rho) = Ma/(1-Ka) < b,$$

or

$$a < b/(M+Kb).$$

Since $b/(M+Kb) \leq 1/K$, we require only

$$a < b/(M+Kb).$$

Theorem III.1.1 is now directly applicable. In the present context it is
as follows:

III.1.6. <u>Theorem</u>. Let $f(\cdot,:)$ be a continuous function defined on a
rectangle R. Further let f satisfy a Lipschitz condition in the second
variable. Then for each (ξ,τ) in R there exists an interval I with

τ at its center such that the equation

$$(V) \qquad x(t) = \xi + \int_{\tau}^{t} f(s,x(s))ds$$

has a unique solution in C[I] for each   t   in   I.

When we discuss systems of equations, we will need a generalization of theorem III.1.6.   The essential ideas are retained.   Only minor changes in notation are needed.

III.1.7.   Definition.   Let $I = [a,b]$, $-\infty < a < b < \infty$,   and let   X   be a Banach space.   By C[I,X] we mean the Banach space of continuous functions with domain   I   and range in   X   with norm

$$\|g\| = \sup_{t \in I} \|g(t)\|_X,$$

where $\|\cdot\|_X$ denotes the norm in   X.

We shall assume that there exists an integral $\int_{\tau}^{t} g(s)ds$ for each element   g   in C[I,X] which satisfies

1. $\left\| \int_{\tau}^{t} g(s)ds \right\|_X \leq \int_{\tau}^{t} \|g(s)\|_X \, ds.$

2. $\int_{\tau}^{t} [g(s)+h(s)]ds = \int_{\tau}^{t} g(s)ds + \int_{\tau}^{t} h(s)ds.$

3. $\int_{\alpha}^{t} \alpha g(s)ds = \alpha \int_{\tau}^{t} g(s)ds,$ where α is a complex (real) number.

It is also possible for elements g in C[I,X] to have derviatives. We mean
in that case that there exists an element g'(t) in C[I,X] such that

$$\lim_{h \to 0} \|g'(t) - [g(t+h) - g(t)]/h\|_X = 0.$$

III.1.8. Theorem. Let X be a Banach space. Let f(·,·) be an X valued,
continuous, bounded function defined on D = I' × S, where S is an open
sphere in X. (That is, the first variable lies in the interval I', the
second in S, which is in X). Further let f satisfy

$$\|f(t,x) - f(t,y)\|_X \leq K\|x-y\|_X$$

for all (t,x), (t,y) in D. Then for each (τ,ξ) in D, there exists
an interval I ⊂ I' with τ at its center such that the equation

$$(V) \qquad x(t) = \xi + \int_\tau^t f(s,x(s))ds,$$

has a unique solution in C[I,X] for each t in I.

The proof is very similar to that of III.1.6. We define the operator A by

$$Ax(t) = \int_\tau^t f(s,x(s))ds$$

with domain $D_A = \{x:x \in C[I,X], \|x-\xi\|_{C[I,X]} < b\}.$

III.1.9. <u>An Example</u>. Let $X = R^n$, n-dimensional space, whose elements

$x = (x_1,\ldots,x_n)$ are normed by $\|x\| = \sum_{i=1}^{n} |x_i|$. Let F be a vector valued

function,

$$F(t,x) = (f_1(t,x_1,\ldots,x_n),\ldots,f_n(t,x_1,\ldots,x_n)),$$

defined on $I' \times R^n$, where I' is an interval of real numbers. Then the
system of integral equations

$$x_1(t) = \xi_1 + \int_{\tau}^{t} f_1(s,x_1(s),\ldots,x_n(s))ds,$$

$$\cdots$$

$$x_n(t) = \xi_n + \int_{\tau}^{t} f_n(s,x_1(s),\ldots,x_n(s))ds$$

has a unique solution, valid over a subinterval $I \subset I'$, for each
$(\tau,(\xi_1,\ldots,\xi_n))$ in $I' \times R^n$.

III.2. <u>Existence and Uniqueness of Solutions for Ordinary Differential</u>
<u>Equations</u>. We apply the results of the previous section to ordinary
differential equations to prove what are called local existence theorems.
The results obtained are then extended to larger regions whenever it is
possible. The key to these theorems is the Lipschitz condition.

III.2.1. <u>Definition</u>. <u>By a first order ordinary differential equation we mean</u>

the following problem. Let D be an open region in the plane R^2, and let

$f(\cdot,\cdot)$ be a real valued function which is continuous on D.   The problem is

to find a continuously differentiable (C^1) function x,   defined on a real

interval I   such that

1. (t,x(t)) is in D, when t is in I.

2. x'(t) = f(t,x(t)), when t is in I.

We use the following abbreviated notation to describe the first order

ordinary differential equation:

$$\text{DE:} \quad x' = f(t,x), \quad (t,x) \in D.$$

If such an interval I   and continuously differentiable function x

exist, then x is a solution of DE on I.

Of course, there may be many or no solutions for a particular differ-

ential equation. Frequently, however, many solutions exist because a con-

stant of integration is introduced. As long as there is some freedom in the

choice of this parameter, an entire family of solutions can be found.   In

order to eliminate this possibility, we specify what is called an initial value

problem.

III.2.2. Definition. By an initial value problem for DE we mean the

following problem. Let D be an open region in the plane R^2, and let

$f(\cdot,\cdot)$ be a real valued function, which is continuous on D. Let (τ,ξ)

be a point in D. The problem is to find a continuously differentiable

(C^1) function x which satisfies DE and also which satisfies $x(\tau) = \xi$.

We use the following abbreviated notation to describe the first order

initial value problem:

$$\text{IVP:} \qquad x' = f(t,x), \quad x(\tau) = \xi.$$

If such a continuously differentiable function x is defined over an

interval I, if τ is in I, $x(\tau) = \xi$, and (τ,ξ) is in D, then x

is a solution of IVP.

In geometric language DE describes a collection of paths, called

trajectories, through D. The slope at each point (t,x) of D is given

by f(t,x). On the other hand IVP specifies that the path should go

through a specific point (τ,ξ).

The following graph uses isoclines to show the solutions of the

DE: $x' = x^2$. The solid line shows the solution to the IVP: $x' = x^2$,

$x(-1) = 1$, which is $x = -1/t$.

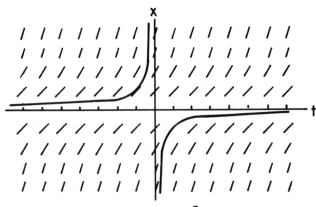

ISOCLINES: $x' = x^2$

We next show that solutions to IVP do exist with some regularity. The theorem we present is associated with the French mathematician Picard.

III.2.3. <u>Theorem (Picard's Existence Theorem).</u> <u>Let D be an open region</u> <u>in the plane R^2, let $f(\cdot,\cdot)$ be a real valued function which is continuous</u> <u>on D, and let f satisfy a Lipschitz condition in the second variable.</u> <u>Then there exists an interval I with τ in I such that</u>

$$\text{IVP: } x' = f(t,x), \quad x(\tau) = \xi$$

<u>has a unique, continuously differentiable (C^1) solution x(t) for all t in</u> <u>I.</u>

Proof. The IVP is equivalent to the integral equation

$$(V) \qquad x(t) = \xi + \int_{\tau}^{t} f(s,x(s))ds.$$

The result follows immediately from Theorem III.1.6.

III.2.4. <u>Corollary.</u> <u>Let D be an open region in the plane R^2, let</u> <u>$f(\cdot,\cdot)$ be a real valued function which is continuous on D and which</u> <u>possesses a continuous partial derivative in the second variable.</u> <u>Then there</u> <u>exists an interval I with τ in I such that</u>

$$\text{IVP: } \qquad x' = f(t,x), \quad x(\tau) = \xi$$

has a unique continuously differentiable (C^1) solution x(t) for all t

in I.

Proof. The mean value theorem states that when the partial derivative with

respect to the second variable exists,

$$f(t,x) - f(t,y) = \int_{y}^{x} \frac{\partial f}{\partial \xi} (t,\xi) d\xi .$$

Since $\frac{\partial f}{\partial \xi}$ is continuous, it is bounded, $\left|\frac{\partial f}{\partial \xi}\right| < M,$ and

$$\left| f(t,x) - f(t,y) \right| < M \left| x-y \right|,$$

Thus f satisfies a Lipschitz condition in the second variable, and the

result follows from Theorem III.2.3.

III.2.5. Examples. 1. We have already briefly glanced at $x' = x^2,$

x(-1) = 1. In this problem $f(t,x) = x^2.$ Since

$$\left| f(t,x) - f(t,y) \right| = \left| x^2 - y^2 \right|,$$

$$= \left| x+y \right| \left| x-y \right|,$$

if x and y are bounded such that $\left| x+y \right| < M,$ then f will satisfy a

Lipschitz condition with Lipschitz constant M. For example since x(-1) = 1,

we might restrict our interval of attention to that when $\left| x \right| < 2.$ Thus in a neighborhood of t = -1, M = 4 is a suitable Lipschitz constant. The

solution x = -1/t is unique near t = -1.

2. The IVP x′ = sin tx, x(0) = 1 has a unique solution, since

$f(t,x) = \sin tx$ is differentiable with respect to x. $\dfrac{\partial f}{\partial x}$ = t cos tx, which

is bounded as long as t is constrained. In this case, however, a formula

for the solution in terms of the elementary functions in not available.

3. The IVP $x′ = x^{1/3}$, x(0) = 0 has many solutions. x(t) = 0 is

one.

$$
x(t) = \begin{cases} 0 & , \quad t < c \\[2ex] (\tfrac{2}{3})^{3/2}(t-c)^{3/2}, & t \geq c \end{cases}
$$

is another whenever c ≥ 0. The lack of uniqueness is due to the failure

of $f(t,y) = x^{1/3}$ to satisfy a Lipschitz condition. In fact,

$|x^{1/3}-y^{1/3}| < K|x-y|$ implies $|x^{1/3}| < K|x|$, when y = 0. If $x = 1/n^3$,

then $(1/n) < K(1/n^3)$, or $1 < (K/n^2)$. If $n \geq K^{1/2}$, this leads to a

contradiction.

We now turn our attention to the extension of these results to systems

of equations.

III.2.6. <u>Definition</u>. <u>Let n be a positive integer. By a system of n</u>

<u>ordinary differential equations of the first order we mean the following</u>

<u>problem. Let D be an open region in n+1 dimensional space R^{n+1}, and let</u>

$f_1(\cdot,\cdot),\ldots,f_n(\cdot,\cdot)$ be real valued functions which are continuous on D, where the first variable in $f_1(\cdot,\cdot),\ldots,f_n(\cdot,\cdot)$ is one dimensional, and the second is n-dimensional. The problem is to find n continuously differentiable (C^1) functions $x_1,\ldots,x_n$, defined on a real interval I such that

1. $(t,x_1(t),\ldots,x_n(t))$ is in D when t is in I.

2. $x_i'(t) = f_i(t,x_1(t),\ldots,x_n(t))$ when t is in I, $i = 1,\ldots,n$.

We use the following abbreviated notation to describe the system of first order ordinary differential equations: In matrix notation, we have

$$\text{DS:} \qquad x' = f(t,x), \quad (t,x) \in D,$$

where x is an n-dimensional vector $(x_1,\ldots,x_n)$, and $f = (f_1,\ldots,f_n)$.

If an interval I and continuously differential functions $x_1,\ldots,x_n$ exist, then $x = (x_1,\ldots,x_n)$ is a solution of DS on I.

As before, precisely how many solutions exist is difficult to say. We would expect to encounter an n-parameter family of solutions because of the n indicated integrations.

III.2.7. Definition. By an initial value problem for DS we mean the following problem. Let D be an open region in n+1 dimensional space R^{n+1}, and let $f_1(\cdot,\cdot),\ldots,f_n(\cdot,\cdot)$ be real valued functions, which are continuous on D, where the first variable in $f_1(\cdot,\cdot),\ldots,f_n(\cdot,\cdot)$ is one dimensional, and the second is n-dimensional. Let $\xi = (\xi_1,\ldots,\xi_n)$, and let (τ,ξ) be a point in D. The problem is to find n continuously differentiable (C^1) functions $x_1,\ldots,x_n$ such that $x = (x_1,\ldots,x_n)$ satisfies DS, and which also satisfies $x(\tau) = \xi$.

We use the following notation to describe the initial value problem for the nth order differential system:

$$\text{IVS:} \qquad x' = f(t,x), \quad x(\tau) = \xi .$$

If such continuously differentiable functions $x_1,\ldots,x_n$ are defined over an interval I, if τ is in I, $x(\tau) = \xi$, and (τ,ξ) is in D, then x is a solution of IVS.

III.2.8. Theorem (Picard's Existence Theorem for Systems). Let D be an open region in n+1 dimensional space R^{n+1}, let $f_1(\cdot,\cdot),\ldots,f_n(\cdot,\cdot)$ be real valued functions which are continuous on D, and let $f_1,\ldots,f_n$ satisfy the Lipschitz condition

$$\sum_{j=1}^{n} |f_j(t,x_1,\ldots,x_n) - f_j(t,y_1,\ldots,y_n)| \le K \sum_{j=1}^{n} |x_j - y_j|$$

for $(t,x_1,\ldots,x_n)$, $(t,y_1,\ldots,y_n)$ in D. Then there exists an interval I with τ in I such that

$$\text{IVS:} \qquad x' = f(t,x), \quad x(\tau) = \xi$$

has a unique, continuously differentiable (C^1) solution $x(t) = (x_1(t), \ldots, x_n(t))$ for all t in I.

Proof. We refer the reader to Theorem III.1.8. The Banach space X referred to therein is in this case R^n.

We note in passing that in Theorem III.2.8 we have used the norm $\|x\| = \sum_{j=1}^{n} |x_j|$. Other norms such as $[\sum_{j=1}^{n} |x_j|^p]^{1/p}$, $p > 1$, or $\sup_j |x_j|$ will also work. The norm with $p = 1$ is merely the most convenient. Of course, they are all equivalent.

The solutions given in Theorems III.2.3 and III.2.8 are _local_ existence theorems. That is, the interval I exists only locally about the point τ. We know nothing about the solution at a distance. Under certain conditions the local solutions can be extended.

III.2.9. Theorem. Let D be an open region in n+1 dimensional space R^{n+1}, let $f_1(\cdot, \cdot), \ldots, f_n(\cdot, \cdot)$ be real valued functions which are continuous on D and which satisfy the Lipschitz condition

$$\sum_{j=1}^{n} |f_j(t, x_1, \ldots, x_n) - f_j(t, y_1, \ldots, y_n)| \leq K \sum_{j=1}^{n} |x_j - y_j|$$

for $(t, x_1, \ldots, x_n)$, $(t, y_1, \ldots, y_n)$ in D. Further let $f_1, \ldots, f_n$ be uniformly bounded.

1. If $x_1, \ldots, x_n$ is a solution of IVS on $I = (a, b)$, then $\lim_{t \to a+} x_j(t)$ exists for $j = 1, \ldots, n$ and $\lim_{t \to b-} x_j(t)$ exists for $j = 1, \ldots, n$.

2. <u>If $(a,x_1(a),\ldots,x_n(a))$ or $(b,x_1(b),\ldots,x_n(b))$ is in D, then</u>
<u>the solution may be continued to the left of a or the right of b.</u>

Proof. Let $x = (x_1,\ldots,x_n)$, and consider

$$x(t) = \xi + \int_\tau^t f(s,x(s))ds,$$

where t is in $I = (a,b)$, and x is in R^n. We shall show that $\lim_{t \to b} x(t)$
exists. Let $\{t_n\}_{n=1}^\infty$ be a (Cauchy) sequence, which approaches b. Then

$$\|x(t_n) - x(t_m)\|_{R^n} = \|\int_{t_m}^{t_n} f(s,x(s))ds\|_{R^n} ,$$

$$\leq \int_{t_m}^{t_n} \|f(s,x(s))\|ds.$$

If $\|f\| < M$ for all points in D, then

$$\|x(t_n) - x(t_m)\| < M|t_n - t_m|.$$

Thus as $\{t_n\}_{n=1}^\infty$ approaches b, $\{x(t_n)\}_{n=1}^\infty$ is also a Cauchy sequence.
Since R^n is a Banach space, $\{x(t_n)\}_{n=1}^\infty$ also converges to a limit we call
$x(b-)$.

We then consider

$$y(t) = x(b-) + \int_b^t f(s,y(s))ds.$$

By Theorem III.2.8 this has a solution in a neighborhood of b, which we
denote by y(t). Let

$$x_{ext}(t) = \begin{cases} x(t), & a < t \leq b, \\ \\ y(t), & b \leq t. \end{cases}$$

Then x_{ext} is a solution of IVS: $x' = f(t,x)$, $x(\tau) = \xi$ in a region which
extends beyond b. In a similar way x can be extended to the left of a.

The proof of the previous theorem is so easy largely because the assumption that f be bounded is so strong. For example, when considering the
IVP: $x' = x^2$, $x(-1) = 1$, the solution $x(t) = -1/t$, and hence $f(t,x) =
1/t^2$, which is not bounded as t approaches 0. This is not obvious at a
glance. On the other hand, IVP's such as $x' = \cos x$, $x(0) = 1$ do have
f's which are bounded, and solutions which are extendable indefinitely. We
emphasize that the assumption that the f's be bounded is more subtle than
might be first expected.

III.3. First Order Linear Systems. Within this section we denote special
attention to linear systems. We show that the results concerning existence,
uniqueness, and extensions are applicable to such systems. Hence in the
future the solutions we encounter will exist uniquely not merely locally,
but throughout the entire range of our interest.

The following theorem is quite widely used in a number of circumstances.
It is best known as Gronwall's inequality.

III.3.1. Theorem (Gronwall's Inequality). Let x, y and z be continuous
functions on I = [τ,b], −∞ < τ < b < ∞. Let z > 0 on I and

$$x(t) \leq y(t) + \int_\tau^t z(s)x(s)\,ds$$

for all t in I. Then

$$x(t) \leq y(t) + \int_\tau^t z(s)y(s)e^{\int_s^t z(\xi)\,d\xi}\,ds.$$

Proof. Let $R(t) = \int_\tau^t z(s)x(s)\,ds$. Then $R(\tau) = 0$, and $R' = zx$. Multiplying
the first inequality by z, we have

$$R' - zR \leq zy\ .$$

We multiply by the positive integrating factor $e^{-\int_\tau^t z(\xi)\,d\xi}$ and integrate
to find

$$R(t) \leq e^{\int_\tau^t z(\xi)\,d\xi} \int_\tau^t e^{-\int_\tau^s z(\xi)\,d\xi} z(s)y(s)\,ds.$$

Thus

$$R(t) \leq \int_\tau^t e^{\int_s^t z(\xi)\,d\xi} z(s)y(s)\,ds.$$

Since $x \leq y+R$, the proof is complete.

We next apply Gronwall's inequality to the differential system DS.
Afterwards we specialize to consider linear systems.

III.3.2. Definition. The vector valued function y in n-dimensional space R^n is an ε-approximate solution for DS: $x' = f(t,x)$, t in I, if

$$\|y'(t) - f(t,y(t))\|_{R^n} < \varepsilon$$

for all t in I.

III.3.3. Theorem. Let y and z be ε_1 and ε_2-approximate solutions of DS respectively, and let f satisfy a Lipschitz condition, such as in Theorem III.2.8, with Lipschitz constant K.   Then

$$\|y(t) - z(t)\|_{R^n} \leq \|y(\tau) - z(\tau)\|_{R^n} e^{K|t-\tau|} + \frac{\varepsilon_1 + \varepsilon_2}{K} [e^{K|t-\tau|} - 1].$$

Proof. By integrating the assumptions,

$$y(t) = y(\tau) + \int_\tau^t f(s,y(s))ds + \int_\tau^t \delta_1(s)ds,$$

$$z(t) = z(\tau) + \int_\tau^t f(s,z(s))ds + \int_\tau^t \delta_2(s)ds,$$

where $\|\delta_1(s)\|_{R^n} < \varepsilon_1$, and $\|\delta_2(s)\|_{R^n} < \varepsilon_2$. Subtracting,

$$\|y(t) - z(t)\|_{R^n} \leq \|y(\tau) - z(\tau)\|_{R^n} + \|\int_\tau^t [f(s,y(s)) - f(s,z(s))]ds\|_{R^n}$$

$$+ \|\int_\tau^t \delta_1(s)ds\|_{R^n} + \|\int_\tau^t \delta_2(s)ds\|_{R^n}.$$

If we let $F(t) = \|y(t) - z(t)\|_{R^n}$ and employ the Lipschitz condition on f, we find

$$F(t) \leq [F(\tau) + (\varepsilon_1 + \varepsilon_2)|t - \tau|] + K \int_{\tau}^{t} F(s)ds.$$

We are now in a position to employ Gronwall's inequality. Doing so yields the desired result.

III.3.4. <u>Definition.</u> <u>Let n be a positive integer. By a linear system of n ordinary differential equations of the first order, we mean the following problem. Let $I = [\alpha,\beta]$ be a closed interval, and let $\{a_{ij}(t)\}_{i,j=1}^{n}$ and $\{b_j(t)\}_{j=1}^{n}$ be real valued continuous functions on I. The problem is to find n continuously differentiable (C^1) functions $x_1,\ldots,x_n$ such that when t is in I</u>

$$x_1'(t) = a_{11}(t)x_1(t)+\ldots+a_{1n}(t)x_n(t)+b_1(t)$$

$$\cdot \quad \cdot \quad \cdot$$

$$x_n'(t) = a_{n1}(t)x_1(t)+\ldots+a_{nn}(t)x_n(t)+b_n(t).$$

<u>In matrix form this can be represented as</u>

$$\text{LDS:} \quad x' = A(t)x + b(t), \quad \underline{t \text{ in } I,}$$

<u>where $x = (x_j)$, $A(t) = (a_{ij}(t))$ and $b(t) = (b_j(t))$.</u>

III.3.5. Definition. By an initial value problem for LDS, we mean the following problem. Let $\{a_{ij}\}_{i,j=1}^{n}$ and $\{b_j(t)\}_{j=1}^{n}$ be real valued continuous functions on an interval $I = [\alpha, \beta]$. Let $\{\xi_j\}_{j=1}^{n}$ be arbitrary constants, and let $\xi = (\xi_j)$. The problem is to find n continuously differentiable functions $x_1, \ldots, x_n$, such that $x = (x_j)$ satisfies LDS, and also satisfies $x(\tau) = \xi$, where τ is a point in I.

This is written in abbreviated form as

$$\text{LIVS:} \qquad x' = A(t)x + b(t), \quad x(\tau) = \xi.$$

III.3.6. Theorem. Let $\{a_{ij}(t)\}_{i,j=1}^{n}$ and $\{b_j(t)\}_{j=1}^{n}$ be continuous on an interval $I = [\alpha, \beta]$. Then for each τ in I there exists a sub-interval I such that $\tau \in I_\tau \subset I$, and LIVS possesses a unique solution on I_τ.

Proof. We have in this case $f(t,x) = A(t)x + b(t)$, using the matrix notation. We now notice that

$$\|f(t,x) - f(t,y)\|_{R^n} = \|\{\sum_{k=1}^{n} a_{jk}(t)(x_k - y_k)\}\|_{R^n},$$

$$= \sum_{j=1}^{n} |\sum_{k=1}^{n} a_{jk}(t)(x_k - y_k)|,$$

$$\leq [\sup_{t \in I} \sum_{j=1}^{n} |a_{jk}(t)|] \sum_{k=1}^{n} |x_k - y_k|,$$

$$= K\|x_k - y_k\|_{R^n}$$

Thus f(t,x) satisfies a Lipschitz condition, and Theorem III.2.8 is directly

applicable.

III.3.7. Theorem. Let $\{a_{ij}(t)\}_{i,j=1}^{n}$ and $\{b_j\}_{j=1}^{n}$ be real and continuous

on the interval $I = [\alpha, \beta]$. Then for each τ in I there exists a unique

solution for LIVS, which is valid throughout all of I.

Proof. Let y be an extension of the solution as given by Theorem III.3.6.

Then by Theorem III.3.3, with z = 0,

$$\|y(t)\|_{R^n} \leq \|y(\tau)\|_{R^n}\, e^{K|t-\tau|} + \frac{\|b(t)\|_{R^n}}{K}\, [e^{K|t-\tau|}-1].$$

Thus for all t in I, $\|y(t)\|_{R^n}$ is bounded. As t approaches any point

within I, the bound remains, and y can be further extended. Thus y can

be extended over all of I.

III.3.8. Corollary. If $\{a_{ij}(t)_{i,j=1}^{n}$ and $\{b_j(t)\}_{j=1}^{n}$ are real and

continuous throughout an interval I, and τ is in I, then the solution

to LIVS: $x' = A(t)x + b(t)$, t in I, satisfies the following inequality.

$$\|x(t)\|_{R^n} \leq \|x(\tau)\|_{R^n}\, e^{K|t-\tau|} + \frac{\|b(t)\|_{R^n}}{K}\, [e^{K|t-\tau|}-1].$$

III.3.9. An Example. Consider the first order equation $x' = ax + 1$,

$x(0) = 2$, where a is constant and positive. The inequality of Theorem

III.3.8 shows that when $t \geq 0$,

$$x \leq (2 - 1/a)e^{at} - 1/a,$$

since $\|x(0)\|_{R^n} = 2,$ $K = a,$ and $\|b(t)\|_{R^n} = 1.$ In this simple case the equation is easily solved. The right side is, in fact, equal to x.

III.4. **n-th Order Differential Equations.** So far we have discussed only first order equations and systems. However, in most physical applications the problems are of second order or higher. We must accordingly discuss such problems in order to be able later to handle the standard equations of mathematical physics.

We first define such problems. We then show that they can be reduced to problems of the first order.

III.4.1. **Definition.** By an n-th order ordinary differential equation we mean the following problem. Let D be an open region in n+1 dimensional space R^{n+1}. Let $f(\cdot,\ldots,\cdot)$ be a real valued function which is continuous on D. The problem is to find an n times continuously differentiable (C^n) function x, defined on a real interval I, such that

1. $(t,x(t),x'(t),\ldots,x^{(n)}(t))$ is in D when t is in I,

2. $x^{(n)}(t) = f(t,x(t),\ldots,x^{(n-1)}(t))$ when t is in I.

We use the following abbreviated notation to describe the n-th order ordinary differential equation:

$$\text{DE(n):} \qquad x^{(n)} = f(t,x,\ldots,x^{(n-1)}), \ (t,x,x',\ldots,x^{(n-1)}) \in D.$$

If such an interval I and an n times continuously differentiable function x exist, then x is a solution of DE(n) on I.

III.4.2. Definition. By an initial value problem for DE(n) we mean the following problem. Let D be an open region in n+1 dimensional space R^{n+1}, and let $f(\cdot,\ldots,\cdot)$ be a real valued function, which is continuous on D. Let $(\tau,\xi_1,\ldots,\xi_n)$ be a point in D. The problem is to find an n times continuously differentiable (c^n) function x which satisfies DE(n), and which also satisfies $x^{(j-1)}(\tau) = \xi_j$, $j = 1,\ldots,n$.

We use the following abbreviated notation to describe the n-th order initial value problem:

$$\text{IVP}(n): \qquad x^{(n)} = f(t,x,\ldots,x^{(n-1)}), x^{(j-1)}(\tau) = \xi_j, \; j = 1,\ldots,n.$$

If such an n times continuously differentiable function x is defined over an interval I, if τ is in I, $x^{(j-1)}(\tau) = \xi_j$, $j = 1,\ldots,n$, and $(\tau,\xi_1,\ldots,\xi_n)$ is in D, then x is a solution of IVP(n).

There are also similar definitions for n-th order vector and n-th order linear systems. We invite the reader to write them out.

We now show that n-th order problems are reducible through matrix notation to first order problems.

III.4.3. Theorem. The n-th order problem IVP(n): $x^{(n)} = f(t,x,\ldots,x^{(n-1)})$, $x^{(j-1)}(\tau) = \xi_j$, $j = 1,\ldots,n$ is reducible to a problem which is in the form of a first order system (IVS).

Proof. We define $x_1 = x$, $x_2 = x'$,...,$x_n = x^{(n-1)}$. Then we have

$$x_1' = x_2,$$

$$x_2' = x_3,$$

$$\ldots$$

$$x_{n-1}' = x_n,$$

$$x_n' = f(t,x_1,x_2,\ldots,x_{n-1}).$$

The initial values take the form

$$x_1(\tau) = \xi_1,$$

$$x_2(\tau) = \xi_2,$$

$$\ldots$$

$$x_n(\tau) = \xi_n.$$

This is clearly equivalent to IVS when $f_1 = x_2$, $f_2 = x_3,\ldots,f_{n-1} = x_n$, and $f_n = f(t,x_1,\ldots,x_{n-1})$.

III.4.4. Theorem. Let D be an open region in n+1 dimensional space R^{n+1}, and let $f(\cdot,\ldots,\cdot)$ be a real valued function which is continuous on D.

Further let f satisfy the Lipschitz condition

$$\left| f(t,x_1,\ldots,x_n) - f(t,y_1,\ldots,y_n) \right| \le K \sum_{j=1}^{n} \left| x_j - y_j \right|$$

for all $(t,x_1,\ldots,x_n)$, $(t,y_1,\ldots,y_n)$ in D. Then the problem

$$\text{IVP}(n): \qquad x^{(n)} = f(t,x,\ldots,x^{(n-1)}), \ x^{(j-1)}(\tau) = \xi_j, \ j = 1,\ldots,n,$$

has a unique solution over some interval I containing τ.

Proof. We apply Theorem III.2.8, where $f_1 = x_1$, $f_2 = x_2,\ldots,f_n = f(t,x_1,\ldots,x_n)$.
The Lipschitz condition, required in Theorem III.2.8, is

$$\left| x_1 - y_1 \right| + \left| x_2 - y_2 \right| + \ldots + \left| f(t,x_1,\ldots,x_n) - f(t,y_1,\ldots,y_n) \right|$$

$$\le \sum_{j=1}^{n} \left| x_j - y_j \right| + K \sum_{j=1}^{n} \left| x_j - y_j \right|,$$

$$= (K+1) \sum_{j=1}^{n} \left| x_j - y_j \right|,$$

and the theorem is immediately applicable.

The preceding theorem is, of course, applicable to n-th order linear equations. Since we can say more, however, the n-th order linear equations is worth separate consideration.

III.4.5. Definition. Let n be a positive integer. By an n-th order linear differential equation we mean the following problem. Let $I = [\alpha,\beta]$

be a closed interval, let $a_0(t),\ldots,a_n(t)$ and $b(t)$ be real valued

continuous functions on I, and let $a_0(t) > 0$ on I. The problem is to

find an n times continuously differentiable (C^n) function x, such that

when t is in I

$$a_0(t)x^{(n)}(t) + \ldots + a_n(t)x(t) = f(t).$$

We denote this problem by LDE(n).

III.4.6. Definition. By an initial value problem for LDE(n), we mean

the following problem. Let $a_0(t),\ldots,a_n(t)$ and $b(t)$ be real valued

continuous functions on an interval $I = [\alpha,\beta]$, let $a_0(t) > 0$ on I, and

let $\{\xi_j\}_{j=1}^n$ be arbitrary constants. The problem is to find an n times

continuously differentiable (C^n) function x such that x satisfies

LDE(n), and which also satisfies $x^{(j-1)}(\tau) = \xi_j$, $j = 1,\ldots,n$, where τ

is a point of I.

We denote the n-th order initial value problem by

$$\text{LIVP}(n): \qquad a_0(t)x^{(n)}(t) + \ldots + a_n(t)x(t) = f(t),$$

$$x^{(j-1)}(\tau) = \xi_j, \; j = 1,\ldots,n.$$

III.4.7. Theorem. The n-th order initial value problem LIVP(n) is

reducible to a problem which is in the form of a first order linear system

(LIVS).

Proof. We define $x_1 = x$, $x_2 = x'$, $\ldots, x_n = x^{(n-1)}$. Then we have

$$x_1' = x_2,$$

$$x_2' = x_3,$$

$$\cdots$$

$$x_{n-1}' = x_n,$$

$$x_n' = -(1/a_0)(a_n x_1 + \cdots + a_1 x_n) + (1/a_0)b.$$

The initial values take the form

$$x_1(\tau) = \xi_1,$$

$$x_2(\tau) = \xi_2,$$

$$\cdots$$

$$x_n(\tau) = \xi_n.$$

If we employ matrix notation, these become

$$
\begin{pmatrix} x_1 \\ x_2 \\ \cdots \\ x_n \end{pmatrix}'
=
\begin{pmatrix}
0 & 1 & \cdots & \\
0 & 0 & 1 & \\
\cdots & & & \\
-\dfrac{a_n}{a_0} & -\dfrac{a_{n-1}}{a_0} & \cdots & -\dfrac{a_1}{a_0}
\end{pmatrix}
\begin{pmatrix} x_1 \\ x_2 \\ \cdots \\ x_n \end{pmatrix}
+
\begin{pmatrix} 0 \\ 0 \\ \cdots \\ \dfrac{b}{a_0} \end{pmatrix} ,
$$

$$\begin{pmatrix} x_1(\tau) \\ \cdot \\ \cdot \\ \cdot \\ x_n(\tau) \end{pmatrix} = \begin{pmatrix} \xi_1 \\ \cdot \\ \cdot \\ \cdot \\ \xi_n \end{pmatrix},$$

which is in the form of LIVS.

III.4.8. Theorem. Let $a_0(t),\dots,a_n(t)$ and $b(t)$ be continuous on an interval $I = [\alpha,\beta]$, let $a_0(t) > 0$ on I, and let $\{\xi_j\}_{j=1}^{n}$ be arbitrary constants. Then the problem LIVP(n) has a unique solution which is valid throughout all of I.

We refer to Theorem III.3.7 for the proof.

III.5. Some Extensions. Sometimes it is convenient to eliminate the independent variable t from the function $f(t,x)$ in IVP or IVS, or, for that matter, the other initial value problems. At other times a parameter appears, and it is useful to know the existence theorems still apply. Yet another situation occurs when the dependent variables in question are complex in nature. We will now show how each of these may be reduced to a problem previously considered.

III.5.1. Theorem. The problem IVS: $x' = f(t,x)$, $x(\tau) = \xi$, may be reduced to a similar IVS in which the variable t does not appear in the function f.

Proof. IVS is of the form $x_j' = f_j(t,x_1,\dots,x_n)$, $x_j(\tau) = \xi_j$, $j = 1,\dots,n$. We define a new vector system as follows. We let $y_1 = t$, $y_{j+1} = x_j$, j = 1,...,n.   Then

$$y_1' = 1,$$

$$y_2' = f_1(y_1,\ldots,y_{n+1}),$$

$$\ldots$$

$$y_{n+1}' = f_n(y_1,\ldots,y_{n+1}).$$

If $y = (y_j)$, $F_1 = 1$, $F_{j+1} = f_j$, $j = 1,\ldots,n$,   and $F = (F_j)$,   then $y' = F(y)$. The initial value for y is given by $y_1(\tau) = \xi$, $y_{j+1}(\tau) = \xi_j$, $j = 1,\ldots,n$.

Clearly F is continuous and Lipschitzian if and only if f is.   So the existence and uniqueness theorems may be easily applied.

III.5.2.   <u>Theorem</u>.   <u>Let $f(t,x)$ of IVS be continuously dependent upon a parameter λ.   Thus $f(t,x) = f(t,x,\lambda)$.   Then the initial value problem</u>

$$x' = f(t,x,\lambda),\quad x(\tau) = \xi$$

<u>is reducible to one of the form of IVS involving no parameter.</u>

Proof.   We have $x_j' = f_j(t,x_1,\ldots,x_n,\lambda)$, $x_j(\tau) = \xi_j$, $j = 1,\ldots,n$.   We define a new vector system by letting $y_j = x_j$, $j = 1,\ldots,n$,   and $y_{n+1} = \lambda$.   Our system then has the form

$$y_1' = f_1(t,y_1,\ldots,y_{n+1}),$$

...

$$y_n' = f_n(t,y_1,\ldots,y_{n+1}),$$

$$y_{n+1}' = 0,$$

with initial values $y_j(\tau) = \xi_j$, $j = 1,\ldots,n$, $y_{n+1}(\tau) = \lambda$.

In this case Picard's existence theorem is applicable if f satisfies a Lipschitz condition with respect to $x_1,\ldots,x_n$ and λ.

Finally we indicate what happens when the functions involved are complex valued.

III.5.3. Theorem. Consider the problem IVP: $x' = f(t,x)$, $x(\tau) = \xi$, where x, f and ξ are complex valued. IVP is reducible to an IVS with real valued elements.

Proof. Let $x = y + iz$, $f = g + ih$, and $\xi = \mu + i\nu$. Then IVP is equivalent to

$$y' = g(t,y,z),$$

$$z' = h(t,y,z),$$

...

$$y(\tau) = \mu,$$

$$z(\tau) = \nu,$$

which is a 2-dimensional IVS.

By using the same technique, vector systems with complex elements may be reduced to systems with real elements.

By using various combinations of these reductions a great variety of problems may be similarly reduced.

III.   Exercises

1. Prove Theorem III.1.3.

2. In the proof of Theorem III.1.4 we indirectly used the inequality

$$\| \sum_{n=0}^{\infty} \|x_n\| \le \sum_{n=0}^{\infty} \|x_n\|,$$

where $\{x_n\}_{n=1}^{\infty}$ are elements in a Banach space. Prove that this inequality is valid.

3. Give a direct proof of Theorem III.1.5. Compare it with the proof of Theorem III.1.1.

4. Give some additional examples where Theorem III.1.8 is applicable. In particular, in example III.1.9 how vital is the particular norm which is used there?

5. Consider the IVP: $x' = |x|^a$, $x(0) = 0$. Show that it has at least two solutions when $0 < a < 1$, and exactly one solution when $a = 0$ or $a = 1$. Why is this so?

6. Consider the IVP: $x' = x^2$, $x(-1) = 1$. Calculate the first few approximations $x_0, x_1, \ldots$ as illustrated in Example 2 of Theorem III.1.

7. Find some additional examples if IVP's which (1) have unique solutions, (2) do not have unique solutions.

8. State a corollary to Theorem III.2.8 similar to that of Corollary III.2.

9. Solve the system

$$\begin{pmatrix} x_1 \\ x_2 \end{pmatrix}' = \begin{pmatrix} 1 & 1 \\ 1 & 1 \end{pmatrix} \begin{pmatrix} x_1 \\ x_2 \end{pmatrix}, \quad \begin{pmatrix} x_1(0) \\ x_2(0) \end{pmatrix} = \begin{pmatrix} 1 \\ 1 \end{pmatrix}$$

by successive approximations. Compute the first three approximations. Compute the estimates as given by Theorem III.3.8.

10. Specify definitions for problems for n-th order ordinary differential systems and n-th order initial value systems.

11. Show that an n-th order linear system can be reduced to a first order linear system.

12. Write out precisely a Lipschitz condition necessary for Picard's existence theorem to be applicable to $x' = f(t,x,\lambda)$, $x(\tau) = \xi$.

References

1. R. Bellman, "Stability Theory of Differential Equations," McGraw-Hill, New York, 1953.

2. G. Birkhoff and G.C. Rota, "Ordinary Differential Equations," Blaisdell, Waltham, Mass., 1959.

3. E.A. Coddington and N. Levinson, "Theory of Ordinary Differential Equations," McGraw-Hill, New York, 1955.

4. R.H. Cole, "Theory of Ordinary Differential Equations," Appleton-Century-Crofts, New York, 1968.

5. E. Hewitt and K. Stromberg, "Real and Abstract Analysis," Springer-Verlag, New York, 1965.

6. W. Hurewicz, "Lectures on Ordinary Differential Equations," M.I.T. Press, Cambridge, Mass., 1958.

7. E.L. Ince, "Ordinary Differential Equations," Dover Publications, New York, 1956.

8. A.E. Taylor, "Introduction to Functional Analysis," John Wiley and Sons, New York, 1958.

IV. LINEAR ORDINARY DIFFERENTIAL EQUATIONS

This is the first of two chapters concerning linear systems and equations.
We have already established the unique existence of solutions to initial value
problems concerning them. However, because of their linear nature we can
say much more by using the framework of the linear space.

We shall begin by considering first order systems. Next we shall
consider the n-th order equation.   Lastly we shall restrict ourselves to
the important case of constant coefficients, both for first order systems and
for the n-th order equation. We make free use of matrix notation and matrix
manipulation throughout. We point out necessary facts as they are needed.

IV.1. <u>First Order Linear Systems.</u> We begin by considering the vector system

$$\text{LDS:} \quad x' = A(t)x + b(t),$$

where $A(t) = (a_{ij}(t))$ and $b(t) = (b_j(t))$ are, respectively, $n \times n$ and
$n \times 1$ matrices with complex valued, continuous elements, defined over an
interval $I = [\alpha, \beta]$.

We associate with LDS the operator L, where

$$Lx = x' - A(t)x$$

for all vectors x in $C^1(I)$, the linear space of continuously differentiable
vectors, defined on I.

We shall first consider homogeneous systems.

IV.1.1. Definition. The system LDS is said to be homogeneous if $b(t) \equiv 0$.
LDS then has the form $Lx = 0$, or $x' - A(t)x = 0$.

We note that $x \equiv 0$ is a solution of the homogeneous system. It is
given the name of the null or trivial solution. Conversely, if a given
solution vanishes at some point τ in I, then the uniqueness part of
Picard's Theorem guarantees that the solution is the null solution.

We next show that the solutions of a homogeneous LDS have many properties
in common with linear spaces. In fact, they form a linear space of vector
valued functions.

IV.1.2. Theorem. The set of all solutions of the linear homogeneous system
$Lx = 0$ forms an n-dimensional linear manifold in $C[I]$.

Proof. Let $\xi_1,\ldots,\xi_n$ be a basis for R^n (real n-dimensional space) or
K^n (complex n-dimensional space) depending upon the circumstances. Let
$x_1,\ldots,x_n$ be solutions of $Lx = 0$ such that $x_j(\tau) = \xi_j$, $j = 1,\ldots,n$, for
some point τ in I. We shall show that $\{x_j\}_{j=1}^n$ is a linearly independent
set and spans the space of solutions.

Suppose there exist coefficients $\alpha_1,\ldots,\alpha_n$ such that $\sum\limits_{j=1}^n \alpha_j x_j = \theta$,
the zero element in R^n or K^n. Then, letting $t = \tau$, $\sum\limits_{j=1}^n \alpha_j \xi_j = \theta$. Since
$\{\xi_j\}_{j=1}^n$ are linearly independent, this implies that $\alpha_j = 0$, $j = 1,\ldots,n$.
Thus $\{x_j\}_{j=1}^n$ are linearly independent.

Now let x be an arbitrary solution of $Lx = 0$. Let $\xi = x(\tau)$. Then since ξ is in R^n or K^n, there exist coefficients $\alpha_1, \ldots, \alpha_n$ such that $\xi = \sum_{j=1}^{n} \alpha_j \xi_j$. Then $x - \sum_{j=1}^{n} \alpha_j x_j$ is a solution to $Lx = 0$ and equals θ at $t = \tau$. From the remarks made previously, we see that $x - \sum_{j=1}^{n} \alpha_j x_j = \theta$ for all t, and $x = \sum_{j=1}^{n} \alpha_j x_j$.

We remark that the previous theorem does not hold for nonlinear equations. For a general nonlinear equation, knowledge about any group of solutions is usually not too much help in finding others.

IV.2. <u>Fundamental Matrices</u>. Before we can seriously discuss fundamental matrices, by means of which all the solutions for homogeneous problems can be found, we need to know something about differentiating matrices themselves.

Let $X = L(K^n, K^n)$, the space of $n \times n$ (complex) matrices. By an elementary inspection of their make-up it is immediately apparent that X has n^2-dimensions. For each $A = (a_{ij})$ in X, we define the norm $|\cdot|$ by

$$|A| = \sum_{i=1}^{n} \sum_{j=1}^{n} |a_{ij}|.$$

With this norm X is easily shown to be a Banach space. In fact, since

$$(a_{ij})(b_{ij}) = (\sum_{k=1}^{n} a_{ik} b_{kj}),$$

an easy calculation shows

$$|AB| \leq |A||B|,$$

so X is a Banach algebra. Finally if A is in X, and x is in K^n,

then

$$\|Ax\|_{K^n} \leq |A| \|x\|_{K^n} ,$$

and

$$\|A\|_{op} \leq |A| \quad (\text{see II.4.3.}).$$

Now let A = A(t) be a function with domain an interval I and range

in X. If for a point τ in I there exists a matrix, denoted by $A'(\tau)$,

such that

$$\lim_{t \to \tau} |\{[A(t) - A(\tau)]/(t-\tau)\} - A'(\tau)| = 0,$$

we say that A is differentiable at τ with derivative $A'(\tau)$. If A is

differentiable for each τ in I, then it is differentiable on I.

IV.2.1. Definition. 1. Let $\{x_i\}_{i=1}^n$ be a set of n linearly independent

solutions of Lx = 0 on an interval I. Then $\{x_i\}_{i=1}^n$ is called a funda-

mental set of solutions.

2. Let X(t) be an n × n matrix valued function, defined over an

interval I, whose columns are linearly independent solutions of Lx = 0.

Then X(t) is called a fundamental matrix for Lx = 0 on I.

3. Let $\phi(t)$ be a $C^1(I)$ matrix valued function such that $\phi' = A(t)\phi$.

Then ϕ is called a solution matrix for Lx = 0 on I.

Of course, the fundamental matrix X equals $(x_1,\ldots,x_n)$, the matrix formed from the adjoining one after the other the column vectors x_i, $i = 1,\ldots$ in the fundamental set. We shall see shortly that the order in which $x_1,\ldots,$ appear is unimportant.

IV.2.2. Theorem. Let $X(t)$ be a fundamental matrix for $Lx = 0$. Then $X(t)$ is also a solution matrix.

Proof. Let $X = (x_1,\ldots,x_n)$. Then the j-th column of x is x_j. Thus

$$X' = (x_1',\ldots,x_n'),$$

$$= (Ax_1,\ldots,Ax_n),$$

$$= A(x_1,\ldots,x_n),$$

$$= Ax.$$

Let us again briefly diverge to state some necessary facts concerning matrices and determinants. Let $A = (a_{ij})$ and $B = (b_{ij})$ be $n \times n$ matrices Then

1. The trace of A, tr A, is $\sum_{j=1}^{n} a_{jj}$, the sum along the main diagonal. It is possible to show that $\text{tr}(AB) = \text{tr}(BA)$.

2. The determinant of A is by definition

$$\det A = \sum_{j_1=1}^{n} \cdots \sum_{j_n=1}^{n} \varepsilon\,(j_1,\ldots,j_n) a_{1j_1} \cdots a_{nj_n},$$

where

$$\varepsilon(j_1,\ldots,j_n) = \begin{cases} 1 & \text{if } (j_1,\ldots,j_n) \text{ is a cyclic permutation of } (1,\ldots, \\ 0 & \text{if } j_\alpha = j_\beta, \quad \alpha \neq \beta, \\ -1 & \text{if } (j_1,\ldots,j_n) \text{ is an acyclic permutation of } (1,\ldots \end{cases}$$

3. $\det(AB) = \det A \det B$.

4. $\det A = \det A^t$, or $\det(a_{ij}) = \det(a_{ji})$.

5. $\det I = 1$.

6. $\det A \neq 0$ if and only if A^{-1} exists.

7. Let $A = (A_1,\ldots,A_n)$, where $A_1,\ldots,A_n$ are the columns of A and are in K^n, n-dimensional complex space, then

$$\det(A_1,\ldots,A_j,\ldots,A_k,\ldots,A_n) = -\det(A_1,\ldots,A_k,\ldots,A_j,\ldots,A_n).$$

Thus if two columns (or rows) are equal, $\det A = 0$.

8. $\det(A_1,\ldots,A_j+B,\ldots,A_n) =$

$$\det(A_1,\ldots,A_j,\ldots,A_n) + \det(A_1,\ldots,B,\ldots,A_n).$$

9. $\det(A_1,\ldots,\alpha A_j,\ldots,A_n) = \alpha \det(A_1,\ldots,A_n)$, when α is a complex (real) number.

10. $\det(A_1,\ldots,A_n) = 0$ if and only if $A_1,\ldots,A_n$ are linearly dependent.

11. $\frac{d}{dt} \det(A_1(t),\ldots,A_n(t)) = \det(A_1',A_2,\ldots,A_n) +$

$$+ \det(A_1,A_2',\ldots,A_n) + \ldots + \det(A_1,A_2,\ldots,A_n').$$

IV.2.3. Definition. Let X(t) be a solution matrix for Lx = 0. Then

det X(t) is called the Wronskian of X(t).

IV.2.4. Theorem. Let X(t) be a solution matrix for Lx = 0. Then the

Wronskian of X(t) is either identically zero or never zero. If τ is

in I, then

$$\det X(t) = [\det X(\tau)] \exp \int_{\tau}^{t} \mathrm{tr}A(s)ds.$$

Proof. Let $X(t) = (x_{ij}(t))$. Then

$$[\det X(t)]' = \begin{vmatrix} x'_{11} & \cdots & x'_{1n} \\ \cdot & \cdot & \cdot \\ x_{n1} & \cdots & x_{nn} \end{vmatrix} + \ldots + \begin{vmatrix} x_{11} & \cdots & x_{1n} \\ \cdot & \cdot & \cdot \\ x'_{n1} & & x'_{nn} \end{vmatrix},$$

where each determinant on the right only one row is differentiated. Since

$x'_{ij} = \sum_{k=1}^{n} a_{ik}x_{kj}$, we find after substitution,

$$[\det X(t)]' = \sum_{k=1}^{n} a_{1k} \begin{vmatrix} x_{k1} & \cdots & x_{kn} \\ \cdot & \cdot & \cdot \\ x_{n1} & \cdots & x_{nn} \end{vmatrix} + \ldots + \sum_{k=1}^{n} a_{nk} \begin{vmatrix} x_{11} & \cdots & x_{1n} \\ \cdot & \cdot & \cdot \\ x_{k1} & \cdots & x_{kn} \end{vmatrix},$$

$$= \sum_{k=1}^{n} a_{kk} \begin{vmatrix} x_{11} & \cdots & x_{1n} \\ \cdot & \cdot & \cdot \\ x_{n1} & \cdots & x_{nn} \end{vmatrix},$$

Thus

$$[\det X(t)]' = \text{tr}A(t)[\det X(t)].$$

Solving this elementary differential equation completes the derivation of the formula. From it the first statement is obvious.

IV.2.5. <u>Corollary</u>. 1. <u>If the Wronskian of a solution matrix is nonzero</u> <u>at any point of an interval I, then it is nonzero at every point of I.</u>

2. <u>If the Wronskian of a solution matrix is zero at any point of an</u> <u>interval I, then it is zero at every point of I.</u>

3. <u>A necessary and sufficient condition that a solution matrix be a</u> <u>fundamental matrix is that its Wronskian be nonzero at one point of the</u> <u>interval I.</u>

4. <u>If $X(\tau)^{-1}$ exists for some point τ in I, and $X(t)$ is a</u> <u>solution matrix, then $X(t)^{-1}$ exists for all t in I.</u>

Before we proceed to the next theorem, let us recall that matrix differential calculus is similar to ordinary differential calculus with the exception that the order in which terms appear is important. This is so, because the commutative law of multiplication fails to hold for matrices. Thus we have

$$(AB)' = A'B + AB',$$

$$(A^{-1})' = -A^{-1}A'A^{-1},$$

when A and B are differentiable matrices, and in the second case when
A is invertible.

We conclude this section with a statement that it really doesn't matter
which fundamental matrix one works with. They are all equivalent.

IV.2.6. Theorem. Let X(t) be a fundamental matrix for Lx = 0, and let
C be a constant matrix. Then X(t)C is a solution matrix. If C is nonsingular, then X(t)C is a fundamental matrix. Conversely, if X and Y
are fundamental matrices for Lx = 0, then there exists a nonsingular constant matrix C such that Y = XC.

Proof. Since $X' = AX$, $(XC)' = X'C = A(XC)$. Since $\det(XC) = \det X \det C$,
XC is nonsingular if and only if C is nonsingular. To show the converse
we consider $X^{-1}Y$.

$$(X^{-1}Y)' = (X^{-1})'Y + X^{-1}Y',$$

$$= -X^{-1}X'X^{-1}Y + X^{-1}Y',$$

$$= -X^{-1}AXX^{-1}Y + X^{-1}AY,$$

$$= 0.$$

Thus $X^{-1}Y = C$ a constant and $Y = XC$.

IV.3. Nonhomogeneous Systems.

IV.3.1. Definition. The system LDS is said to be nonhomogeneous if b(t) is not identically zero. LDS then has the form $Lx = b(t)$, or $x' - A(t)x = b(t)$.

It is a rather remarkable fact that if the solutions to the homogeneous system are known so that a fundamental matrix can be formed, then the solution to the nonhomogeneous system can be given in closed form. The technique is called variation of parameters.

IV.3.2. Theorem (Variation of Parameters). Let X be fundamental matrix for $Lx = 0$ satisfying $X(\tau) = I$, the identity matrix, for some point τ in $[\alpha, \beta]$. Then the unique solution to the initial value problem

$$x' = A(t)x + b(t), \quad x(\tau) = \xi$$

is given by

$$x(t) = X(t)\xi + X(t)\int_{\tau}^{t} X(s)^{-1}b(s)ds.$$

Proof. We can verify directly that the formula given works. It is more instructive, however to derive it. Let us assume that $x = Xy$, where y is an unknown vector valued function. Assuming we can differentiate,

$$x' = Xy' + X'y.$$

Substituting for x' and X', we have

$$Xy' + A(t)Xy = A(t)Xy + b(t),$$

or

$$y' = X^{-1}b(t).$$

This can be integrated from τ to t. We use $y(\tau) = \xi$ to find

$$y(t) = \xi + \int_{\tau}^{t} X(s)^{-1}b(s)ds,$$

and

$$x(t) = X(t)\xi + X(t) \int_{\tau}^{t} X(s)^{-1}b(s)ds.$$

This knowledge of a fundamental matrix enables us to solve not only the homogeneous initial value problem (by letting $b(t) \equiv 0$), but the non-homogeneous problem as well.

This completes the theoretical solution of linear differential systems. We next turn our attention to n-th order one dimensional problems. We finall consider both situations when the coefficients are constants.

IV.4. **n-th Order Equations.** We consider the n-th order equation

$$\text{LDE}(n): \qquad A_0(t)x^{(n)} + \ldots + a_n(t)x = b(t),$$

where $a_0(t), \ldots, a_n(t)$ and $b(t)$ are continuous and complex valued over an interval I, and $a_0(t) > 0$ on I.

We associate with LDS(n) the operator L_n, where

$$L_n x = a_0(t)x^{(n)} + \ldots + a_n(t)x$$

for all functions x in $C^n(I)$, the linear space of n times continuously differentiable functions, defined on I.

As in the case of systems, we shall first consider homogeneous equations.

IV.4.1. **Definition.** **The equation LDE(n) is said to be homogeneous if**
$b(t) \equiv 0$. LDE(n) then has the form $L_n x = 0$, or $a_0(t)x^{(n)} + \ldots + a_n(t)x = 0$.

As with the first order linear system, $x \equiv 0$ is a solution of LDE(n), and is called the null or trivial solution. Conversely if any solution, together with its first n-1 derivatives, vanishes at any point τ in I, then the uniqueness part of Picard's Theorem guarantees that that solution is the null solution.

IV.4.2. **Theorem.** **The set of all solutions of the linear homogeneous equation**
$L_n x = 0$ forms an n-dimensional linear manifold in $C^n[I]$.

Proof. Let x_j be that solution of $L_n x = 0$ satisfying $x_j^{(i-1)}(\tau) = \delta_{ij}$, where $\delta_{ij} = \begin{cases} 0 & \text{if } i \neq j \\ 1 & \text{if } i = j \end{cases}$ is the Kronecker delta function, and τ is a point of I. Let x be any solution, and let $x^{(i-1)}(\tau) = \xi_i$. Then consider the function

$$y = x - \sum_{j=1}^{n} \xi_j x_j .$$

For any i, $i = 1,\ldots,n,$

$$y^{(i-1)}(\tau) = x^{(i-1)}(\tau) - \sum_{j=1}^{n} \xi_j x_j^{(i-1)}(\tau),$$

$$= \xi_i - \sum_{j=1}^{n} \xi_j \delta_{ij},$$

$$= 0.$$

Clearly y is a solution of $L_n x = 0$. Since it and its first $n-1$ derivatives vanish at τ in I, y is identically zero. Thus $\{x_j\}_{j=1}^{n}$ forms a basis for the solutions. The space of solutions is an n-dimensional linear manifold.

We are again faced with the problem of determining whether a given set of solutions spans the space of solutions. To do so, we reintroduce the concept of a Wronskian.

IV.4.3. Definition. Let $x_1, \ldots, x_n$ be a set of solutions for $L_n x = 0$. Then the determinant

$$W[x_1, \ldots, x_n] = \begin{vmatrix} x_1 & \cdots & x_n \\ x_1' & \cdots & x_n' \\ \cdot & \cdot & \cdot \\ x_1^{(n-1)} & \cdots & x_n^{(n-1)} \end{vmatrix}$$

is called the Wronskian of $x_1, \ldots, x_n$.

IV.4.4. Theorem. Let $x_1, \ldots, x_n$ be solutions of $L_n x = 0$. Then the Wronskian $W[x_1, \ldots, x_n]$ is either identically zero or never zero. If τ is in I, then

$$W[x_1,\ldots x_n](t) = W[x_1,\ldots x_n](\tau)\exp[-\int_{\tau}^{t}[a_1(s)/a_0(s)]ds].$$

Proof. We recall from the previous chapter that $L_nx = 0$ is equivalent to

the vector system $\hat{L}\hat{x} = 0$, where $\hat{L}\hat{x} = \hat{x}' - \hat{A}(t)\hat{x}$,

$$\hat{A} = \begin{pmatrix} 0 & 1 & & 0 \\ 0 & 0 & 1 & 0 \\ \cdot & \cdots & \cdots & \ddots \\ 0 & 0 & \cdots & 1 \\ \dfrac{a_n}{a_0} & -\dfrac{a_{n-1}}{a_0} & \cdots & -\dfrac{a_1}{a_0} \end{pmatrix}, \qquad \hat{x} = \begin{pmatrix} x \\ x' \\ \cdot \\ \cdot \\ \cdot \\ x^{(n-1)} \end{pmatrix},$$

By making the appropriate substitutions in the formula of Theorem IV.2.4,

the result follows immediately.

IV.4.5. Corollary. 1. If the Wronskian of a set of solutions is nonzero at
any point of an interval I, then it is nonzero at every point of I.

2. If the Wronskian of a set of solutions is zero at any point of an
interval I, then it is zero at every point of I.

3. A necessary and sufficient condition that a set of n solutions be
a basis for the space of solutions for $L_nx = 0$ is that its Wronskian be
nonzero at one point of I. Such a set is called a fundamental set.

Part 3 is not necessarily true if the elements in question are not solutions of an appropriate ordinary differential equation. For example, let ϕ and ψ represent two continuously differentiable functions, defined over an interval I, but such that $\phi \cdot \psi \equiv 0$ on I. This will occur when one of them is always 0, but neither is identically 0. Then $W[\phi, \psi] = \phi\psi' - \psi\phi' \equiv 0$. It is clear, however, they are independent. What is needed, of course, is a formula such as that in Theorem IV.4.4. It is missing here.

We conclude this section with a theorem which is rather remarkable, but which is probably more useful to instructors making out examinations than to others.

IV.4.6. Theorem. Let $x_1, \ldots, x_n$ be in $C^n(I)$ and let the Wronskian $W[x_1, \ldots, x_n]$ be nonzero for all t in I. Then there exists a unique homogeneous linear differential equation of order n, $L_n x = 0$, such that $x_1, \ldots, x_n$ forms a fundamental set of solutions (that is, such that $x_1, \ldots, x_n$ forms a basis for the space of solutions) and such that the coefficient of the highest derivative is 1. It is

$$(-1)^n W[x, x_1, \ldots, x_n] / W[x_1, \ldots, x_n] = 0.$$

This is easily proved by inspection.

IV.5. Nonhomogeneous n-th Order Equations. Before we can discuss the solution to the nonhomogeneous n-th order equation, we need to know a bit about the formula for inverting a matrix, since we intend to use the formula for first order linear systems, which employed the inverse of the fundamental matrix.

Let $A = (a_{ij})$ be an $n \times n$ nonsingular matrix. Let (c_{ij}) denote the $n-1 \times n-1$ matrix obtained from A by deleting row i and column j, and let $A_{ij} = (-1)^{i+j} \det (c_{ij})$. A_{ij} is the ij-th cofactor of A. Then, when $\det A \neq 0$,

$$A^{-1} = [1/\det A](A_{ji}).$$

As an example let

$$A = \begin{pmatrix} a_{11} & a_{12} \\ a_{21} & a_{22} \end{pmatrix}$$

be nonsingular. Then

$$A^{-1} = [1/(a_{11}a_{22} - a_{12}a_{21})] \begin{pmatrix} a_{22} & -a_{12} \\ -a_{21} & a_{11} \end{pmatrix}$$

We are now in a position to solve the nonhomogeneous n-th order initial value problem. The result depends upon the same technique and for first order systems, and is again called variation of parameters.

IV.5.1. Theorem (Variation of Parameters). Let $x_1,\ldots,x_n$ be a fundamental set of solutions for $L_n x = 0$, over the interval $I = [\alpha,\beta]$. Let x_0 be the solution to the homogeneous initial value problem $L_n x = 0$, $x^{(j-1)}(\tau) = \xi_j$, $j = 1,\ldots,n$, where τ is in I. Then the solution to the initial value problem

LIVP(n): $a_0(t)x^{(n)} + \ldots + a_n(t)x = b(t)$, $a_0(t) > 0$,

$$x^{(j-1)}(\tau) = \xi_j, \quad j = 1,\ldots,n,$$

is given by

$$x(t) = x_0(t) + \sum_{j=1}^{n} x_j(t) \int_\tau^t \frac{W_j[x_1,\ldots,x_n](s)b(s)}{W[x_1,\ldots,x_n](s)a_0(s)} \, ds,$$

where $(W_j[x_1,\ldots,x_n]$ is the nj-th cofactor of the matrix $\hat{x} = (x_j^{(i-1)})$.

Proof.   We recall that the n-th order initial value problem is equivalent
to the first order system

$$\hat{L}\hat{x} = \hat{x}' - \hat{A}(t)\hat{x} = \hat{b}(t),$$

$$\hat{x}(\tau) = \hat{\xi},$$

where

$$\hat{x} = \begin{pmatrix} x \\ x' \\ \cdot \\ \cdot \\ \cdot \\ x^{(n-1)} \end{pmatrix}, \quad \hat{A} = \begin{pmatrix} 0 & 1 & 0 & \cdot\cdot \\ 0 & 0 & 1 & \cdot\cdot \\ & & & \cdot \\ & & & 1 \\ -\dfrac{a_1}{a_0} & & -\dfrac{a_n}{a_0} \end{pmatrix}, \quad \hat{b} = \begin{pmatrix} 0 \\ 0 \\ \cdot \\ \cdot \\ \dfrac{b}{a_0} \end{pmatrix}, \quad \hat{\xi} = \begin{pmatrix} \xi_1 \\ \xi_2 \\ \cdot \\ \cdot \\ \xi_n \end{pmatrix}.$$

The system has as its solution

$$
\begin{pmatrix} x_1 \\ x_2 \\ \vdots \\ x_n \end{pmatrix} = \begin{pmatrix} x_0 \\ x_0' \\ \vdots \\ x_0^{(n-1)} \end{pmatrix} + \begin{pmatrix} x_1 & \cdots & x_n \\ x_1' & & x_n' \\ \vdots & & \vdots \\ x_1^{(n-1)} & \cdots & x_n^{(n-1)} \end{pmatrix} \int_\tau^t \frac{\begin{pmatrix} A_{11} & \cdots & A_{n1} \\ \cdot & & \\ \cdot & & \\ \cdot & & \\ A_{1n} & \cdots & A_{nn} \end{pmatrix} \begin{pmatrix} 0 \\ 0 \\ \vdots \\ \frac{b(s)}{a_0(s)} \end{pmatrix}}{W[x_1,\ldots,x_n](s)} \, ds
$$

where A_{ij} is the ij-th cofactor of $\hat{X}$. The first component is all that is
needed.

IV.5.2. <u>Corollary</u>. <u>Let $x_1,\ldots,x_n$ be a fundamental set of solutions for</u>
<u>$L_n x = 0$ over the interval $I = [\alpha,\beta]$. Let x_0 be the solution to the</u>
<u>homogeneous initial value problem $L_n x = 0$, $x^{(j-1)}(\tau) = \xi_j$, $j = 1,\ldots,n$,</u>
<u>where τ is in I. Then the solution to the initial value problem</u>

$$\text{LIVP}(n): \qquad a_0(t)x^{(n)} + \ldots + a_n(t)x = b(t), \quad a_0(t) > 0,$$

$$x^{(j-1)}(\tau) = \xi_j, \quad j = 1,\ldots,n,$$

<u>is also given by</u>

$$
x(t) = x_0(t) + \int_\tau^t \frac{\det \begin{vmatrix} x_1(s) & \cdots & x_n(s) \\ x_1'(s) & & x_n'(s) \\ \cdots & & \cdots \\ x_1^{(n-2)}(s) & & x_n^{(n-2)}(s) \\ x_1(t) & & x_n(t) \end{vmatrix} \, b(s)}{W[x_1,\ldots,x_n](s) a_0(s)} \, ds.
$$

The proof is an exercise.

IV.5.3. **An Example.** If $n = 2$, then the solution to the nonhomogeneous

second order initial value problem

$$LIVP(n): \quad a_0(t)x'' + a_1(t)x' + a_2(t)x = b(t), \quad a_0(t) > 0,$$

$$x(\tau) = \xi_1, \quad x'(\tau) = \xi_2,$$

is given by

$$x(t) = x_0(t) + \int_\tau^t \frac{[x_1(s)x_2(t) - x_2(s)x_1(t)]}{[x_1(s)x_2'(s) - x_2(s)x_1'(s)]} \frac{b(s)}{a_0(s)} \, ds,$$

where x_0 satisfies the corresponding homogeneous equation.

IV.6   <u>Reduction of Order.</u> The following discussion is inserted here primarily

because it doesn't seem to fit as well elsewhere. It is concerned with the

rather unusual fact that given a homogeneous n-th order linear equation and

m solutions, it is possible to reduce the order of the equation to $n - m$.

In particular, if one solution of a second order equation is known, then

the differential equation may be reduced to a linear first order equation,

which can be completely solved.

IV.6.1. <u>Theorem.</u> Let $L_n x = a_0(t)x^{(n)} + \dots + a_n(t)x = 0$, $a_0 > 0$, be a

homogeneous linear differential equation of order n with continuous coefficients, and let $x_1, \ldots, x_m$ be m linearly independent solutions of $L_n x = 0$. Then the equation $L_n x = 0$ is reducible to a homogeneous linear differential equation of order $n - m$. If the solutions of this equation are known, then the remaining solutions of $L_n x = 0$ may also be recovered.

Proof. We reduce the equation to one of n-1st order and proceed by induction. We let $x = x_n y$. Then since

$$x^{(j)} = \sum_{k=0}^{n} \binom{j}{k} x_n^{(j-k)} y^{(k)} \ ,$$

$$\sum_{j=0}^{n} a_{n-j}(t) x^{(j)} = 0$$

is equivalent to

$$\sum_{j=0}^{n} \sum_{k=0}^{j} a_{n-j}(t) \binom{j}{k} x_n^{(j-k)} y^{(k)} = 0.$$

The coefficient of y in this series is given when $k = 0$ and is $\sum_{j=0}^{n} a_{n-j}(t) x_n^{(j)}$. Since x_n is a solution of $L_n x = 0$, this is zero. Thus the series is reduced to

$$\sum_{j=0}^{n} \sum_{k=1}^{j} a_{n-j}(t) \binom{j}{k} x_n^{(j-k)} y^{(k)} = 0.$$

Letting $k = i+1$ and $y' = z$, we have

$$\sum_{j=0}^{n} \sum_{i=0}^{j-1} a_{n-j} \binom{j}{i+1} x_n^{(j-i-1)} z^{(i)} = 0.$$

This is a homogeneous linear equation of order $n-1$ in z. Now since $x_1, \ldots, x_{n-1}$ are solutions to $L_n x = 0$, there exist corresponding $y_1, \ldots, y_{n-1}$, given by $y_j = x_j/x_n$, $j = 1, \ldots, n-1$. Then $z_j = y_j'$, $j = 1, \ldots$ are solutions to the reduced equation.

The induction proceeds until we are left with an equation of order $n - m$ and no solutions given.

If we are able at any stage to somehow produce a solution to a reduced equation z, then a corresponding y is found by an integration, and a corresponding additional solution x is found by multiplying by x_n. Again by induction we ultimately recover an additional solution x.

IV.6.2. <u>Examples</u>. 1. Let x_1 be one solution of

$$x'' + a_1(t)x' + a_2(t)x = 0.$$

Then if $x = x_1 y$ this becomes

$$x_1 y'' + (2x_1' + a_1(t)x_1)y' = 0.$$

If we let $y' = z$, and divide by zx_1, then

$$\frac{z'}{z} + 2\frac{x_1'}{x_1} + a_1(t) = 0.$$

Integrating

$$\ln z + \ln x_1^2 + \int^t a_1(s)ds = \ln c_1,$$

or

$$y' = c_1 e^{-\int^t a_1(s)ds}/x_1^2.$$

Thus a second solution of the differential equation has the form

$$x = c_1 \int^t [e^{-\int^r a_1(s)ds}/x_1(r)^2]dr + c_2 x_1(t),$$

where c_1 and c_2 are arbitrary constants of integration.

2. Consider the equation

$$t^3 x''' - 6t^2 x'' + 15tx' - 15x = 0,$$

with known solutions $x_1 = t^3$ and $x_2 = t$. If we let $x = ty$, then the equation is reduced to

$$t^2 y''' - 3ty'' + 3y' = 0.$$

Letting $z = y'$, we have a second order equation in z,

$$t^2 z'' - 3tz' + 3z = 0.$$

Now one of the solutions, z, of the above equation must yield $x_1 = t^3$.

Thus $t^3 = ty_1$, where y_1 is the corresponding y. Dividing and differentiat

we find $z_1 = y_1' = 2t$ is a solution to the equation in z.

If we let $z = 2tw$, the equation in z is reduced to

$$tw'' - w' = 0.$$

Letting $s = w'$, we have

$$ts' - s = 0,$$

which is easily solved to yield $s = c_1 t$. Then $w = c_1 t^2/2 + c_2$,

$z = c_1 t^3 + 2c_2 t$, $y = c_1 t^4/4 + c_2 t^2 + c_3$, and finally $x = c_1 t^5/4 + c_2 t^3 + $

$c_3 t$. The constants c_1, c_2, c_3 are constants of integration. By choosing

$c_1 = 4$, $c_2 = 0$, $c_3 = 0$, we find a third solution $x_3 = t^5$.

We remark that these examples have been chosen because the operations
involved are particularly elementary. This is not usually the case,
especially when considering higher order equations.

IV.7. **Constant Coefficients.** The final section in this chapter is concerned
with the more elementary, but very important problems of linear differential
systems and equations with constant coefficients. Central to their discussion
is the role of the exponential matrix. Therefore, we proceed by first defining
what we mean by an exponential matrix, and exploring its properties. Next we
discuss linear differential systems with constant coefficients. Finally we

consider the n-th order linear differential equation with constant co-

efficients.

IV.7.1. <u>Definition.</u> <u>Let A be an n × n matrix. Then by e^A we mean</u>

the series $\sum\limits_{n=0}^{\infty} A^n/n!$

It is easy to show that the exponential matrices have the following

properties.

1. $e^{At} = \sum\limits_{n=0}^{\infty} A^n t^n/n!$.

2. $|e^{At}| \leq e^{|A||t|} + (n-1)$, where $|\cdot|$ is the norm defined in

Section IV.2.

3. $e^{At} e^{As} = e^{A(t+s)}$ for all complex numbers s, t.

4. e^{At} represents a regular transformation from K^n into K^n.

$(e^{At})^{-1} = e^{-At}$.

5. $(e^{At})' = A e^{At}$.

The first follows from the definition. To show the second, we note

$$|e^{At}| = |\sum_{n=0}^{\infty} A^n t^n/n!|,$$

$$\leq \sum_{n=0}^{\infty} |A|^n |t|^n/n!,$$

$$\leq |I| + \sum_{n=1}^{\infty} |A|^n |t|^n/n!,$$

$$\leq n + e^{|A||t|} - 1.$$

To show the third, we resort to the infinite series.

$$e^{At} e^{As} = \sum_{n=0}^{\infty} A^n t^n / n! \sum_{m=0}^{\infty} A^m s^m / m!$$

$$= \sum_{n=0}^{\infty} \sum_{m=0}^{\infty} A^{n+m} t^n s^m / n! m! \; .$$

If we let $m + n = p$ and eliminate m,

$$e^{At} e^{As} = \sum_{p=0}^{\infty} \sum_{n=0}^{\infty} A^p t^n s^{p-n} / n! (p-n)! \; ,$$

$$= \sum_{p=0}^{\infty} [A^p / p!] \sum_{n=0}^{p} \binom{p}{n} t^n s^{p-n} ,$$

$$= \sum_{p=0}^{\infty} A^p (t+s)^p / p! \; ,$$

$$= e^{A(t+s)} .$$

The fourth now follows from part 3 by letting $s = -t$. The fifth follows immediately from the definition.

We are now prepared to consider the first order linear system with constant coefficients in both homogeneous and nonhomogeneous from.

IV.7.2 Theorem. Let A be a constant $n \times n$ matrix. Then

1. e^{At} is a fundamental matrix for $x' - Ax = 0$.

2. $\det(e^{At}) = \exp \int_0^t (\text{tr}A)ds$, so

   $\det e^A = e^{\text{tr}A}$.

3. The solution to the homogeneous linear initial value system with

   constant coefficients

   $$x' - Ax = 0, \quad x(\tau) = \xi,$$

   is $x(t) = e^{A(t-\tau)}\xi$.

4. The solution to the nonhomogeneous linear initial value system

   with constant coefficients

   $$x' - Ax = b(t), \quad x(\tau) = \xi$$

is

   $$x(t) = e^{A(t-\tau)}\xi + \int_\tau^t e^{A(t-s)}b(s)ds.$$

The proof of these statements follows immediately from the remarks concerning exponential matrices and the previous formulae.

IV.7.3. An Example. Let us find the fundamental matrix for

$$x_1' = x_1 + x_2$$

$$x_2' = x_1 - x_2,$$

so that we may solve the initial value problem

$$x_1' = x_1 + x_2 + b_1(t), \quad x_1(0) = \xi_1$$

$$x_2' = x_1 - x_2 + b_2(t), \quad x_2(0) = \xi_2.$$

In this case $A = \begin{pmatrix} 1 & 1 \\ 1 & -1 \end{pmatrix}$. In order to calculate e^{At}, we first note that $A^2 = 2I$, where I is the identity matrix. We then have

$$\sum_{n=0}^{\infty} A^n t^n / n! = \sum_{m=0}^{\infty} 2^m I t^{2m}/(2m)! + \sum_{m=0}^{\infty} 2^m A t^{2m+1}/(2m+1)!,$$

$$= (A/\sqrt{2}) \sum_{m=0}^{\infty} (\sqrt{2}t)^{2m+1}/(2m+1)! + I \sum_{m=0}^{\infty} (\sqrt{2}t)^{2m}/(2m)!,$$

$$= (A/\sqrt{2}) \sinh \sqrt{2}t + I \cosh \sqrt{2}t.$$

Thus

$$e^{At} = \begin{pmatrix} \dfrac{\sinh \sqrt{2}t}{\sqrt{2}} + \cosh \sqrt{2}t & \dfrac{\sinh \sqrt{2}t}{\sqrt{2}} \\[4mm] \dfrac{\sinh \sqrt{2}t}{\sqrt{2}} & -\dfrac{\sinh \sqrt{2}t}{\sqrt{2}} + \cosh \sqrt{2}t \end{pmatrix}$$

The solution to the nonhomogeneous problem is

$$\begin{pmatrix} x_1(t) \\[2mm] x_2(t) \end{pmatrix} = \begin{pmatrix} \dfrac{\sinh \sqrt{2}t}{\sqrt{2}} + \cosh \sqrt{2}t & \dfrac{\sinh \sqrt{2}t}{\sqrt{2}} \\[4mm] \dfrac{\sinh \sqrt{2}t}{\sqrt{2}} & -\dfrac{\sinh \sqrt{2}t}{\sqrt{2}} + \cosh \sqrt{2}t \end{pmatrix} \begin{pmatrix} \xi_1 \\[2mm] \xi_2 \end{pmatrix} +$$

$$+ \int_0^t \begin{pmatrix} \dfrac{\sinh \sqrt{2}\ (t-s)}{\sqrt{2}} + \cosh \sqrt{2}\ (t-s) & \sinh \sqrt{2}\ (t-s) \\[2ex] \dfrac{\sinh \sqrt{2}\ (t-s)}{\sqrt{2}} & -\dfrac{\sinh \sqrt{2}\ (t-s)}{\sqrt{2}} + \cosh \sqrt{2}\ (t-s) \end{pmatrix} \begin{pmatrix} b_1(s) \\[2ex] b_2(s) \end{pmatrix} ds$$

In this problem, as well as all similar ones, the real problem is to find e^{At}. This is not usually as easy a task as in the problem just completed. In fact it can be an extremely tedious chore to find it directly, even for a 2×2 matrix. Rather it is easier to "diagonalize" A first. The computation is then substantially reduced. We describe without proof just what does happen.

IV.7.4. Definition. Let A be an $n \times n$ matrix. The characteristic polynomial for A (in λ) is

$$\det (A - \lambda I).$$

The characteristic equation for $x' - Ax = 0$ is

$$\det (A - \lambda I) = 0.$$

It is well known that this characteristic equation possesses n roots $\lambda_1, \ldots, \lambda_n$, which may not necessarily be distinct. For each root λ_j, called an eigenvalue, there exists at least one pair of vectors, called eigenvectors, a row vector $v_j = (v_j^1, \ldots, v_j^n)$ satisfying $v_j A = \lambda_j v_j$, and a column vector $u_j = \begin{pmatrix} u_j^1 \\ \ldots \\ u_j^n \end{pmatrix}$ satisfying $Au_j = \lambda_j u_j$. These vectors can be appropriately adjusted so that the dot product $v_j u_k = \sum_{i=1}^{n} v_j^i u_k^i = \delta_{ij}$, the Kronecker function.

Let us assume for the moment that all the eigenvalues $\lambda_1,\ldots,\lambda_n$ are distinct. Then there exist eigenvectors $u_1,\ldots,u_n$ and $v_1,\ldots,v_n$ with the properties just described. We then define

$$V = \begin{pmatrix} v_1 \\ \cdots \\ v_n \end{pmatrix} = (v_j^i),$$

which has as its rows the row vectors v_j, and

$$U = (u_1,\ldots,u_n) = (u_i^j),$$

which has as its columns the column vectors u_j. It is then an easy calculation to show

$$VU = UV = I,$$

and

$$VAU = V(\lambda_1 u_1,\ldots,\lambda_n u_n) = D,$$

where

$$D = \begin{pmatrix} \lambda_1 & & 0 \\ & \ddots & \\ 0 & & \lambda_n \end{pmatrix}$$

is a diagonal matrix with the eigenvalues $\lambda_1, \ldots, \lambda_n$ along the diagonal. Then, since $U^{-1} = V$ and $V^{-1} = U$, we find

$$A = UDV.$$

A is said to be diagonalized. It is then an elementary task to see that

$$(At)^n = U[Dt]^n V.$$

Inserting this in the expansion for the exponential function,

$$e^{At} = Ue^{Dt}V,$$

$$= U \begin{pmatrix} e^{\lambda_1 t} & & 0 \\ & \ddots & \\ 0 & & e^{\lambda_n t} \end{pmatrix} V.$$

The situation when $\lambda_1, \ldots, \lambda_n$ are not all distinct is somewhat more complicated. In this case the matrices U and V are found in a similar manner, but are not exactly the same as before. There exists an additional matrix N which satisfies $N^m = 0$, $m \geq n$, and $ND = DN$. In fact, the elements of N are all zero with the exception of the diagonal lying just above the main diagonal, where a number of ones may be found. Then

$$A = U[D + N]V,$$

$$e^{At} = Ue^{Dt}e^{Nt}V.$$

But since $N^m = 0$, $m \geq n$,

$$e^{Nt} = \sum_{j=0}^{m-1} N^j t^j / j! \ ,$$

and

$$e^{At} = U \begin{pmatrix} e^{\lambda_1 t} & & 0 \\ & \ddots & \\ 0 & & e^{\lambda_n t} \end{pmatrix} [I + Nt + \ldots + \frac{N^{m-1} t^{m-1}}{(m-1)!}] V .$$

The components of e^{At} consist of sums of the form $c_{ij} e^{\lambda_j t} t^i$, $i = 1, \ldots, m-$ $j = 1, \ldots, n$.

IV.7.5. __Theorem.__ Let A be an $n \times n$ matrix with eigenvalues $\lambda_1, \ldots, \lambda_n$. Then there exist matrices U, V satisfying $UV = VU = I$ and matrices D, N satisfying $N^m = 0$, $m \geq n$, and $D = \begin{pmatrix} \lambda_1 & 0 \\ & \ddots & \\ 0 & \lambda_n \end{pmatrix}$, such that

$$e^{At} = U \begin{pmatrix} e^{\lambda_1 t} & & 0 \\ & \ddots & \\ 0 & & e^{\lambda_n t} \end{pmatrix} [\sum_{j=0}^{m-1} (Nt)^j / j!] V .$$

For a proof of these statements we recommend any of the many books on linear algebra.

IV.7.6. __An Example.__ Let us consider the initial value system

$$\begin{pmatrix} x_1 \\ x_2 \end{pmatrix}' = \begin{pmatrix} 2 & 1 \\ 1 & 2 \end{pmatrix} \begin{pmatrix} x_1 \\ x_2 \end{pmatrix} + \begin{pmatrix} b_1(t) \\ b_2(t) \end{pmatrix}, \quad \begin{pmatrix} x_1(0) \\ x_2(0) \end{pmatrix} = \begin{pmatrix} 1 \\ 0 \end{pmatrix} .$$

Here $A = \begin{pmatrix} 2 & 1 \\ 1 & 2 \end{pmatrix}$. To compute e^{At} directly is quite tedious, and requires

the solution of a nonhomogeneous linear difference equation. However, if

we attempt to diagonalize, we find that $\lambda_1 = 1$, $\lambda_2 = 3$, $u_1 = \begin{pmatrix} 1 \\ -1 \end{pmatrix}$,

$v_1 = (1/2, -1/2)$, $u_2 = \begin{pmatrix} 1 \\ 1 \end{pmatrix}$, $v_2 = (1/2, 1/2)$. Thus

$$U = \begin{pmatrix} 1 & 1 \\ -1 & 1 \end{pmatrix}, \qquad V = \begin{pmatrix} 1/2 & -1/2 \\ -1/2 & 1/2 \end{pmatrix},$$

and

$$e^{\begin{pmatrix} 2 & 1 \\ 1 & 2 \end{pmatrix}t} = \begin{pmatrix} 1 & 1 \\ -1 & 1 \end{pmatrix} \begin{pmatrix} e^t & 0 \\ 0 & e^{3t} \end{pmatrix} \begin{pmatrix} 1/2 & -1/2 \\ -1/2 & 1/2 \end{pmatrix},$$

$$= \begin{pmatrix} \dfrac{e^t + e^{3t}}{2} & \dfrac{-e^t + e^{3t}}{2} \\ \dfrac{-e^t + e^{3t}}{2} & \dfrac{e^t + e^{3t}}{2} \end{pmatrix}.$$

The solution to the original problem is

$$\begin{pmatrix} x_1(t) \\ x_2(t) \end{pmatrix} = \begin{pmatrix} 1/2[e^t + e^{3t}] & 1/2[-e^t + e^{3t}] \\ 1/2[-e^t + e^{3t}] & 1/2[e^t + e^{3t}] \end{pmatrix} \begin{pmatrix} 1 \\ 0 \end{pmatrix}$$

$$+ \int_0^t \begin{pmatrix} 1/2[e^{(t-s)} + e^{3(t-s)}] & 1/2[-e^{(t-s)} + e^{3(t-s)}] \\ 1/2[-e^{(t-s)} + e^{3(t-s)}] & 1/2[e^{(t-s)} + e^{3(t-s)}] \end{pmatrix} \begin{pmatrix} b_1(s) \\ b_2(s) \end{pmatrix} ds$$

As the final subject of this chapter, we now turn our attention to the

discussion of n-th order linear differential equations with constants coefficients.

IV.7.7. Definition. Let $a_0, \ldots, a_n$ be constants $a_0 > 0$. The characteristic equation for the differential equation

$$a_0 x^{(n)} + \ldots + a_n x = 0$$

is

$$a_0 \lambda^n + \ldots + a_n = 0.$$

IV.7.8. Theorem. The characteristic equation for $a_0 x^{(n)} + \ldots + a_n x = 0$, $a_0 > 0$, is equivalent to the characteristic equation of its matrix representation $\hat{x}' - \hat{A}\hat{x} = 0$.

Proof. In its matrix representation,

$$A = \begin{pmatrix} 0 & 1 & & & \\ 0 & 0 & 1 & & \\ & & & \ddots & \\ & & & & 1 \\ -\dfrac{a_n}{a_0} & -\dfrac{a_{n-1}}{a_0} & \cdots & & -\dfrac{a_1}{a_0} \end{pmatrix}.$$

The characteristic equation is, therefore,

$$\det \begin{pmatrix} -\lambda & 1 & 0 & & & \cdot & \cdot \\ 0 & -\lambda & 1 & 0 & & \cdot & \cdot \\ & \cdot & \cdot & \cdot & \cdot & & \\ 0 & & & & \cdot & & 1 \\ -\dfrac{a_n}{a_0} & -\dfrac{a_{n-1}}{a_0} & & \cdots & & -\lambda & -\dfrac{a_1}{a_0} \end{pmatrix} = 0 .$$

We multiply the next to last row by $\frac{a_1}{a_0} + \lambda$ and add it to the last. We

multiply the second from the last by $\frac{a_2}{a_0} + \frac{a_1}{a_0}\lambda + \lambda^2$ and add it to the

last. Finally we multiply the first row by $\frac{a_{n-1}}{a_0} + \frac{a_{n-2}}{a_0}\lambda + \ldots + \frac{a_1}{a_0}\lambda^{n-2}$

$+ \lambda^{n-1}$ and add it to the last. This results in

$$\det \begin{pmatrix} -\lambda & 1 & & & \\ 0 & -\lambda & \ddots & & \\ & \ddots & \ddots & \ddots & \\ -\sum_{j=0}^{n}\frac{a_j}{a_0}\lambda^{n-j} & 0 & \cdots & 0 \end{pmatrix} = 0,$$

which, when expanded, gives us the scalar characteristic equation.

IV.7.9. <u>Theorem</u>. <u>Let</u> $L_n x = a_0 x^{(n)} + \ldots + a_n x = 0,$ <u>where</u> $a_0, \ldots, a_n$ <u>are</u>

<u>coefficients, and</u> $a_0 > 0.$ <u>Let the characteristic equation be</u> $p(\lambda) = 0,$

<u>where</u> $p(\lambda) = a_0\lambda^n + \ldots + a_n.$ <u>Finally let</u> $p(\lambda)$ <u>have the factored form</u>

$$p(\lambda) = (\lambda - \lambda_1)^{n_1} \ldots (\lambda - \lambda_s)^{n_s},$$

<u>where</u> $n_1 + \ldots + n_s = n.$ <u>Then</u> $t^j e^{\lambda_i t},$ $j = 0, \ldots, n_i - 1,$ $i = 1, \ldots, s$ <u>are</u>

<u>all linearly independent solutions of</u> $L_n x = 0.$

Proof. We have already essentially proved this theorem, based on our remarks

concerning the diagonalization of the matrix $\hat{A}.$ However, since those remarks

were not verified, we will give an independent proof.

We note that if λ_i is a root of $p(\lambda) = 0$ of order $n_i,$ then λ_i

is also a root of $p^{(j)}(\lambda) = 0$, $j = 0,\ldots,n_i-1$. Now for any

$$L_n(e^{\lambda t}) = p(\tfrac{d}{dt})e^{\lambda t},$$

$$= p(\lambda)e^{\lambda t}.$$

In general,

$$L_n(t^j e^{\lambda t}) = L_n(\frac{\partial^j}{\partial \lambda^j} e^{\lambda t}),$$

$$= \frac{\partial^j}{\partial \lambda^j} L_n(e^{\lambda t}),$$

$$= \frac{\partial^j}{\partial \lambda^j} (p(\lambda)e^{\lambda t}),$$

$$= \sum_{k=0}^{j} \binom{j}{k} p^{(k)}(\lambda) t^{j-k} e^{\lambda t}.$$

If $\lambda = \lambda_i$, all the derivatives of $p(\lambda)$ through n_i-1 vanish. Thus if $j \le n_i-1$, $t^j e^{\lambda_i t}$ is a solution.

To show linear independence, we assume the contrary. Then there exist constants c_{ij} such that

$$\sum_{i=1}^{s} \sum_{j=0}^{n_i-1} c_{ij} t^j e^{\lambda_i t} = 0.$$

We divide by λ_s, let $\mu_i = \lambda_n - \lambda_i$, $i = 1,\ldots,s-1$, to find

$$\sum_{i=1}^{s-1} \sum_{j=0}^{n_i-1} c_{ij} t^j e^{\mu_i t} + \sum_{j=0}^{n_s-1} c_{sj} t^j = 0.$$

We differentiate n_s times to find

$$\sum_{i=1}^{s-1} \sum_{j=0}^{n_i-1} d_{ij} t^j e^{\mu_i t} = 0,$$

where the coefficients d_{ij} are linear combinations of the c_{ij}'s. We next divide by λ_{s-1} and differentiate n_{s-1} times. We eventually find that

$$\sum_{j=0}^{n_1-1} e_{ij} t^j e^{\lambda_1 t} = 0.$$

Dividing by $e^{\lambda_1 t}$, we have $\sum_{j=0}^{n_1-1} e_{1j} t^j = 0$ for all t. Thus $e_{1j} = 0$ for all $j = 0,\ldots,n_1-1$. This in turn forces $c_{1j} = 0$ for $j = 0,\ldots,n_1-1$. We then repeat this process to find all $c_{ij} = 0$.

Having found a fundamental set of solutions to the homogeneous equation $L_n x = 0$, we could insert them in the formula of Theorem IV.5.1 or Corollary IV.5.2. There is nothing to be gained at this point, however, by doing so.

IV. Exercises

1. Show that the two representations for the n-th order initial value problem LIVP$(\hat{n})$ $\hat{L}\hat{x} = \hat{b}(\hat{t})$, $\hat{x}(\tau) = \hat{\xi}$ and $L_n x = b(t)$, $x^{j-1}(\tau) = \xi_j$,

$j = 1,...,n$ are equivalent to each other.

2. In the proof of Theorem IV.1.2 why can one let $t = \tau$ and have

$$\sum_{j=1}^{n} \alpha_j x_j(\tau) = \theta?$$ That is, if $\sum_{j=1}^{n} \alpha_j x_j(\tau) = \theta$, what does this imply

about $\sum_{j=1}^{n} \alpha_j x_j$ for other values of t?

3. Let $X = L(K^n, K^n)$, the space of complex $n \times n$ matrices. Show X

has n^2 dimensions. For $A = (a_{ij})$ in X let $|A| = \sum_{i=1}^{n} \sum_{j=1}^{n} |a_{ij}|$.

With $|\cdot|$ as a norm show X is a Banach space. For A, B in X,

show $|AB| \le |A||B|$. Further if A is in X and x is in K^n,

show $\|Ax\|_{K^n} \le |A| \|x\|_{K^n}$, so $\|A\| \le |A|$.

4. If $A(t) = (a_{ij}(t))$ is a differentiable matrix valued function, show
   $A'(t) = (a'_{ij}(t))$.

5. For $n \times n$ matrices A and B show $\text{tr}(AB) = \text{tr}(BA)$.

6. Can $\begin{pmatrix} t & t^2 \\ 0 & 0 \end{pmatrix}$ be a solution matrix for some linear homogeneous system
   $Lx = 0$?

7. Consider the matrix initial value problem $X' = A(t)X$, $X(\tau) = I$.
   Assuming that X exists, show that the solution to the LIVS
   $x' = A(t)x$, $x(\tau) = \xi$ is $x = X\xi$.

8. Prove that two different homogeneous linear systems cannot have the
   same fundamental matrix.

9. If A is a differentiable nonsingular $n \times n$ matrix, show
   $(A^{-1})' = -A^{-1}A'A^{-1}$.

10. If $A = \begin{pmatrix} a_{11} & a_{12} & a_{13} \\ a_{21} & a_{22} & a_{23} \\ a_{31} & a_{32} & a_{33} \end{pmatrix}$ is nonsingular, find A^{-1}.

11. Verify Corollary IV.5.2.

12. Simplify the formula in Corollary IV.5.2 when $n = 1$.

13. If, in the section on reduction of order, the original differential

equation is nonhomogeneous, can the procedure still be used? What does

the reduced equation look like?

14. Calculate e^{At} for several different matrices A.

15. Give a proof of Theorem IV.7.5.

16. Recalculate e^{At} for the examples used in problem 14 by first

diagonalizing A.

References

1. G. Birkhoff and G.C. Rota, "Ordinary Differential Equations," Blaisdell,
       Waltham, Mass. 1959.

2. E.A. Coddington and N. Levinson, "Theory of Ordinary Differential
       Equations," McGraw-Hill, New York, 1955.

3. R.H. Cole, "Theory of Ordinary Differential Equations," Appleton-
       Century-Crofts, New York, 1968.

4. W. Hurewicz, "Lectures on Ordinary Differential Equations," M.I.T. Press,
       Cambridge, Mass. 1958.

5. E.L. Ince, "Ordinary Differential Equations," Dover Publications,
       New York, 1956.

V.   SECOND ORDER ORDINARY DIFFERENTIAL EQUATIONS

This chapter is devoted to the study of second order differential operato

which will be used extensively throughout the later portions of the book.

Extensions to higher order operators, which we will not need, are straigh

forward, and can be found in the references at the end of the chapter.

The chapter begins with a very brief review of the results of the previou

chapter as applied to second order operators. We next introduce the concept

of an adjoint operator, and derive Green's formula. We then diverge briefly

to prove an oscillation theorem. The regular Sturm-Liouville problem is

next discussed, followed by a discussion of the inverse problem and a deriva-

tion of the Green's function associated with it.

V.1.   A Brief Review.   The results of Chapter IV, applied to second order

equations, are stated in their special form.

V.1.1.   Definition.   Let a_0, a_1, a_2 be continuous, complex valued functions,

defined on an interval $I = [\alpha, \beta]$, and let $a_0 > 0$ on I. We define the

linear differential operator L by stating that the domain of L, D_L, con-

sists of all functions x in $C^2[I]$. For each x in D_L,

$$Lx = a_0(t)x'' + a_1(t)x' + a_2(t)x.$$

We recall that Picard's Existence Theorem states that the initial value

problem

$$Lx = b, \quad x(\tau) = \xi_1, \quad x'(\tau) = \xi_2,$$

has a unique solution in $C^2[I]$ for each τ in I, each pair of complex

numbers ξ_1, ξ_2, and b in $C[I]$. The set of solutions of the differential

equation $Lx = 0$ forms a two dimensional linear manifold in $C[I]$.

V.1.2. Definition. A particular integral of $Lx = b$ is any function x_0

in $C^2[I]$ which satisfies $Lx_0 = b$. The general solution of $Lx = b$ is a two

parameter family of solutions of $Lx = b$ such that any solution may be

obtained by a suitable choice of parameters.

V.1.3. Theorem. Let x_0 be a particular integral of $Lx = b$, and let

x_1, x_2 be linearly independent solutions of $Lx = 0$. Then the general

solution of $Lx = b$ has the form

$$x = c_1 x_1 + c_2 x_2 + x_0,$$

where c_1, c_2 are arbitrary constants.

V.1.4. An Example. The general solution for $x'' + x = t$ is given by

$x = c_1 \sin t + c_2 \cos t + t$.

Finally we recall that when x_1, x_2 are solutions of $Lx = 0$, the

Wronskian $W[x_1, x_2] = x_1 x_2' - x_1' x_2$ satisfies

$$W[x_1, x_2](t) = W[x_1, x_2](\tau) \exp[- \int_\tau^t (a_1(s)/a_0(s)) ds]$$

for each t, τ in I.

If x_1 is a known solution of Lx = 0 which is nonzero on I, then
an additional solution x_2 is given by

$$x_2(t) = x_1(t) \int_\tau^t x_1^{-2}(s) \exp[- \int_\tau^s (a_1(\xi)/a_0(\xi)) d\xi] ds.$$

V.2. The Adjoint Operator. Associated with every linear differential
operator is another similar, but usually different operator called the adjoint
operator. Originally introduced to provide integrating factors for the related differential equations, adjoints have been formed to possess a number
of additional properties, not only in mathematical theory, but also in other
fields such as quantum mechanics and calculus of variations as applied to
certain problems in mechanics.

V.2.1. Definition. Let I = [α,β], let a_0 be in $C^2[I]$, let a_2 be
in C[I], and let Lx = $a_0(t)x'' + a_1(t)x' + a_2(t)x$ be a linear differential
operator defined for all functions x in $C^2[I]$. The FORMAL Lagrange adjoint
operator, L^+, of L is defined by letting $D_{L^+} = C^2[I]$ and

$$L^+x = (\overline{a_0}(t)x)'' - (\overline{a_1}(t)x)' + (\overline{a_2}(t)x).$$

This, of course, can be expanded, so

$$L^+x = \overline{a_0}(t)x'' + [2\overline{a_0'}(t) - \overline{a_1}(t)]x'$$

$$+ [\overline{a_0''}(t) - \overline{a_1'}(t) + \overline{a_2}(t)]x.$$

We note that in the formulation of L, differentiation is first carried out upon x. The various derivatives are then multiplied by the coefficients. In the construction of the adjoint, multiplication by the coefficients is first, with differentiation following. To complete the picture, the coefficients are replaced by their complex conjugates and the signs are alternated so odd derivatives change sign.

V.2.2. Theorem (Lagrange Identity). Let $I = [\alpha,\beta]$, let a_0 be in $C^2[I]$, let a_1 be in $C^1[I]$, and let a_2 be in $C[I]$. Let L be the differential operator defined for all x in $C^2[I]$ by $Lx = a_0(t)x'' + a_1(t)x' + a_2(t)x$. Then for all x, y in $C^2[I]$,

$$Lx \cdot \overline{y} - x \cdot \overline{L^+ y} = \frac{d}{dt}[x,\overline{y}],$$

where $[x,\overline{y}] = a_0[x'\overline{y} - x\overline{y}'] - (a_1 - a_0')x\overline{y}$.

We can see at this point the original use of the adjoint. If y satisfies $L^+ y = 0$, the adjoint equation, then $Lx \cdot \overline{y}$ is the exact derivative of some function, and an integration can be performed. Knowing y, therefore we can move one step closer to finding the unknown x.

V.2.3. Corollary (Green's Formula). Let $I = [\alpha,\beta]$, let a_0 be in $C^2[I]$, let a_1 be in $C^1[I]$, and let a_2 be in $C[I]$. Let L be the differential operator defined for all x in $C^2[I]$ by $Lx = a_0(t)x'' + a_1(t)x' + a_2(t)x$. Then for all x, y in $C^2[I]$ and all t_1, t_2 in I,

$$\int_{t_1}^{t_2} [Lx \cdot \overline{y} - x \cdot \overline{L^+ y}]dt = [x,\overline{y}](t_2) - [x,\overline{y}](t_1),$$

where $[x,\bar{y}] = a_0[x'\bar{y} - x\bar{y}'] - (a_1 - a_0)x\bar{y}$.

Green's formula is used most often when the right side vanishes. In this case the left side is a statement concerning inner products, which we will properly define in Chapter VII.

The formula is of primary importance when L is self-adjoint.

V.2.4. <u>Definition. Let I = [α,β], let a_0 be in $C^2[I]$, let a_1 be in</u>
<u>$C^1[I]$, and let a_2 be in C[I]. Let L be the differential operator defined</u>
<u>for all x in $C^2[I]$ by $Lx = a_0(t)x'' + a_1(t)x' + a_2(t)x$. L is formally</u>
<u>self-adjoint if $Lx = L^+x$ for all x in $C^2[I]$.</u>

<u>L is real when a_0, a_1, and a_2 are all real valued functions.</u>

V.2.5. <u>Theorem. Let L be real. Then</u>

1. <u>L is formally self-adjoint if and only if $a_0' = a_1$.</u>

2. <u>If L is formally self-adjoint, then L has the form</u>

$$Lx = (a_0(t)x')' + a_2(t)x,$$

<u>where a_0 is in $C^2[I]$, and a_2 is in C[I].</u>

3. <u>If L is formally self-adjoint and real, and has the form expressed</u>
<u>in part 2, then for all t_1, t_2 in I and x, y in $C^2[I]$, Green's formula</u>
<u>is</u>

$$\int_{t_1}^{t_2} [Lx \cdot \bar{y} - x \cdot \overline{Ly}] dt = [x,\bar{y}](t_2) - [x,\bar{y}](t_1),$$

where $[x,\bar{y}] = a_0[x'\bar{y} - x\bar{y}'] = -a_0 W[x,\bar{y}]$. Thus $[x,\bar{y}]$ is constant when $Lx = 0$ and $Ly = 0$. $[x,\bar{y}] = 0$ if and only if x and y are dependent.

Proof. If x is in $C^2[I]$, and L is real, then

$$L^+ x = a_0 x'' + (2a_0' - a_1)x' + (a_0'' - a_1' + a_2)x.$$

when $L = L^+$, this implies

$$a_1 = 2a_0' - a_1, \quad a_2 = a_0'' - a_1' + a_2.$$

The first implies $a_1 = a_0'$. The second follows from differentiating the first. The converse is similar.

To show the converse we note that if L is self-adjoint, then $a_1 = a_0'$. Inserting this in L,

$$Lx = a_0 x'' + a_0' x' + a_2 x$$

$$= (a_0 x')' + a_2 x.$$

The third follows easily from Corollary V.2.3.

With the obvious simplification present in self-adjoint operators we may ask: Can differential operators be modified in some way so they will all be self-adjoint? The answer for real second order operators is yes. Unfor-

tunately the answer in general is no.

V.2.6. Theorem. Any real, second order linear differential operator L,

defined by

$$Lx = a_0(t)x'' + a_1(t)x' + a_2(t)x,$$

where a_0 is in $C^2[I]$, a_1 is in $C^1[I]$, a_2 is in $C[I]$, can be converted

into a formally self-adjoint operator by multiplying L by

$$\mu = \exp[- \int_\tau^t \{(a_0 - a_1)/a_0\}ds].$$

Proof. We consider

$$\mu Lx = \mu a_0 x'' + \mu a_1 + \mu a_2 x$$

and require that it equal its adjoint. We find by Theorem V.2.5, part 1,

that

$$(\mu a_0)' = \mu a_1.$$

This is a first order equation in μ, which, when solved, yields the formula

for μ.

From now on we shall write formally self-adjoint operators in the form
$Lx = (px')' + qx$.

V.2.7. <u>Examples</u>. 1. The Legendre operator is

$$Lx = (1 - t^2)x'' - 2tx'$$

$$= ((1 - t^2)x')',$$

and is already formally self-adjoint.

2. The Laguerre operator is

$$Lx = tx'' + (1 - t)x'.$$

The factor $\mu = \exp[- \int_0^t [\{1-(1 - s)\}/s]ds] = e^{-t}$. $e^{-t}Lx = (te^{-t}x')'$.

3. The Hermite operator is

$$Lx = x'' - 2tx'.$$

The factor $\mu = \exp[- \int_0^t [\{0-(-2s)\}/1]ds = e^{-t^2}$. $e^{-t^2}Lx = (e^{-t^2}x')'$.

V.3. <u>An Oscillation Theorem</u>. We diverge briefly to present an oscillation theorem in which the self-adjoint form is a necessity. There are several important applications for which the theorem is used. We shall see two examples involving sines and cosines. Bessel functions, which we shall also use as an example, provides another. The classical orthogonal poly-

nomials.offer still others.

V.3.1. Theorem. Let $I = [\alpha,\beta]$, let $p_1 > 0$, $p_2 > 0$ be in $C^2[I]$, and

let q_1, q_2 be in $C[I]$. Further suppose that x is nontrivial, real

valued, and satisfies

$$(p_1 x')' + q_1 x = 0,$$

and y is nontrivial, real valued, and satisfies

$$(p_2 y')' + q_2 y = 0.$$

If   1. $p_1 > p_2$, $q_2 \geq q_1$,

      2. $p_1 \geq p_2$, $q_2 > q_1$,

or   3. $p_1 \geq p_2$, $q_2 \geq q_1$, and x, y are linearly independent on I,

then, when t_1, t_2 are successive zeros of x, there is a zero of y at

some point τ in $[t_1, t_2]$.

Proof. We first observe that the zeros of the nontrivial solution x are

isolated. For if $\{t_n\}_{n=1}^{\infty}$ are zeros of x and $\lim_{n \to \infty} t_n = \tau$, then $x(\tau) = 0$,

and $x'(\tau) = \lim_{n \to \infty} (x(t_n) - x(\tau))/(t_n - \tau) = 0$. This is impossible when x

is nontrivial.

The proof consists of using an identity due to Picone:

$$[\frac{x}{y}(p_1yx' - p_2xy')]' =$$

$$(q_2 - q_1)x^2 + (p_1 - p_2)(x')^2 + p_2 [\frac{yx' - xy'}{y}]^2 .$$

If y fails to vanish on the interval $[t_1,t_2]$, then both sides are integrable. The integral of the left is zero, since x vanishes at t_1 and t_2. But the integral on the right is not zero, and we have a contradiction.

V.3.2. Examples. 1. Let the equations be

$$4x'' + x = 0,$$

$$y'' + y = 0.$$

Then $x = A \sin(t/2 + \theta)$, $y = B \sin(t + \phi)$, where A, θ, B, ϕ are arbitrary constants. Between each pair of zeros of x, y must vanish.

2. Consider the equations

$$(tx')' + (t - n^2/t)x = 0,$$

$$(ty')' + (t - m^2/t)y = 0,$$

where $n > m$. Let $x = J_n(t)$, $y = J_m(t)$, the Bessel functions of the first

kind of order   n   and   m,   respectively.   Between each pair of zeros of J_n
there must be a zero of J_m.

3.   Let the equations be

$$x'' + \omega^2 x = 0,$$

$$y'' + \omega^2 y = 0,$$

and let $x = \cos \omega t$, $y = \sin \omega t$.   In this case the zeros interlace.

V.4.   The Regular Sturm-Liouville Problem.   We now begin the study of the
regular Sturm-Liouville problem, so named after the two nineteenth century
mathematicians who first considered it.   The results we shall derive have
considerable application to the partial differential equations of mathematica
physics.

V.4.1.   Definition.   Let $I = [\alpha, \beta]$, $-\infty < \alpha < \beta < \infty$,   and let   p, q, w   be
real valued functions,   p   in $C^1[I]$,   q, w   in C[I], and let $p > 0$,
$w > 0$ on   I.   We define the operator   A   by letting its domain D_A be
those functions   x   which satisfy

1.   x   is in $C^2[I]$,

2.   x   satisfies

$$\alpha_1 x(\alpha) + \alpha_2 x'(\alpha) = 0,$$

$$\beta_1 x(\beta) + \beta_2 x'(\beta) = 0,$$

where $\alpha_1, \alpha_2, \beta_1, \beta_2$ are real numbers which satisfy

$$\alpha_1^2 + \alpha_2^2 \neq 0, \quad \beta_1^2 + \beta_2^2 \neq 0.$$

For all x in D_A we let

$$Ax = [(px')' + qx]/w.$$

By the regular Sturm-Liouville problem we mean the problem of finding the eigenvalues and eigenfunctions of A (which we will next define), and the problem of representing certain functions as a series of these eigenfunctions.

V.4.2. **Definition.** Let x be a nontrivial function in D_A which also satisfies

$$Ax = \lambda x$$

for some complex number λ. Then x is an eigenfunction of A. λ is its associated eigenvalue.

Therefore x is an eigenfunction, with eigenvalue λ, if x satisfies the differential equation

$$(px')' + qx = \lambda wx$$

as well as the boundary conditions at α and β. The word _regular_ means

that the interval $I = [\alpha, \beta]$ is finite, and that p and w are nonzero on
I.

V.4.3. Examples. 1. If the operator A is defined by $Ax = -x''$ and the
boundary conditions are $x(0) = 0$ and $x(1) = 0$. Then the eigenfunctions
satisfy $-x'' = \lambda x$, $x(0) = 0$, $x(1) = 0$. An elementary calculation shows that
the eigenvalues are $\lambda_n = n^2 \pi^2$, $n = 1, \ldots$. The eigenfunctions are
$x_n = \sin n \pi t$, $n = 1, \ldots$.

2. Let the operator A be defined by

$$Ax = -[(e^{2t}x')' + e^{2t}x]/e^{2t},$$

where x satisfies the boundary conditions $x(0) = 0$, $x(\pi) = 0$. Then
another elementary calculation shows that the eigenvalues are $\lambda_n = n^2$,
$n = 1, \ldots$. The eigenfunctions are $x_n = e^{-t} \sin nt$, $n = 1, \ldots$.

V.5. The Inverse Problem, Green's Functions. The eigenvalue problem just
discussed was concerned with solving the equation $Ax = \lambda x$. As we saw in
the two examples, this is only possible for certain special values of λ,
the eigenvalues. There is another problem we can consider: Is the operator
A invertible? This is, when can we solve the equation $Ax = f$, when f
is in C[I], so that x will also satisfy the auxiliary boundary conditions?
It turns out that this is possible only when 0 is not an eigenvalue. The
device which produces this solution is called a Green's function.

V.5.1. Definition. The Green's function for the regular Sturm-Liouville problem is a function $G(\cdot,\cdot)$ which is continuous on the square $I \times I$, and which also has the following properties.

1. $G(\cdot,\cdot)$ is in $C^2[I \times I - \{(t,t); t \in I\}]$.

2. For fixed s in I, $G(\cdot,s)$ satisfies the boundary conditions for the regular Sturm-Liouville problem.

3. $\dfrac{\partial}{\partial t} [p(t)\dfrac{\partial G(t,s)}{\partial t}] + q(t)G(t,s) = 0$ for all $t \neq s$ and s, t in I.

4. $\left.\dfrac{\partial G(t,s)}{\partial t}\right|_{t=s+} - \left.\dfrac{\partial G(t,s)}{\partial t}\right|_{t=s-} = 1/p(s)$, for all s in I.

V.5.2. Examples. 1. Let the operator A be defined by $Ax = -x''$, defined for those functions x in $C^2[0,1]$ which satisfy $x(0) = 0$, $x(1) = 0$. In order to invert A we must solve

$$-x'' = f,$$

$$x(0) = 0, \quad x(1) = 0.$$

After two integrations we find

$$x(t) = \int_0^t \int_0^u f(s)ds\,du + c_1 t + c_2.$$

Reversing the order of integration, one integration can be performed to yield.

$$x(t) = - \int_0^t (t-s)f(s) \, ds + c_1 t + c_2.$$

Substituting in the boundary conditions, we find $c_2 = 0$ and $c_1 = \int_0^1 (1-s)f(s$

When these are inserted we find

$$x(t) = \int_0^t (1-t)sf(s) \, ds + \int_t^1 (1-s)tf(s) \, ds,$$

or

$$x(t) = \int_0^1 G(t,s)f(s) \, ds,$$

where

$$G(t,s) = \begin{cases} (1-t)s, & 0 \le s \le t \le 1, \\ \\ (1-s)t, & 0 \le t \le s \le 1. \end{cases}$$

Note that if s and t are reversed, G remains unchanged. It is easy to verify that it satisfies the requirements of Definition V.5.1.

2. Let the operator A be defined by $Ax = -[(e^{2t}x')' + e^{2t}x]/e^{2t}$, defined for those functions x in $C^2[0,\pi]$ which also satisfy $x(0) = 0$, $x(\pi) = 0$. In order to invert A we must solve

$$-[(e^{2t}x')' + e^{2t}x]/e^{2t} = f,$$

$$x(0) = 0, \quad x(\pi) = 0.$$

The differential equation is equivalent to

$$x'' + 2x' + x = f.$$

If we let $x = e^{-t}y$, this is reducible to

$$e^{-t}y'' = -f,$$

which has as its general solution

$$y(t) = - \int_0^t (t-s)e^s f(s)ds + c_1 t + c_2.$$

Thus

$$x(t) = - \int_0^t (t-s)e^{-(t-s)} f(s)ds + c_1 t e^{-t} + c_2 e^{-t}.$$

Substituting in the boundary conditions shows $c_2 = 0$ and $c_1 = (1/\pi)\int_0^\pi (\pi-s)e^s f(s)ds$. With these values inserted,

$$x(t) = (1/\pi) \int_0^t (\pi-t)se^{-(s+t)}e^{2s} f(s)ds$$

$$+ (1/\pi) \int_t^\pi (\pi-s)te^{-(s+t)}e^{2s} f(s)ds,$$

or

$$x(t) = \int_0^\pi G(t,s)e^{2s} f(s)ds,$$

where

$$
G(t,s) = \begin{cases} (1/\pi)(\pi-t)se^{-(s+t)}, & 0 \leq s \leq t \leq \pi, \\ \\ (1/\pi)(\pi-s)te^{-(s+t)}, & 0 \leq t \leq s \leq \pi. \end{cases}
$$

As in example 1, if s and t are reversed, G is unchanged. Again it is
easy to show G satisfies the requirements of Definition V.5.1. We now
show these properties always hold.

V.5.3. <u>Theorem</u>. <u>Suppose 0 is not an eigenvalue for the regular Sturm-
Liouville problem. Then A has a unique Green's function G(t,s) which
has the following additional properties.</u>

   1. <u>G is symmetric. That is, $G(t,s) = G(s,t)$ for all s, t in I.</u>

   2. <u>G is real valued.</u>

   3. <u>If x_1 and x_2 are (linearly independent) solutions of $(px')' +$
$qx = 0$, which satisfy $x_1(\alpha) = \alpha_2$, $x_1'(\alpha) = -\alpha_1$, $x_2(\beta) = \beta_2$, $x_2'(\beta) = -\beta_1$,
then</u>

$$
G(t,s) = \begin{cases} (1/c)x_1(t)x_2(s), & \alpha \leq t \leq s \leq \beta, \\ \\ (1/c)x_1(s)x_2(t), & \alpha \leq s \leq t \leq \beta, \end{cases}
$$

<u>where $c = p(t)W[x_1,x_2](t) \neq 0$ and is constant.</u>

Proof. We note that x_1 and x_2 are linearly independent, since, if not, 0 would be an eigenvalue of A. Now let us construct G. Because of requirements 2 and 3 in Definition V.5.1,

$$G(t,s) = \begin{cases} c_1(s)x_1(t), & \alpha \le t < s \le \beta, \\ \\ c_2(s)x_2(t), & \alpha \le s < t \le \beta. \end{cases}$$

G is continuous, so when $s = t$,

$$c_2(s)x_2(s) - c_1(s)x_1(s) = 0.$$

Using requirement 4 of Definition V.5.1,

$$c_2(s)x_2'(s) - c_1(s)x_1'(s) = 1/p(s).$$

Solving these simultaneously, we find

$$c_1(s) = \frac{\begin{vmatrix} 0 & -x_2(s) \\ -1/p(s) & -x_2'(s) \end{vmatrix}}{\begin{vmatrix} x_1(s) & -x_2(s) \\ x_1'(s) & -x_2'(s) \end{vmatrix}} \qquad c_2(s) = \frac{\begin{vmatrix} x_1(s) & 0 \\ x_1'(s) & -1/p(s) \end{vmatrix}}{\begin{vmatrix} x_1(s) & -x_2(s) \\ x_1'(s) & -x_2'(s) \end{vmatrix}}.$$

Thus $c_1(s) = x_2(s)/c$, and $c_2(s) = x_1(s)/c$. This verifies the formula for G.

We now verify that the Green's function G does indeed generate the inverse of A.

V.5.4. Theorem. Suppose 0 is not an eigenvalue of A. Let K be the integral operator defined by

$$Kf(t) = \int_\alpha^\beta G(t,s)f(s)w(s)ds$$

for all f in $C[I]$. Then $A^{-1} = K$. That is,

1. If f is in $C[I]$, then Kf is in D_A, and $AKf = f$.

2. If x is in D_A, then $KAx = x$.

Proof. We recall the following theorem due to Leibnitz: If $f(\cdot,\cdot)$ maps $I \times I$ into the complex numbers, where both $f(t,s)$ and $\frac{\partial f}{\partial t}(t,s)$ are continuous for all s and t in I, then

$$\frac{d}{dt} \int_\alpha^t f(t,s)ds = f(t,t) + \int_\alpha^t \frac{\partial f}{\partial t}(t,s)ds,$$

$$\frac{d}{dt} \int_t^\beta f(t,s)ds = -f(t,t) + \int_t^\beta \frac{\partial f}{\partial t}(t,s)ds.$$

This is shown in the following way: Let $\varepsilon > 0$. Since f is continuous in s and t, there exists a $\delta_1 > 0$ such that if $|h| < \delta_1$ and $|s-t| < \delta_1$, then

$$|f(t+h,s) - f(t,t)| < \epsilon/2,$$

or

$$f(t+h,s) = f(t,t) + e_1(t,s,h),$$

where $|e_1| < \epsilon/2$. Since $\frac{\partial f}{\partial t}$ (t,s) is continuous, there is a $\delta_2 > 0$ such that if $|h| < \delta_2$, then

$$|(f(t+h,s) - f(t,s))/h - \frac{\partial f}{\partial t}(t,s)| < \epsilon/2(\beta-\alpha),$$

or

$$(f(t+h,s) - f(t,s))/h = \frac{\partial f}{\partial t}(t,s) + e_2(s,t,h),$$

where $|e_2| < \epsilon/2(\beta-\alpha)$. Now let

$$F(t) = \int_{\alpha}^{t} f(t,s)ds.$$

Then

$$\left|\frac{F(t+h)-F(t)}{h} - f(t,t) - \int_{\alpha}^{t}\frac{\partial f}{\partial t}(t,s)ds\right| =$$

$$\left|\frac{1}{h}\int_{t}^{t+h}[f(t+h) - f(t,t)]ds + \int_{\alpha}^{t}[(f(t+h,s)-f(t,s))/h - \frac{\partial f}{\partial t}(t,s)]ds\right|,$$

$$\leq \left|\frac{1}{h}\frac{\epsilon}{2}h\right| + \left|\frac{\epsilon}{2(\beta-\alpha)}(t-\alpha)\right| \leq \epsilon.$$

The other formula is similarly verified. We now proceed with the proof of

the theorem.

 1. Let   f   be in C[I], let

$$x(t) = \int_{\alpha}^{\beta} G(t,s)f(s)w(s)ds.$$

Then

$$x(t) = \int_{\alpha}^{t} G(t,s)f(s)w(s)ds + \int_{t}^{\beta} G(t,s)f(s)w(s)ds,$$

where both integrands satisfy the conditions of Leibnitz$'$ theorem.   Then

$$x'(t) = G(t,t)f(t)w(t) + \int_{\alpha}^{t} \frac{\partial G}{\partial t}(t,s)f(s)w(s)ds$$

$$- G(t,t)f(t)w(t) + \int_{t}^{\beta} \frac{\partial G}{\partial t}(t,s)f(s)w(s)ds,$$

$$= \int_{\alpha}^{t} \frac{\partial G}{\partial t}(t,s)f(s)w(s)ds + \int_{t}^{\beta} \frac{\partial G}{\partial t}(t,s)f(s)w(s)ds.$$

$$x''(t) = \frac{\partial G}{\partial t}(t,t-)f(t)w(t) + \int_{\alpha}^{t} \frac{\partial^2 G}{\partial t^2}(t,s)f(s)w(s)ds$$

$$- \frac{\partial G}{\partial t}(t,t+)f(t)w(t) + \int_{t}^{\beta} \frac{\partial^2 G}{\partial t^2}(t,s)f(s)w(s)ds,$$

$$= [1/p(t)]f(t)w(t) + \int_{\alpha}^{\beta} \frac{\partial^2 G}{\partial t^2}(t,s)f(s)w(s)ds.$$

Since G(t,s) satisfies the boundary conditions at α and β, x(t) also does. Thus x is in D_A. From the formulas above, we conclude AKf = f for all f in C[I].

2. By assumption the range of A is in C[I]. By part 1 C[I] is in the range of A. Thus the range of A is C[I]. Since 0 is not an eigenvalue, Ax = 0 implies x = 0, so by Theorem II.2.16 A^{-1} exists, and by Theorem II.2.11 A^{-1} = K.

As the final subject of this chapter we explore the relationship between the eigenvalues of A and those of K.

V.5.5. <u>Theorem</u>. <u>Suppose 0 is not an eigenvalue of A. Then</u>

1. <u>x is an eigenfunction of A with eigenvalue λ if and only if</u> <u>x is an eigenfunction of K with eigenvalue 1/λ.</u>

2. <u>0 is not an eigenvalue of K.</u>

Proof. 1. If Ax = λx and λ ≠ 0, then x = KAx = Kλx, so Kx = (1/λ)x. Conversely, if Kx = (1/λ)x, then AKx = (1/λ)Ax, and Ax = λx.

2. If Kx = 0x, then x = AKx = Aθ = θ.

We have now two equivalent representations of the regular Sturm–Liouville problem

$$(px')' + qx = wx,$$

$$\alpha_1 x(\alpha) + \alpha_2 x'(\alpha) = 0,$$

$$\beta_1 x(\beta) + \beta_2 x'(\beta) = 0,$$

and

$$(1/\lambda)x(t) = \int_\alpha^\beta G(t,s)x(s)w(s)ds$$

where $G(t,s) = G(s,t)$.

We have throughout made the BASIC ASSUMPTION: <u>0 is not an eigenvalue of A.</u>

V.5.6. <u>Theorem.</u> <u>If x_1 and x_2 are eigenfunctions of A corresponding to the same eigenvalue, then they are linearly dependent. The eigenvalues of A have multiplicity 1.</u>

Proof. Let $Ax_1 = \lambda x_1$, $Ax_2 = \lambda x_2$,

$$\alpha_1 x_1(\alpha) + \alpha_2 x_1'(\alpha) = 0, \quad \beta_1 x_1(\beta) + \beta_2 x_1'(\beta) = 0,$$

$$\alpha_1 x_2(\alpha) + \alpha_2 x_2'(\alpha) = 0, \quad \beta_1 x_2(\beta) + \beta_2 x_2'(\beta) = 0.$$

From Theorem V.2.5, part 3, we find

$$\int_{t_1}^{t_2} [Lx_1 \cdot x_2 - x_1 \cdot Lx_2]dt = 0.$$

Thus $W[x_1, x_2]$ is constant. By using either of the boundary conditions, this constant is seen to be 0. Hence x_1 and x_2 are dependent.

V.   Exercises

1. What conditions must a complex differential operator $Lx = a_0 x'' + a_1 x' + a_2 x$ satisfy in order to be formally self-adjoint.

2. Show that if y satisfies the adjoint equation $L^+ y = 0$, then $Lx \cdot \bar{y}$ is integrable, and $Lx = 0$ can be completely solved.

3. Can a complex second order differential operator be made formally self-adjoint by multiplication by an appropriate function?

4. Show that $(L^+)^+ = L$ when L is a second order ordinary differential operator. Is this true for higher order operators? What would L^+ look like for higher order operators?

5. How would one define L^+ if the coefficients in L were not sufficiently differentiable? (Hint: Restrict the domain of L. Check Green's formula, since it essentially depends upon integration by parts.)

6. Assume that x, y are solutions of

$$(px')' + q_1 x = 0,$$

$$(py')' + q_2 y = 0,$$

where $p > 0$ on $I = [\alpha, \beta]$, and

1. when $q_1 = q_2$, x and y are linearly independent, or

2. when $q_2 > q_1$, there exists a $\delta > 0$ such that $x > 0$, $y > 0$ on $(\alpha, \alpha+\beta)$. Let $\lim\limits_{t \to \alpha+} p(t)(x'(t)y(t) - x(t)y'(t)) > 0$ and $x(t_1) = 0$ for some t_1 in (α, β). Show there exists t_2 in (α, t_1) such that $y(t_2) = 0$.

7. Find the eigenvalues, eigenfunctions and Green's functions (if possible) for $Ax = -x''$, $x'(0) = 0$, $x'(1) = 0$. For $Ax = -x''$, $x'(0) = 0$, $x(1) = 0$. For $Ax = -x''$, $x(0) = 0$, $x'(1) = 0$.

References

1. G. Birkhoff and G.C. Rota, "Ordinary Differential Equations," Blaisdell, Waltham, Mass. 1959.

2. M. Bocher, "Lecons sur les Methodes de Sturm," Gauthier-Villars, Paris, 1917.

3. E.A. Coddington and N. Levinson, "Theory of Ordinary Differential Equations," McGraw-Hill, New York, 1955.

4. R.H. Cole, "Theory of Ordinary Differential Equations," Appleton-Century-Crofts, New York, 1968.

5. W. Hurewicz, "Lectures on Ordinary Differential Equations," M.I.T. Press, Cambridge, Mass. 1958.

6. E.L. Ince, "Ordinary Differential Equations," Dover Publications, New York, 1956.

VI.   THE STONE–WEIERSTRASS THEOREM

The following chapter is concerned with a rather remarkable and extremely powerful theorem first proved for polynomials by Weierstrass and then extended to its present form by Stone. Although mathematicians have long recognized its importance, most of those people in applications have for some reason failed to appreciate its significance. It is fundamental to certain arguments later on. We shall point out precisely where at that point. For the moment, we hope the reader will have faith enough to wait.

VI.1. <u>Preliminary Remarks</u>. What follows involves a bit of special notation which might be unfamiliar to the reader. So we briefly diverge to explain the language to be used. We list in succession five definitions, followed by a theorem from topology, which we shall not prove.

The setting which we employ is called a Hausdorff space. It is, roughly speaking, a space in which sets behave as one might näively expect.

VI.1.1. <u>Definition</u>. A Hausdorff space X is one in which for each x, y in X there exist open sets N_x, N_y such that x is in N_x, y is in N_y, and their intersection $N_x \cap N_y$ is empty.

A picture of such a space might look like this.

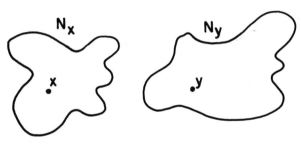

Of course, the Euclidean spaces, in fact, all the Banach spaces, are all Hausdorff spaces.

We also need the notion of compactness.

VI.1.2. Definition. A compact set is one in which every infinite subset contains at least one limit point.

It is easy to see that when the concept of distance is valid, a compact set is closed and bounded.

VI.1.3. Definition. If S is a set of elements, then by the closure $\overline{S}$ of S we mean the set of all points in S together with the set of all limit points of S.

For instance in the Cartesian plane R^2 if $S = \{(x,y): x^2 + y^2 < 1$, x and y are both rational\}$, then $\overline{S} = \{(x,y): x^2 + y^2 \leq 1\}$.

We will also need the idea of a maximum and a minimum of numbers of functions.

VI.1.4. Definition. Let a and b be real numbers. Then the maximum or supremum of a and b is

$$\sup(a,b) = a \vee b = \begin{cases} a, & \text{if } a \geq b, \\ \\ b, & \text{if } b \geq a, \end{cases}$$

The minimum or infimum of a and b is

$$
\inf(a,b) = a \wedge b = \begin{cases} a, & \text{if} \quad a \leq b, \\[2ex] b, & \text{if} \quad b \leq a. \end{cases}
$$

Let f and g be two real valued functions. Then

$$
\sup(f,g) = f \vee g = \begin{cases} f, & \text{whenever,} \quad f(t) \geq g(t), \\[2ex] g, & \text{whenever,} \quad g(t) \geq f(t). \end{cases}
$$

$$
\inf(f,g) = f \wedge g = \begin{cases} f, & \text{whenever,} \quad f(t) \leq g(t), \\[2ex] g, & \text{whenever,} \quad g(t) \leq f(t). \end{cases}
$$

For example if $f(t) = t$ and $g(t) = t^2$, then

$$
f \vee g = \begin{cases} t^2, & -\infty < t \leq 0, \\[1.5ex] t, & 0 \leq t \leq 1, \\[1.5ex] t^2, & 1 \leq t < \infty, \end{cases}
$$

$$
f \wedge g = \begin{cases} t, & -\infty < t \leq 0, \\[1.5ex] t^2, & 0 \leq t \leq 1, \\[1.5ex] t, & 1 \leq t < \infty. \end{cases}
$$

Finally, we will need the idea of a covering.

VI.1.5. Definition. Let X be a set of elements. Suppose for each x in X, there is an open set N_x such that x is in N_x. The collection of sets $\{N_x\}_{x \in X}$ is said to cover X, since for each x in X there is a set N_x in $\{N_x\}_{x \in X}$ such that x is in N_x.

We now state a theorem which is fundamental in basic topology and will be necessary for what is to follow. Although we will not give it, the proof is not difficult. We invite the reader to attempt it.

VI.1.6. Theorem. If X is a compact Hausdorff space and $\{N_x\}_{x \in X}$ is an open covering of X, then there exists a FINITE collection $\{N_{x_i}\}_{i=1}^n$ which also covers X. That is, for each x in X there exists an open set $\{N_{x_i}\}_{i=1}^n$ such that x is in N_{x_i}.

VI.2. Algebras and Subalgebras. Central to the statement of the Stone-Weierstrass theorem is the concept of an algebra, which we have already briefly encountered. We redefine it for the sake of completeness.

VI.2.1. Definition. An algebra is a set of elements denoted by Y, together with a scalar field F, which is closed under the binary operators of + (addition between elements of Y), × (multiplication between elements of Y), · (multiplication of elements in Y by elements from the scalar field F), such that

1. Y together with F, + and · forms a linear space,

2. If f, g, h are in Y, α is in F, then

    a. f × g is in Y,

    b. f × (g × h) = (f × g) × h,

    c. f × (g + h) = f × g + f × h,

    d. (f + g) × h = f × h + g × h,

    e. α(f × g) = (αf) × g = f × (αg).

If a norm $\|\cdot\|$ is defined on Y such that the linear space formed by Y, F, + and · is a Banach space, and if $\|f × g\| \leq \|f\|\|g\|$ for all f, g in Y, then Y, together with its scalar field, binary operations, and norm, is called a Banach algebra.

We shall be primarily concerned with the case where X is a compact Hausdorff space, such as a finite interval, a rectangle, a circle, a sphere, etc., and Y = C[X,R], the space of real valued continuous functions with domain X, and with norm $\|f\| = \sup_{x \in X} |f(x)|$ for all f in C[X,R]. It is an elementary calculation to show C[X,R] is a Banach algebra.

VI.2.2. Definition. A set S is a subalgebra of the algebra A if

1. S is a linear subspace of A,

2. S is closed under the operation ×. That is, if f, g are in S, then f × g is also in S.

In passing we note that if S is a subalgebra of a Banach algebra A, then the closure of S, $\overline{S}$, is also a subalgebra of A.

We now pose the following question: Given a subset T of $C[X,R]$, we consider the set (in $C[X,R]$) generated by taking sums, scalar products, products, uniform limits of elements of T. When is this class of generated functions equal to $C[X,R]$?

Clearly we may assume that T is already a subalgebra of $C[X,R]$, so we will call it S. Our problem is this: When is $\overline{S} = C[X,R]$? The Stone-Weierstrass Theorem provides an answer to this question.

VI.3.   The Stone-Weierstrass Theorem.

VI.3.1.   Lemma. $\sqrt{1-t}$ has a power series representation

$$\sqrt{1-t} = 1 - \sum_{n=1}^{\infty} \frac{(2n-2)!\,t^n}{2^{2n-1}(n-1)!\,n!} \,,$$

which is uniformly convergent on the interval $-1 \leq t \leq 1$.

Partial proof. This is just the Taylor's series for $\sqrt{1-t}$. The ratio test verifies that the interval of convergence is $-1 \leq t \leq 1$. To show that the series also converges at $t = \pm 1$ requires a bit of delicate and uninteresting analysis, which we prefer to avoid. Those who are interested can find it in Hewitt and Stromberg, "Real and Abstract Analysis," page 91.

VI.3.2.   Theorem.   Let S be a subalgebra of $C[X,R]$ containing $f \equiv 1$. Then if f, g are in $\overline{S}$, so are $|f|$, $f \vee g$, and $f \wedge g$.

Proof. Let f be in $\bar{S}$. Then since $\|f\| = \sup_{x \in X} |f(x)|$, $|f(x)| \le \|f\|$

for all x in X. Let us assume temporarily that $\|f\| \le 1$, since this can

always be obtained by multiplying f by $1/\|f\|$, when f has a greater norm.

Now let $\varepsilon > 0$. There exists an N_ε such that

$$\left| \sqrt{1-t} - (1 - \sum_{n=1}^{m} \frac{(2n-2)!\, t^n}{2^{2n-1}(n-1)!\, n!}) \right| < \varepsilon$$

when $m \ge N_\varepsilon$ for all t in $-1 \le t \le 1$. Thus letting $t = (1-f^2)$

$$\left| \sqrt{1-(1-f^2)} - (1 - \sum_{n=1}^{m} \frac{(2n-2)!\,(1-f^2)^n}{2^{2n-1}(n-1)!\, n!}) \right| < \varepsilon$$

for all x in X.

Since $1-f^2 \le 1$, and $|f| = \sqrt{1-(1-f^2)}$,

$$|f| = 1 - \sum_{n=1}^{\infty} \frac{(2n-2)!\,(1-f^2)^n}{2^{2n-1}(n-1)!\, n!} \, ,$$

which is in $\bar{S}$. When $\|f\| > 1$, we replace f by $f/\|f\|$. We then multiply

by $\|f\|$ to find

$$|f| = \|f\| \, [1 - \sum_{n=1}^{\infty} \frac{(2n-2)!\,[1 - [f/\|f\|]^2)^n}{2^{2n-1}(n-1)!\, n!}] \, ,$$

which is in $\bar{S}$.

Since

$$1/2[f+g + |f-g|] = f \vee g,$$

$$1/2[f+g - |f-g|] = f \wedge g,$$

these are also in $\overline{S}$.

VI.3.3. Definition. A subset S of C[X,R] separates points of X if, for every pair of distinct elements x_1, x_2 in X, there exists f in S such that $f(x_1) \neq f(x_2)$.

As an example of a subset S which does not separate points, let $X = [0,1]$, and let S be generated by sums, products and limits of the functions $f_1 \equiv 1$ and $f_2 = \begin{cases} 3t, & 0 \le t \le 1/3 \\ 1, & 1/3 \le t \le 1 \end{cases}$. Then $\overline{S}$ only contains functions which are constant on $[1/3,1]$, and, therefore, fails to separate points.

On the other hand the set S generated by 1 and t does separate points, since t varies from point to point.

VI.3.4. Lemma. Let S be a subalgebra of C[X,R] such that

1. S contains $f \equiv 1$,

2. S separates points of X.

Then

1. If x_1, x_2 are distinct elements of X and α and β are real numbers, then there exists f in S such that $f(x_1) = \alpha$, $f(x_2) = \beta$.

2. If f is in C[X,R], x is in X, and $\varepsilon > 0$, then there exists

g in $\overline{S}$ such that $g(x_1) = f(x_1)$, and $g(x) \leq f(x)+\varepsilon$ for all x in X.

Proof. Since there exists g in S such that $g(x_1) \neq g(x_2)$, let

$$f(x) = \alpha + (\beta-\alpha) \frac{g(x) - g(x_1)}{g(x_2) - g(x_1)}.$$

To show the second part, for each $y \neq x_1$, let $h_y(x)$ be a continuous

function such that $h_y(x_1) = f(x_1)$, and $h_y(y) \leq f(y)+\varepsilon$. For $y = x_1$, let

$h_{x_1}(x) = f(x)$. Since these functions are all continuous, for each y in

X there exists a neighborhood N_y of y and function $h_y(x)$ such that

$h_y(x_1) = f(x_1)$, and $h_y(x) \leq f(x)+\varepsilon$, when y is in N_y.

Now $\{N_y\}_{y \in X}$ covers X. Since X is compact, there exists a <u>finite</u>

subcovering $\{N_{y_i}\}_{i=1}^{n}$ which also covers X. Let

$$g = h_{y_1} \wedge h_{y_2} \cdots \wedge h_{y_n}.$$

Then $g(x_1) = f(x_1)$, and $g(x) \leq f(x)+\varepsilon$ for all x in X.

VI.3.5. <u>Theorem.(The Stone-Weierstrass Theorem)</u>. Let X be a compact

Hausdorff space. If a subalgebra S of C[X,R] contains the constant

functions and separates points of X, then S is dense in C[X,R]. That is,

$\overline{S} = C[X,R]$.

Proof. Let f be in C[X,R], $\varepsilon > 0$ and y in X. Then Lemma VI.3.4

shows the existence of a function g_y in $\overline{S}$ such that $g_y(y) = f(y)$,

$g_y(x) \le f(x)+\varepsilon$ for all x in X. For all x in a neighborhood N_y of

y we can also require $g_y(x) > f(x)-\varepsilon$, since f and y are continuous.

Now $\{N_y\}_{y \in X}$ covers X. Since X is compact, a finite subcovering

$\{N_{y_i}\}_{i=1}^n$ also covers X.

Let

$$g = g_{y_1} \vee g_{y_2} \cdots \vee g_{y_n} .$$

Then $g(x) \le f(x)+\varepsilon$ and $g(x) \ge f(x)-\varepsilon$ for all x in X. Thus

$|f(x) - g(x)| \le \varepsilon$ for all x in X, and

$$\|f-g\| = \sup_{x \in X} |f(x) - g(x)| \le \varepsilon,$$

By choosing a sequence of functions g_i, i = 1,... such that

$\|f-g_i\| < 1/i$, i = 1,..., we see that since all the functions g_i, i = 1,...

are in S, f is in $\bar{S}$, and $\bar{S} = C[X,R]$.

Thus a subalgebra of functions which separates points and which contains

the function 1 can be used to uniformly approximate any continuous function,

so long as the domain of those functions X is compact.

VI.4. Extensions and Special Cases. The Stone-Weierstrass Theorem is false

if the range R is replaced by K, the complex numbers, as we easily see

by the following example. Let X = {z: $|z = x + iy$, $|z| \le 1$}. Let S be

the subalgebra generated by 1 and z. It is well known then that $\bar{S}$ con-

sists of all bounded analytic functions, defined for $|z| \le 1$. But $z^* = x-iy$

is in C[X,K] and is not analytic. Thus $\overline{S} \neq$ C[X,K]. We need therefore stronger assumptions concerning S. in order to have $\overline{S} = $ C[X,K].

VI.4.1. <u>Theorem</u>. <u>Let X be a compact Hausdorff space. If a subalgebra</u> <u>S of C[X,K] contains the constant functions, separates points of X, and is</u> <u>such that when f is in S, f^{*}, its complex conjugate, is also in S,</u> <u>then S is dense in C[X,K]. That is, $\overline{S} = $ C[X,K].</u>

Proof. Let f be in S. Then

$$\text{Re } f = (1/2)(f + f^{*}),$$

$$\text{Im } f = (1/2i)(f - f^{*})$$

are also in S. Let S_{R} be the real subalgebra of S consisting of all real valued functions in S. Then S_{R} separates points of X, since if x_{1}, x_{2} are in X, there exists f in S such that $f(x_{1}) \neq f(x_{2})$. Thus Re $f(x_{1}) \neq$ Re $f(x_{2})$, or Im $f(x_{1}) \neq$ Im $f(x_{2})$, and these are both in S_{R}. Clearly S_{R} also contains 1. Hence $\overline{S}_{R} = $ C[X,R] and $\overline{S}_{R} + i\overline{S}_{R} = $ C[X,R] + iC[X,R] = C[X,K].

VI.4.2. <u>Examples</u>. 1. Let X = [α,β], $-\infty < \alpha < \beta < \infty$. Let S be the algebra generated by 1 and t. That is, S consists of polynomials in t. Then $\overline{S} = $ C[[α,β],R], or, if complex coefficients are permitted, $\overline{S} = $ C[[α,β],K]. This is the original version due to Weierstrass.

2. More generally, let X be a compact subspace of Euclidean space R^n. If points in R^n have coordinates $(t_1, \ldots, t_n)$, then when S is generated by $1, t_1, \ldots, t_n$, i.e., consists of polynomials in n variables, it is dense in $C[X,R]$, or $C[X,K]$, depending upon the scalar field.

This is easy to show, since the functions $1, t_1, \ldots, t_n$ separate points. (Any two distinct points in X must have at least one coordindate differing. That coordinate function then separates them.) Thus given any real or complex valued continuous function f in n variables and any $\varepsilon > 0$, there exists a polynomial $\sum_{\alpha_1, \ldots, \alpha_n = 0}^{N} C_{\alpha_1 \ldots \alpha_n} t_1^{\alpha_1} \ldots t_n^{\alpha_n}$ such that

$$\left| f(t_1, \ldots, t_n) - \sum_{\alpha_1 \ldots \alpha_n = 0}^{N} C_{\alpha_1 \ldots \alpha_n} t_1^{\alpha_1} \ldots t_n^{\alpha_n} \right| < \varepsilon$$

for all $(t_1, \ldots, t_n)$ within the compact set X.

A word of caution: If a finer approximation is desired and ε is made smaller, all the coefficients will probably change.

$$C_{\alpha_1 \ldots \alpha_n} = C_{\alpha_1 \ldots \alpha_n}(f, \varepsilon).$$

The coefficients depend upon both f and ε.

3. In the Euclidean plane R^2 let $X = \{(x,y): x^2 + y^2 = 1\}$, the unit circle, and consider $C[X,K]$. It is evident that $C[X,K]$ is equivalent to the space of complex valued, continuous functions with period 2π. Each f in $C[X,K]$ can be represented by

$$f = \{f(t), \quad 0 \le t \le 2\pi\}.$$

Now consider the subalgebra S generated by 1, e^{it}, e^{-it}. If $t_1 \neq t_2$, then $e^{it_1} \neq e^{it_2}$. So S separates points. Hence any complex valued, continuous function with period 2π can be uniformly approximated by a trigonometric polynomial $\sum_{n=-N}^{N} C_n e^{int}$.

VI.   Exercises

1.   In the plane, give an example of an open covering. Give an example of a bounded, but not closed set S and an open covering such that no finite subset of this covering also covers S.

2.   Prove Theorem VI.1.6.   Why is compactness needed?

3.   Prove Lemma VI.3.1.   (See Hewitt and Stromberg, "Real and Abstract Analysis," page 91.

4.   Let $X = [\alpha_1, \beta_1] \times [\alpha_2, \beta_2]$, where $-\alpha < \alpha_i < \beta_i < \infty$, $i = 1,2$. Show that any real or complex valued, continuous function $f(\cdot, \cdot)$ can be uniformly approximated over X by continuous functions of the form $\sum_{j=1}^{N} u_j(x) v_j(y)$. Hint: Consider the subalgebra of $C[X,K]$ or $C[X,R]$ generated by products of the form $u(x)v(y)$.

5.   Show that the trigonometric functions

$$(1, \sin t, \cos t, \ldots \sin nt, \cos nt \ldots)$$

are dense in $C[[0,2\pi],R]$.

References

1. N. Dunford and J.T. Schwartz, "Linear Operators, vol. 1," Interscience,
       New York, 1958.

2. E. Hewitt and K. Stromberg, "Real and Abstract Analysis," Springer-Verlag,
       New York, 1965.

VII. HILBERT SPACES

Hilbert spaces were essentially invented by the German mathematician David Hilbert at the beginning of this century while studying quadratic forms. They were first formulated abstractly by John von Neumann during the latter part of the 1920's in order to use them as settings for the study of both ordinary and partial differential equations. We shall follow von Neumann's lead, using them throughout the remainder of this book as the principal setting for our treatment of eigenfunction expansions and their application to partial differential equations.

VII.1. <u>Hermitian Forms</u>. The fundamental concept which distinguish Hilbert spaces from other spaces is that of a hermitian form. Historically it originated with Hilbert.

VII.1.1. <u>Definition</u>. <u>Let X be a complex (real) linear space. A hermitian form on X is a function f whose domain is $X \times X$, whose range is K, the complex numbers (or R, the real numbers), and which satisfies</u>

1. $f(x_1 + x_2, y) = f(x_1, y) + f(x_2, y)$, $f(\alpha x, y) = \alpha f(x, y)$, <u>for all x_1, x_2, x, y in X and α in K (or R)</u>,

2. $f(x, y) = \overline{f(y, x)}$ <u>for all x, y in X.</u>

It is an elementary exercise to show that 1 and 2 imply

$$f(x, y_1 + y_2) = f(x, y_1) + f(x, y_2), \quad f(x, \alpha y) = \overline{\alpha} f(x, y)$$

for all x, y, y_1, y_2 in X and α in K (or $\mathbb{R}$). Thus f is linear in the first variable and conjugate linear in the second.

VII.1.2. Theorem. Let f be a hermitian form on a linear space X. Then

1. $f(x,x)$ is real valued for all x in X.

2. If X is a real linear space, then f is bilinear and symmetric. That is, f is linear in both variables, and is unchanged when the variables are reversed.

3. If $\{x_j\}_{j=1}^n$ and $\{y_k\}_{k=1}^n$ are in X, and $\{\alpha_j\}_{j=1}^n$ and $\{\beta_k\}_{k=1}^n$ are complex (real) numbers, then

$$f\left(\sum_{j=1}^n \alpha_j x_j, \sum_{k=1}^n \beta_k y_k\right) = \sum_{j=1}^n \sum_{k=1}^n \alpha_j \overline{\beta}_k f(x_j, y_k).$$

We leave the proof as an exercise.

VII.1.3. Theorem (Polarization Identity). If x and y are elements in a linear space X, and f is a hermitian form on X, then

1. when X is real,

$$4f(x,y) = f(x+y, x+y) - f(x-y, x-y).$$

2. **when X is complex,**

$$4f(x,y) = f(x+y,x+y) + f(x-y,x-y)$$

$$+ if(x+iy,x+iy) - if(x-iy,x-iy).$$

The proof follows from expanding the terms on the right side of these equations. All terms, except those which appear on the left, cancel.

VII.1.4. **Examples.** 1. Consider ℓ^2, the space of vectors of the form $(x_1,x_2,\ldots,x_n,\ldots)$ which satisfy $\sum_{j=1}^{n} |x_j|^2 < \infty$. If $x = (x_1,x_2,\ldots,x_n,\ldots)$ and $y = (y_1,y_2,\ldots,y_n,\ldots)$ are in X and $\{w_{ij}\}_{i,j=1}^{\infty}$ are complex numbers which satisfy $w_{jk} = \overline{w_{kj}}$, then

$$f(x,y) = \sum_{j=1}^{\infty} \sum_{k=1}^{\infty} w_{jk}x_j\overline{y}_k$$

is a hermitian form.

2. If μ is a countably additive function, defined on the measurable subsets of an interval $[\alpha,\beta]$, and x, y are μ-measurable functions on $[\alpha,\beta]$, then

$$f(x,y) = \int_{\alpha}^{\beta} x(t)\overline{y}(t)\mu(dt)$$

is also a hermitian form. In the special case where $\mu(dt) = dt$, f is defined by ordinary (Lebesgue) integration.

VII.1.5. __Definition.__ A hermitian form f on a linear space X is positive

if $f(x,x) \geq 0$ for all x in X. If $f(x,x) > 0$ for all $x \neq 0$ in X,

and $f(x,x) = 0$ if and only if $x = \theta$, then f is positive definite.

VII.1.6. __Theorem (Schwarz's Inequality).__ If f is a positive hermitian form

on a linear space X, then

$$|f(x,y)|^2 \leq f(x,x)f(y,y)$$

for all x, y in X.

Proof. Let $a = f(x,x)$, $b = f(x,y)$, $c = f(y,y)$. Then $a \geq 0$, $c \geq 0$.

   If $c > 0$, we expand $f(x+\lambda y, x+\lambda y) \geq 0$.

$$f(x,x) + \lambda \overline{f(x,y)} + \overline{\lambda} f(x,y) + |\lambda|^2 f(y,y) \geq 0.$$

so

$$a + \lambda \overline{b} + \overline{\lambda} b + |\lambda|^2 c \geq 0.$$

If $\lambda = -b/c$, then

$$a - |b|^2/c - |b|^2/c + |b|^2 \cdot c/c^2 \geq 0,$$

and $ac \geq b^2$. By interchanging the roles of a and c, we find $ac \geq b^2$

when $a > 0$ and $c = 0$.

If $a = 0$, $c = 0$,   then $\lambda\overline{b} + \overline{\lambda}b \geq 0$.   Then, if $\lambda = -b$,   we have

$-2|b|^2 \geq 0$,   and so $b = 0$ also.

VII.1.7. Theorem (Minkowski's Inequality). If f is a positive hermitian

form on a linear space X,   then

$$f(x+y,x+y)^{1/2} \leq f(x,x)^{1/2} + f(y,y)^{1/2}$$

for all x, y in X.

Proof. From Schwarz's inequality

$$f(x,y) + f(y,x) = 2 \text{ Re } f(x,y) \leq 2 \ f(x,x)^{1/2}f(y,y)^{1/2}.$$

Adding $f(x,x) + f(y,y)$ to both sides of the inequality, we find

$$f(x+y,x+y) \leq [f(x,x)^{1/2} + f(y,y)^{1/2}]^2.$$

We then take a square root.

The following is called the parallelogram law, since in the plane it corresponds to the following picture, with $f(x,x)^{1/2}$ denoting Euclidean Length.

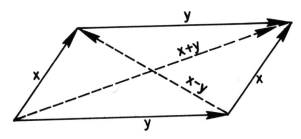

VII.1.8. Theorem (Parallelogram Law). If f is a positive hermitian form
on a linear space X,   then

$$2f(x,x) + 2f(y,y) = f(x+y,x+y) + f(x-y,x-y)$$

for all x,   y   in X.

Proof. Expanding the right side,

$$f(x+y,x+y) + f(x-y,x-y) = f(x,x) + f(x,y) + f(y,x) + f(y,y)$$

$$+ f(x,x) - f(x,y) - f(y,x) + f(y,y) = 2f(x,x) + 2f(y,y).$$

The following theorem again has a pictorial interpretation in the
Euclidean plane. When $f(x,x)^{1/2}$ corresponds to Euclidean length, two
vectors are orthogonal when $f(x,y) = 0$.

VII.1.9. Theorem (Pythagorean Theorem). If f is a positive hermitian form o
a linear space X and $f(x,y) = 0$ for some x, y in X,   then

$$f(x,x) + f(y,y) = f(x+y,x+y).$$

Proof. We expand the right side. The mixed terms cancel.

VII.2. Inner Product Spaces. We implied in the previous section that positi
hermitian forms sometimes correspond to a device for measuring length. This
is always the case when the form is positive definite. Such a form can

always be thought of as generating a distance function.

VII.2.1. Definition. A complex (real) inner product space H is a complex
(real) linear space X together with a positive definite hermitian form f.

   For each x, y in X we call the hermitian form the inner product
between x and y. We denote the inner product by (x,y):

$$(x,y) = f(x,y).$$

For each x in X, the norm, or length, of x is given by

$$\|x\| = (x,x)^{1/2}.$$

It is easy to see that H thus defined is a normed linear space.

VII.2.2. Definition. If an inner product space is complete, that is, a
Banach space, then it is a Hilbert space.

   A Hilbert space, therefore, is a complete, normed, linear space, whose
norm is generated by an inner product or positive definite hermitian form.

VII.2.3. Definition. Let x, y be in an inner product space H. If
$(x,y) = 0,$ then x and y are orthogonal.

   A set A in H is orthogonal if each pair of elements in A is
orthogonal.

For computational purposes, elements with norm 1 are especially useful.

VII.2.4. Definition. An element x in an inner product space H is normal
if $\|x\| = 1$.

A set A in H is normal if each element in A is normal.

A set which is both orthogonal and normal is called an orthonormal
set.

VII.2.5. Examples. For x, y in an inner product space H, the inequaliti
and identities of the previous section have the following form.

Polarization Identity. If H is a real inner product space,

$$(x,y) = (1/4)[\|x+y\|^2 - \|x-y\|^2].$$

If H is a complex inner product space,

$$(x,y) = (1/4)[\|x+y\|^2 - \|x-y\|^2 + i\|x+iy\|^2 - i\|x-iy\|^2].$$

Schwarz's Inequality.

$$|(x,y)| \leq \|x\|\|y\|.$$

Minkowski's Inequality.

$$\|x+y\| \leq \|x\| + \|y\|.$$

Parallelogram Law.

$$2\|x\|^2 + 2\|y\|^2 = \|x+y\|^2 + \|x-y\|^2.$$

Pythagorean Theorem. If x and y are orthogonal, that is $(x,y) = 0$, then

$$\|x\|^2 + \|y\|^2 = \|x+y\|^2.$$

VII.2.6. Examples. We conclude this section with a refinement of those examples previously used to illustrate hermitian forms. The following are positive definite hermitian forms.

1. In ℓ^2 we have seen that

$$f(x,y) = \sum_{j=1}^{\infty} \sum_{i=1}^{\infty} w_{jk} x_j y_k$$

is a hermitian form when $w_{jk} = \overline{w}_{kj}$. We consider here the special case where

$$w_{jk} = \begin{cases} 1 & \text{if } j = k, \\ 0 & \text{if } j \neq k. \end{cases}$$

We leave as an exercise the proof that f generates an inner product for ℓ^2.

2. The second example was defined by

$$f(x,y) = \int_{\alpha}^{\beta} x(t)\overline{y}(t)\mu(dt),$$

where μ is a countably additive function on subsets of $[\alpha,\beta]$, and x, y are μ measurable functions of $[\alpha,\beta]$. To insure that f is positive definite, we require that μ be monotone increasing. That is, if A and B are μ measurable subsets of $[\alpha,\beta]$ and $A \subset B$, then $\mu(A) \leq \mu(B)$.

VII.3. <u>Hilbert Spaces</u>. So far we have been working with an inner product space generated by a positive definite hermitian form. Since such a space is linear, the usual combining processes are valid. We have the same problem we encountered earlier, however, in trying to take limits. This can only be accomplished in a complete inner product space or Hilbert space. We find, however, that just a normed linear space could be isometrically imbedded in a Banach space, so also can an inner product space be isometrically imbedded in a Hilbert space. Our procedure is mathematically the same as before, but the notation is slightly different.

VII.3.1. <u>Definition</u>. <u>Let X be a linear space and M a subspace of X.</u> <u>Then two elements x, y in X are equivalent (modulo M) if x-y is in</u> <u>M. We write</u>

$$x \equiv y \quad (\text{mod } M).$$

VII.3.2. <u>Theorem</u>. <u>The relation $\cdot \equiv \cdot$ (mod M) is an equivalence relation.</u> <u>That is, it is</u>

  1. <u>Reflexive</u>: $x \equiv x$ (mod M) for all x in X,

2. <u>Symmetric</u>: If $x \equiv y$ (mod M), then $y \equiv x$ (mod M) for all x, y in X.

3. <u>Transitive</u>: If $x \equiv y$ (mod M) and $y \equiv z$ (mod M), then $x \equiv z$ (mod M) for all x, y, z in X.

Proof. 1. $x - x = \theta$, which is in every subspace, so $x \equiv x$.

2. If $x - y = m$ is in M, then $y - x = -m$, which is also in M. Thus $x \equiv y$ implies $y \equiv x$.

3. If $x - y = m_1$ and $y - z = m_2$ are in M, then $x - z = m_1 + m_2$, which is also in M. Thus $x \equiv y$ and $y \equiv z$ imply $x \equiv z$.

VII.3.3. <u>Definition</u>. Let X be a linear space and M a subspace of X. We denote by [x] the set of all elements in X equivalent to x(modulo M).

VII.3.4. <u>Theorem</u>. Let X be a linear space and M a subspace of X. Let x be an element in X. Then

1. <u>All elements equivalent to x(modulo M) are equivalent to each other.</u>

2. <u>The sets {[x]:x $\in$ X} are either identical or are mutually disjoint.</u>

3. $X = \cup \{[x]\}$.
       $x \in X$

Proof. 1. If $x \equiv y$ and $x \equiv z$, then $y - x = m_1$ and $x - z = m_2$ are in M. Thus $y - z = m_1 + m_2$ is in M, and $y \equiv z$.

2. Suppose two sets [x] and [y] have an element z in common and w is an arbitrary element in [x]. Then $x-z = m_1$, $y-z = m_2$, $x-w = m_3$ are all in M. Further, $y-w = m_2-m_1-m_3$ is also in M, so w is in [y]. Similarly every element of [y] is in [x], and [x] = [y].

3. It is obvious that $X = \bigcup_{x \in X} \{[x]\}$.

We now show that these sets of equivalent elements can be considered as a linear space themselves.

VII.3.5. Definition. Let X/M be the set of equivalence classes [x] of a linear space X modulo a subspace M. We define addition in X/M by

$$[x] + [y] = [x+y]$$

for all x, y in X. We define scalar multiplication in X/M by

$$\alpha[x] = [\alpha x]$$

for all x in X and complex (real) numbers α.

VII.3.6. Theorem. If X is a linear space and M is a subspace, then X/M is a linear space (called the quotient space).

Proof. This is obvious. It is clear that addition and scalar multiplication are independent of the representatives in [x] and [y].

We need one additional concept, that of a pseudonorm, in order to imbed an inner product space in a Hilbert space.

VII.3.7. Definition. Let X be a linear space. Let $/\cdot/$ be a function with domain X and range R^+, the nonnegative real numbers, which also satisfies

1. $/x+y/ \leq /x/ + /y/$.

2. $/\alpha x/ = |\alpha| \cdot /x/$

for all x, y in X and all complex (real) numbers α. Then $/\cdot/$ is called a pseudonorm.

Only requirement 3 of Definition II.3.1, which defined a norm, is missing. It is possible for an element $x \neq \theta$ to satisfy $/x/ = 0$. This difficulty is eliminated in the following way:

VII.3.8. Theorem. Let X be a linear space with pseudonorm $/\cdot/$. Let M be the subspace of X with pseudonorm equal to 0. (m is in M if and only if $/m/ = 0$). Then X/M is a normed linear space under the norm

$$\| [x] \| = /x/.$$

Proof. There are two steps to the proof. First we must show M is a subspace. Then we must show that $\| \cdot \|$ is, in fact, a norm.

If m_1 and m_2 are in M, then $/m_1/ = 0$ and $/m_2/ = 0$. Since

$/m_1+m_2/ \leq /m_1/ + /m_2/$, $/m_1+m_2/ = 0$ also, and m_1+m_2 is in M. Similarly, the second requirement of Definition VII.3.7 shows $/\alpha m/ = 0$ for all complex (real) numbers α.

We must now show that $\|\cdot\|$ is a norm. Suppose [x] and [y] are two representations for the same set. Then since part 1 of Definition VII.3.7 implies

$$\left| /x/ - /y/ \right| \leq /x-y/,$$

and x-y is in M, $/x/ = /y/$. Again from the Definition, $\|\alpha[x]\| =$ $\|[\alpha x]\| = /\alpha x/ = |\alpha|/x/ = |\alpha|\|[x]\|$. If $\|[x]\| = 0$, then $/x/ = 0$, and x is in M. Since [0] = M is the zero element in X/M and x-0 = x, $x \equiv 0$ (mod M) and $[x] = \theta$. Clearly if $[x] = \theta$, then $\|[x]\| = 0$. So $\|\cdot\|$ is a norm and X/M is a normed linear space.

We have in a sense removed, or factored out, the nonzero elements in X which had 0 pseudonorm.

VII.3.9. <u>Theorem</u>. <u>Let X be a normed linear space with norm denoted by</u> $\|\cdot\|$. <u>Let Y be the pseudonormed linear space of Cauchy sequences in X: if</u> $\{x_n\}_{n=1}^{\infty}$ <u>and</u> $\{y_n\}_{n=1}^{\infty}$ <u>are Cauchy sequences in X and α is a complex</u> <u>(real) number</u>

$$\{x_n\}_{n=1}^{\infty} + \{y_n\}_{n=1}^{\infty} = \{x_n+y_n\}_{n=1}^{\infty},$$

$$\alpha\{x_n\}_{n=1}^{\infty} = \{\alpha x_n\}_{n=1}^{\infty},$$

and

$$\| \{x_n\}_{n=1}^{\infty} \|_Y = \lim_{n \to \infty} \|x_n\|.$$

Finally let M be the subspace of Y with pseudonorm   0 (in Y). Then

1. Y/M is a Banach space.

2. If X is a normed algebra, then Y/M is a Banach algebra.

3. There exists an isometric copy of X in Y/M, which we will also
denote by X. With this identification, X is dense in Y/M, that is every
element in Y/M can be arbitrarily approximated in norm by an element in X.

4. If X is an inner-product space, then Y/M is a Hilbert space.

Proof. Part 1 was proved in Theorem II.3.6. Part 2 is a trivial extension,
as is part 3. To complete the proof we need only show that the inner product
defined on X generates an inner product in Y/M. This follows immediately
from the polarization identities in Example VII.2.5.

VII.3.10. Examples. 1. Let X be the set of all complex valued, continuous
functions, which are defined on n-dimensional space R^n, and which vanish
outside some compact subset of R^n. For each x in X let

$$\|x\|_p = [\int_{R^n} |x|^p dt]^{1/p},$$

where $p \geq 1$. Then X is an incomplete normed linear space. Its completion is denoted by $L^p(R^n)$.

In the special case $p = 2$, $L^2(R^n)$ is a Hilbert space with norm

$$\|x\| = [\int_{R^n} |x|^2 dt]^{1/2}$$

and inner product

$$(x,y) = \int_{R^n} \overline{y} x \, dt$$

for each x, y in $L^2(R^n)$.

2. Euclidean n-dimensional space R^n is a Hilbert space when the norm and inner product are defined in the following way. If $x = (x_1, \ldots, x_n)$ and $y = (y_1, \ldots, y_n)$ are in R^n, then

$$\|x\| = [\sum_{j=1}^{n} |x_j|^2]^{1/2},$$

$$(x,y) = \sum_{j=1}^{n} \overline{y}_j x_j.$$

This is valid in either the case where the components are complex or real valued. In either case we denote the space by E^n. They are called Euclidean spaces.

VII.4. Orthogonal Subspaces. The subject of this section, as its title

implies, is orthogonality. We first find elements orthogonal to subspaces.

Next we examine orthogonal subspaces. Finally we characterize a Hilbert

space in terms óf a subspace and its orthogonal compliment.

VII.4.1. Theorem. Let M be a closed linear manifold in an inner product

space H. Let x be in H and let $\delta = \inf_{y \in M} \|x-y\|$. Then there exists a

unique element y_0 in M such that $\|x-y_0\| = \delta$.

Proof. Let $\{y_n\}_{n=1}^{\infty}$ be in M and satisfy $\lim_{n \to \infty} \|y_n - x\| = \delta$. In the

parallelogram law,

$$2\|x\|^2 + 2\|y\|^2 = \|x+y\|^2 + \|x-y\|^2,$$

replace x by $x-y_n$ and y by $x-y_m$ to find

$$\|y_n - y_m\|^2 + 4\|x - \tfrac{1}{2}(y_n + y_m)\|^2 = 2\|x-y_n\|^2 + 2\|x-y_m\|^2,$$

or

$$\|y_n - y_m\|^2 = 2\|x-y_n\|^2 + 2\|x-y_m\|^2 - 4\|x - \tfrac{1}{2}(y_n + y_m)\|^2.$$

Since M is a linear manifold, $\tfrac{1}{2}(y_n + y_m)$ is in M, and $\|x - \tfrac{1}{2}(y_n + y_m)\| \geq \delta$.

Thus

$$\|y_n - y_m\| \leq 2\|x-y_n\|^2 + 2\|x-y_m\|^2 - 4\delta^2.$$

As n, m approach ∞, the right side approaches $2\delta^2 + 2\delta^2 - 4\delta^2 = 0$.

Thus $\{y_n\}_{n=1}^{\infty}$ is a Cauchy sequence in M. Since M is closed, $\{y_n\}_{n=1}^{\infty}$

approaches a limit y_0 in M. Since

$$\|x-y_0\| \le \|x-y_n\| + \|y_n-y_0\|,$$

and $\|y_n-y_0\|$ approaches 0, $\|x-y_0\| \le \delta$. But y_0 is in M, so $\|x-y_0\| \ge \delta$.

Thus $\|x-y_0\| = \delta$.

If two elements, y_0 and y_0', exist in M and satisfy $\|x-y_0\| = \delta$,

$\|x-y_0'\| = \delta$, then

$$\|y_0-y_0'\|^2 = 2\|x-y_0\|^2 + 2\|x-y_0'\|^2 - 4\|x - \tfrac{1}{2}(y_0+y_0')\|^2,$$

$$\le 2\delta^2 + 2\delta^2 - 4\delta^2,$$

$$= 0.$$

So $y_0 = y_0'$.

We remark that the use of the parallelogram law here is essential. In

fact in a Banach space the theorem is not always true. For example, consider

C[0,1], the space of real valued continuous functions, defined on the interval

[0,1], with the norm

$$\|y\| = \sup_{t\in[0,1]} |y(t)|$$

for all x in C[0,1]. Let M = {y:y ∈ C[0,1], y(0) = 0}. Let x = 1.

Then for any y in M satisfying 0 ≤ y ≤ 1,

$$\|1-y\| = \sup_{t \in [0,1]} |1-y(t)| = 1.$$

Thus $\inf_{y \in M} \|1-y\| = 1$ (see figure). Many functions minimize the distance

from 1 to M.

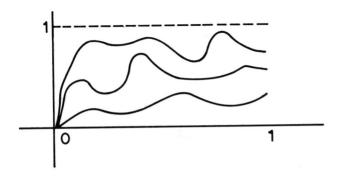

VII.4.2. <u>Theorem.</u> <u>Let M and N be closed linear manifolds in an inner</u>

<u>product space H, M $\subsetneq$ N. Then there exists an element z in N,</u>

<u>z ≠ 0, such that z is orthogonal to M.</u>

Proof. If x is in N − M, and $\delta = \inf_{y \in M} \|y-x\|$, then there exists a unique

y_0 in M such that $\|y_0-x\| = \delta$. Let $z = y_0-x$. If y is in M, then

$$\|z+\lambda y\| = \|y_0-x+\lambda y\| \geq \delta = \|z\|.$$

So

$$0 \leq \|z+\lambda y\|^2 - \|z\|^2$$

for all y in M and any λ. Expanding

$$0 \leq \|z\|^2 + \lambda(y,z) + \overline{\lambda}(z,y) + |\lambda|^2\|y\|^2 - \|z\|^2.$$

If $(y,z) \neq 0$, let $\lambda = -r|(y,z)|/(y,z)$, where $r > 0$. Then

$$0 \leq -r|(y,z)| - r|(y,z)| + r^2\|y\|^2,$$

and

$$2|(y,z)| \leq r\|y\|^2.$$

As r approaches 0, we arrive at a contradiction. Thus $(y,z) = 0$ for
y in M, and z is orthogonal to M.

So far we have considered the possibility of elements in H being
orthogonal to subspaces. This concept can be easily extended.

VII.4.3. Definition. Let M be a subset of an inner product space H.
Then $M^\perp$ is the set of all elements x in H such that x is orthogonal
to M.

While M may be an arbitrary set in H, $M^\perp$ has substantially more
structure.

VII.4.4. Theorem. 1. If H is a Hilbert space, then

1. $M^{\perp}$ is a closed linear manifold.

2. $M \cap M^{\perp} = \{\theta\}$, when $\theta \in M$.

3. $M \subset M^{\perp\perp}$.

4. If $M \subset N$, then $N^{\perp} \subset M^{\perp}$.

5. $M^{\perp} = M^{\perp\perp\perp}$.

Proof. 1. Let $\lim x_n = x$ be in H, and $\{x_n\}_{n=1} \subset M^{\perp}$. For each m in M,

$$|(x,m)| = |(x-x_n,m)| \leq \|x_n-x\| \|m\|,$$

which approaches 0. So $(x,m) = 0$, and x is in $M^{\perp}$. This shows M is closed. It is obviously a linear manifold.

Parts 2 and 3 are easily seen by inspection.

To show part 4, let x be in $N^{\perp}$. Then $(x,y) = 0$ for all y in N. In particular $(x,y) = 0$ for all y in M, since $M \subset N$. Thus x is in $M^{\perp}$.

For part 5, we note $M \subset M^{\perp\perp}$, so by part 4, $M^{\perp\perp\perp} \subset M^{\perp}$. Also by applying part 3 to $M^{\perp}$, $M^{\perp} \subset M^{\perp\perp\perp}$. They are therefore equal.

VII.4.5. Theorem. Let M be a closed linear manifold of a Hilbert space H. Then $M = M^{\perp\perp}$.

Proof. From the previous theorem we know that $M \subset M^{\perp\perp}$. Suppose there

exists z in $M^{\perp\perp} - M$. By Theorem VII.4.2 we may assume that z is

orthogonal to M. Thus z is in $M^{\perp} \cap M^{\perp\perp}$. By Theorem VII.4.4, part 2

$z = \theta$.

We are now in a position to characterize a Hilbert space H in terms

of orthogonal subspaces. We preceed the actual theorem by a lemma.

VII.4.6. <u>Lemma. Let M and N be closed linear manifolds in an inner</u>

<u>product space H, and let M be orthogonal to N. Let M+N = {x+y:x$\in$M,y$\in$N}</u>

<u>Then M+N is a closed linear manifold in H.</u>

Proof. M+N is clearly a linear manifold. Let $\{z_n = x_n + y_n : x_n \in M, y_n \in N\}_{n=1}^{\infty}$

be a Cauchy sequence in M+N. Then, since

$$\|z_n - z_m\|^2 = \|x_n - x_m\|^2 + \|y_n - y_m\|^2,$$

$$\|x_n - x_m\|^2 \leq \|z_n - z_m\|^2,$$

$$\|y_n - y_m\|^2 \leq \|z_n - z_m\|^2.$$

Thus $\{x_n\}_1^{\infty}$ and $\{y_n\}_{n=1}^{\infty}$ are Cauchy sequences in M and N. Since M and

N are closed, $\lim_{n \to \infty} x_n = x_0$ in M, $\lim_{n \to \infty} y_n = y_0$ in N, and $\lim_{n \to \infty} z_n = x_0 + y$

is in M+N.   M+N is closed.

VII.4.7. Theorem (Projection Theorem).   Let M be a closed linear manifold of a Hilbert space H.   Then

$$M + M^{\perp} = H.$$

Proof. Let $M + M^{\perp} = N$.   Since M and $M^{\perp}$ are closed linear manifolds, by Lemma VII.4.6, so is N. Thus $M \subset N$, so $N^{\perp} \subset M^{\perp}$. $M^{\perp} \subset N$, so $N^{\perp} \subset M^{\perp\perp} = M$. Thus $N^{\perp} \subset M^{\perp} \cap M = \{\theta\}$. So $N^{\perp} = \{\theta\}$. Finally, $N = N^{\perp\perp} = \{\theta\}^{\perp} = H$.

We summarize and extend the results of this section in the following theorem:

VII.4.8. Theorem. Let H be a Hilbert space, M a closed linear manifold in H, and x an element in H.   Then

1. There exists exactly one element $y = P_M(x)$ in M such that $\|x - y\| = \inf\limits_{m \in M} \|x - m\|$. Further y is the unique point of M such that x-y is orthogonal to M.

2. Each x in H has the unique decomposition $x = m+n$, where m is in M and n is in $M^{\perp}$.

3. The operator P_M is linear, continuous, and has norm 1, when $M \neq \{\theta\}$.

4. The range of P_M is M. The range of $I - P_M$, where I is the identity transformation, is $M^{\perp}$.

5. $P_M^2 = P_M$.

6. $(P_M x, y) = (x, P_M y)$ <u>for all x, y in H.</u>

Proof. 1 is merely a restatement of Theorem VII.4.1.

To show 2, let $y = P_M(x)$ be chosen as in part 1. Then $x = P_M(x) +$
$(x - P_M(x))$. Clearly $P_M(x)$ is in M, and $x - P_M(x)$ is in $M^\perp$. If
$x = m_1 + n_1$ and $x = m_2 + n_2$ are two such decompositions, then $m_1 - m_2 = n_2 - n_1$.
The left side is in M; the right is in $M^\perp$. Thus both equal θ, and the
representation is unique.

For 3, let $x = m_1 + n_1$, $y = m_2 + n_2$ be arbitrary elements in H, and
α be a complex (real) number. Then $x + y = (m_1 + m_2) + (n_1 + n_2)$, and
$\alpha x = \alpha m_1 + \alpha n_1$. By inspection we find

$$P_M(x+y) = P_M(x) + P_M(x),$$

$$P_M(\alpha x) = \alpha P_M(x).$$

So P_M is linear. Since $\|x\| \geq \|m_1\| = \|P_M(x)\|$, the operator norm $\|P_M\| \leq 1$.
This implies that P_M is continuous. When $M = \{\theta\}$, $P_M x = \theta$ for all x
in H and $\|P_M\| = 0$. If $M \neq \{\theta\}$, then for $m \neq \theta$ in M, $P_M m = m$, and
$\|P_M\| = 1$.

Part 4 is obvious, as is part 5.

To show part 6, let $x = m_1 + n_1$, $y = m_2 + n_2$ be the decompositions of
x and y. Then

$$(P_M x, y) = (m_1, m_2 + n_2) = (m_1, m_2).$$

Similarly,

$$(x, P_M y) = (m_1 + n_1, m_2) = (m_1, m_2).$$

And we are finished. $\underline{P_M}$ is called the projection of $\underline{H}$ onto $\underline{M}$.

VII.5. Continuous Linear Functionals. We recall that a continuous linear functional L on a Hilbert space H is a continuous function whose domain is H, and whose range is the complex (real) numbers. In particular the function defined for all x in H by $L_y(x) = (x, y)$ is a continuous linear functional.

The following remarkable characterization of a continuous linear functional is due to Riesz and Frechet.

VII.5.1. Theorem (Riesz-Frechet). <u>Let L be a continuous linear functional, defined on a Hilbert space H. Then there exists a unique element y_L in H such that</u>

$$L(x) = (x, y_L)$$

<u>for all x in H. Furthermore</u>

$$\|L\| = \sup_{\|x\| \neq 0} |L(x)| / \|x\| = \|y_L\|.$$

Proof. First, if there exist two elements y_1 and y_2 such that

$L(x) = (x,y_1)$ and $L(x) = (x,y_2)$, then $(x,y_1-y_2) = 0$ for all x in H.

Letting $x = y_1-y_2$, we find $\|y_1-y_2\|^2 = 0$, and $y_1=y_2$.

To show the existence of an element y_L such that $L(x) = (x,y_L)$ for

all x in H, we proceed as follows. Let M be the linear manifold of

all elements x such that $L(x) = 0$. If M = H, let $y_L = \theta$. If $M \underset{\neq}{\subset} H$,

let y_0 be in $M^\perp - \{\theta\}$, so $L(y_0) \neq 0$. Now let x be an arbitrary elemen

in H. Then $x - [L(x)y_0/L(y_0)]$ is in M, and is orthogonal to y_0. Thus

$$(x-[L(x)y_0/L(y_0)],y_0) = 0,$$

$$(x,y_0) - L(x)(y_0,y_0)/L(y_0) = 0,$$

and

$$L(x) = L(y_0)(x,y_0)/\|y_0\|^2,$$

$$= (x,\overline{L(y_0)}y_0/\|y_0\|^2),$$

so $y_L = \overline{L(y_0)}y_0/\|y_0\|^2.$

Finally

$$|L(x)| = |(x,y_L)| \leq \|y_L\|\|x\|,$$

so $\|L\| \leq \|y_L\|$. If $x = y_L$,

$$L(y_L) = \|y_L\| \|y_L\|,$$

and so $\|L\| = \|y_L\|.$

VII.5.2. Examples. 1. In n-dimensional Euclidean space E^n, where for
each $x = (x_1,\ldots,x_n)$, $\|x\| = [\sum_{j=1}^{n} |x_j|^2]^{1/2}$, every continuous linear functional
L has the form

$$L(x) = \sum_{j=1}^{n} \overline{y}_j x_j,$$

where $y = (y_1,\ldots,y_n)$ is an element in E^n.

   2. In $L^2(R^n)$, where for each x,

$$\|x\| = [\int_{R^n} |x|^2 dt]^{1/2},$$

every continuous linear functional L has the form

$$L(x) = \int_{R^n} \overline{y}x dt,$$

where y is also an element in $L^2(R^n)$.

VII.6. Fourier Expansions. The reader is certainly familiar with power
series expansions as expressed by Taylor's and MacLaurin's series. As the
Stone-Weierstrass theorem shows, power series expansions can uniformly
approximate any continuous function defined on a compact interval. But

as the Stone-Weierstrass theorem also shows, other functions besides powers of the independent variable may also be used for approximating continuous functions. One type of expansion, especially important in a Hilbert space, is the generalized Fourier expansion, which follows.

VII.6.1. <u>Definition.</u> <u>Let K be an index set, and let $\{e_k\}_{k \in K}$ be an orthonormal set in a Hilbert space H. For any element x in H, the set $\{(x,e_k)\}_{k \in K}$ is the set of Fourier coefficients of x with respect to $\{e_k\}_{k \in K}$. The series $\sum_K (x,e_k)e_k$ is the (formal) Fourier series for x with respect to $\{e_k\}_{k \in K}$.</u>

VII.6.2. <u>An Example.</u> If H is the Hilbert space $L^2(-\pi,\pi)$, generated by the inner product

$$(x,y) = \int_{-\pi}^{\pi} x\bar{y}\,dt,$$

the set $\{e_k\}_{k=-\infty}^{\infty} = \{e^{ikt}/\sqrt{2\pi}\}_{k=-\infty}^{\infty}$ is an orthonormal set. For each x is $L^2(-\pi,\pi)$ the Fourier coefficients are $\{\int_{-\pi}^{\pi} x(s)e^{-iks}/\sqrt{2\pi}\,ds\}_{k=-\infty}^{\infty}$. The Fourier expansion for x is $\sum_{k=-\infty}^{\infty} [\frac{1}{2\pi}\int_{-\pi}^{\pi} x(s)e^{-iks}\,ds]e^{ikt}$. We write

$$x(t) \approx \sum_{k=-\infty}^{\infty} [\frac{1}{2\pi}\int_{-\pi}^{\pi} x(s)e^{-iks}\,ds]e^{ikt}.$$

The following inequality is particularly useful in Hilbert and inner product spaces.

VII.6.3. <u>Theorem (Bessel's Inequality)</u>. <u>Let $\{e_k\}_{k=1}^{\infty}$ be a countable</u>
<u>orthonormal set in an inner product space H, and let x be an element</u>
<u>in H. Then $\sum_{k=1}^{\infty} |(x,e_k)|^2$ converges, and</u>

$$\sum_{k=1}^{\infty} |(x,e_k)|^2 \le \|x\|^2.$$

Proof. We attempt to minimize $\|x - \sum_{k=1}^{n} \lambda_k e_k\|^2$, where $\{\lambda_k\}_{k=1}^{n}$ are arbitrary scalars, and the sum is finite.

$$\|x - \sum_{k=1}^{n} \lambda_k e_k\|^2 = (x - \sum_{k=1}^{n} \lambda_k e_k, x - \sum_{k=1}^{n} \lambda_k e_k),$$

$$= (x,x) - (\sum_{k=1}^{n} \lambda_k e_k, x) - (x, \sum_{k=1}^{n} \lambda_k e_k) + (\sum_{k=1}^{n} \lambda_k e_k, \sum_{k=1}^{n} \lambda_k e_k),$$

$$= \|x\|^2 - \sum_{k=1}^{n} \lambda_k \overline{(x,e_k)} - \sum_{k=1}^{n} \overline{\lambda}(x,e_k) + \sum_{j=1}^{n} \sum_{k=1}^{n} \lambda_j \overline{\lambda}_k (e_j, e_k),$$

$$= \|x\|^2 + \sum_{k=1}^{n} [-\lambda_k \overline{(x,e_k)} - \overline{\lambda}_k (x,e_k) + |\lambda_k|^2],$$

since $(e_j, e_k) = \begin{cases} 1 & \text{when } j = k \\ 0 & \text{when } j \ne k \end{cases}$. Continuing by completing the square
with respect to the scalars $\{\lambda_k\}_{k=1}^{n}$,

$$\|x - \sum_{k=1}^{n} \lambda_k e_k\|^2 = \|x\|^2 - \sum_{k=1}^{n} |(x,e_k)|^2 + \sum_{k=1}^{n} |\lambda_k - (x,e_k)|^2$$

Since the left is nonnegative, so is the right. Further the right side is
clearly minimized when $\lambda_k = (x,e_k)$, $k = 1,\ldots,n$. Thus

$$\|x - \sum_{k=1}^{n} (x,e_k)e_k\|^2 = \|x\|^2 - \sum_{k=1}^{n} |(x,e_k)|^2 \geq 0.$$

Finally, as we permit the number of terms to increase, the equality and inequality both remain valid. The right yields Bessel's inequality.

VII.6.4. **Corollary.** Let $\{e_k\}_{k=1}^{\infty}$ be a countably infinite orthonormal set in an inner product space H, and let x be an element in H. Then

$$\lim_{n \to \infty} (x,e_k) = 0.$$

Proof. Since by Bessel's inequality $\sum_{k=1}^{n} |(x,e_k)|^2$ converges, the terms involved must approach 0.

In particular if $H = L^2(-\pi,\pi)$, then $\lim_{n \to \infty} \int_{-\pi}^{\pi} x(t)e^{-ikt}dt = 0$ for all x in H. This fact when x is piecewise continuous is known as the Riemann-Lebesgue lemma, after the mathematicians who first proved it. It has a large number of uses.

Bessel's inequality also yields another proof of Schwarz's inequality. If x, y are elements in an inner product space H, then $y/\|y\|$ is a one element orthonormal set. Thus

$$\|x\|^2 \geq |(x,y/\|y\|)|^2$$

or

$$\|x\|^2\|y\|^2 \geq |(x,y)|^2.$$

VII.6.5. Theorem. Let $\{e_k\}_{k=1}^n$ be an orthonormal set in an inner product space H, and let M be the linear manifold spanned by $\{e_k\}_{k=1}^n$. Then the projection of each element x in H onto M is given by

$$P_M x = \sum_{k=1}^n (x,e_k)e_k.$$

Proof. In the proof of Bessel's inequality we showed that

$$\inf_\lambda \left\| x - \sum_{k=1}^n \lambda_k e_k \right\| = \left\| x - \sum_{k=1}^n (x,e_k)e_k \right\|.$$

Since all elements in M can be represented by such series,

$$\inf_{m \in M} \|x-m\| = \left\| x - \sum_{k=1}^n (x,e_k)e_k \right\|.$$

The result then follows from Theorem VII.4.8.

VII.6.6. Theorem (Riesz-Fischer Theorem). 1. If $\{e_k\}_{k=1}^\infty$ is a countable orthonormal set in a Hilbert space H, and $\{\lambda_k\}_{k=1}^\infty$ are complex (real) numbers which satisfy $\sum_{k=1}^\infty |\lambda_k|^2 < \infty$, then $\sum_{k=1}^\infty \lambda_k e_k$ converges to an element in H. In particular $\sum_{k=1}^\infty (x,e_k)e_k$ converges for each x in H.

2. Let M be the closure of the linear manifold spanned by $\{e_k\}_{k=1}^\infty$. Then $P_M x = \sum_{k=1}^\infty (x,e_k)e_k$ for all x in H.

Proof. 1. If $\sum_{k=1}^\infty |\lambda_k|^2 < \infty$, then

$$\left\| \sum_{k=m+1}^{n} \lambda_k e_k \right\|^2 = \sum_{k=m+1}^{n} |\lambda_k|^2,$$

Thus $s_n = \sum_{k=1}^{n} (x,e_k)e_k$ is a Cauchy sequence. Since H is complete, s_n converges to an element in H.

2. From part 1 of Theorem VII.4.8 $P_M x = \sum_{k=1}^{\infty} (x,e_k)e_k$ if and only if $x - \sum_{k=1}^{\infty} (x,e_k)e_k$ is orthogonal to all of M, which in this case is equivalent to being orthogonal to $\{e_k\}_{k=1}^{\infty}$. This is obvious.

We come now to a procedure, stated formally in theorem form, which says that from every set of elements in an inner product space H we can generate an orthonormal set which spans the same linear manifold as the original set. Far more important than the statement of this fact, however, is the proof, which actually illustrates the procedure.

VII.6.7. <u>Theorem (Gram-Schmidt Process)</u>. <u>Let $\{x_j\}_{j=1}^{\infty}$ be a countable sub-</u>
<u>set of an inner product space H such that every finite subset of $\{x_j\}_{j=1}^{\infty}$</u>
<u>is linearly independent, and let M_n be the linear manifold spanned by</u>
<u>$\{x_j\}_{j=1}^{n}$. From $\{x_j\}_{j=1}^{\infty}$ we can extract an orthonormal set $\{y_j\}_{j=1}^{\infty}$, such</u>
<u>that the linear manifold spanned by $\{y_j\}_{j=1}^{n}$ is M_n.</u>

Of course, if any of the original sets $\{x_j\}_{j=1}^{n}$ are linearly dependent those x_j which are dependent upon $x_1,\ldots,x_{j-1}$ may be discarded before the Gram-Schmidt process is applied. Hence our assumption concerning linear independence is for convenience only.

Proof. We let

$$y_1 = x_1/\|x_1\|,$$

$$y_2 = [x_2 - (x_2,y_1)y_1]/\|x_2 - (x_2,y_1)y_1\|,$$

and proceed inductively to define

$$y_n = [x_n - \sum_{j=1}^{n-1} (x_n,y_j)y_j]/\|x_n - \sum_{j=1}^{n-1} (x_n,y_j)y_j\|.$$

Each y_n is orthogonal to $M_1,\ldots,M_{n-1}$. Each y_n is a linear combination of $x_1,\ldots,x_n$. Thus the linear manifold spanned by $y_1,\ldots,y_n$ is M_n.

As we stated previously, where every possible we shall use orthonormal sets for computational purposes. With this in mind, we conclude this section with the definitions of completeness and separability and a theorem connecting them.

VII.6.8. <u>Definition.</u> <u>An orthonormal set</u> $\{y_j\}_{j=1}^{\infty}$ <u>is complete in a Hilbert space</u> H <u>if whenever</u> $(x,y_j) = 0,$ $j = 1,\ldots,$ <u>for some</u> x <u>in</u> H, <u>then</u> $x = \theta.$

VII.6.9. <u>Definition.</u> <u>A Hilbert space</u> H <u>is separable if</u> H <u>contains a countable dense set.</u>

VII.6.10. <u>Theorem.</u> <u>A Hilbert space</u> H <u>is separable if and only if</u> H <u>contains a complete orthonormal set.</u>

Proof. If H is separable, then there exists a countable dense set $\{x_j\}_{j=1}^{\infty}$. From Theorem VII.6.7 we know we can extract a countable orthonormal dense

$\{y_j\}_{j=1}^{\infty}$. Let x be an arbitrary element of H. Then for each $\varepsilon > 0$ there exists an n and scalars $\lambda_1, \ldots, \lambda_n$ such that

$$\|x - \sum_{j=1}^{n} \lambda_j y_j\| < \varepsilon.$$

Since the λ_j's which minimize the left are $\lambda_j = (x, y_j)$, $j = 1, \ldots, n$.

$$\|x - \sum_{j=1}^{n} (x, y_j) y_j\| < \varepsilon.$$

If $x - \sum_{j=1}^{\infty} (x, y_j) y_j \neq \theta$, then for all n sufficiently large $\|x - \sum_{j=1}^{n} (x, y_j) y_j\|$ would become bounded away from 0, and a proper choice of ε yields a contradiction. So

$$x - \sum_{j=1}^{\infty} (x, y_j) y_j = \theta.$$

Thus if for some x in H, $(x, y_j) = 0$, $j = 1, \ldots$, then $x = \theta$.

Conversely, if H contains a complete orthonormal set $\{y_j\}_{j=1}^{\infty}$, then finite combinations of $\{y_j\}_{j=1}^{\infty}$ with rational coefficients are dense in H. This set is clearly countable.

VII.7. <u>Isometric Hilbert Spaces</u>. On several occasions to follow it will be convenient to change the notation involved in some calculations. We do so, of course, without changing the fundamental underlying structure of the Hilbert space . The transformation representing this change is called an

isometry. It is primarily in anticipation of the notational change that
this section is included here.

VII.7.1. <u>Definition. Let H_1 and H_2 be Hilbert spaces with the same</u>
<u>scalar field. H_1 and H_2 are isomorphic and isometric if there exists a</u>
<u>linear operator U transforming H_1 onto H_2 such that</u>

$$\|Ux\|_{H_2} = \|x\|_{H_1}$$

<u>for all x in H_1.</u>

VII.7.2. <u>Theorem. If the Hilbert spaces H_1 and H_2 are isomorphic and</u>
<u>isometric through a transformation U, then</u>

$$(Ux_1, Ux_2)_{H_2} = (x_1, x_2)_{H_1}$$

<u>for all x_1, x_2 in H_1. U^{-1} exists and satisfies</u>

$$(U^{-1}y_1, U^{-1}y_2)_{H_1} = (y_1, y_2)_{H_2}$$

<u>for all y_1, y_2 in H_2.</u>

Proof. Since the first inner product is valid when $x_1 = x_2$, the polarization identities show that it also holds for inner products. To show U^{-1}
exists, we note that $Ux = \theta_2$. implies $0 = \|Ux\|_{H_2} = \|x\|_{H_1}$, and so $x = \theta_1$.

By Theorem II.2.16 U^{-1} exists and is linear. It is obvious that

$$(U^{-1}y_1, U^{-1}y_2)_{H_1} = (y_1, y_2)_{H_2} \quad \text{for all} \quad y_1, \; y_2 \; \text{in} \; H_2.$$

VII.7.3. <u>Theorem.</u> <u>Let H be a Hilbert space. Then</u>

1. <u>If H has dimension n and a complex (real) scalar field, then</u>
<u>H is isomorphic and isometric to E^n.</u>

2. <u>If H has infinite dimension and is separable, then H is isomorphi</u>
<u>and isometric to ℓ^2</u>

$$(\ell^2 = \{x = \{x_j\}_{j=1}^{\infty} \; ; \; \|x\| = [\sum_{j=1}^{\infty} \|x_j\|^2]^{1/2} < \infty\}.).$$

<u>If the scalar field of H is complex (real) then the components in ℓ^2 are</u>
<u>complex (real).</u>

Proof. We prove part 2 first. Let $\{e_k\}_{k=1}^{\infty}$ be a complete orthonormal set
in H (See Theorem VII.6.10). If x is in H let $Ux = \{\alpha_k\}_{k=1}^{\infty}$, where
$\alpha_k = (x, e_k)$, $k = 1, \ldots$. U is obviously linear, and from Bessel's inequality
the range of U is in ℓ^2. Since $\sum_{k=1}^{\infty} \lambda_k e_k$ is in H whenever $\sum_{k=1}^{\infty} |\lambda_k|^2 <$
(from the Riesz-Fischer Theorem), the range of U is all of ℓ^2. For all
x in H,

$$\|x\|^2 = \sum_{k=1}^{\infty} |(x, e_k)|^2 = \sum_{k=1}^{\infty} |\alpha_k|^2 = \|Ux\|^2.$$

To prove the first part we merely replace ∞ by n.

VII.7.4. <u>An Example.</u> Let $H = L^2(-\pi,\pi)$, with inner product

$(x,y) = \int_{-\pi}^{\pi} \bar{y}x\,dt$ for all x, y in H. Then $\{e^{ikt}/\sqrt{2\pi}\}_{k=-\infty}^{\infty}$ is a (complete)

orthonormal set in H. We will also later show that for all x in H the

transformation

$$Ux = \{\int_{-\pi}^{\pi} [e^{-ikt}/\sqrt{2\pi}]x(t)\,dt\}_{k=-\infty}^{\infty}$$

is an isometry between $L^2(-\pi,\pi)$ and ℓ^2.

VII. <u>Exercises</u>

1. Verify the statements following Definition VII.1.1.

2. Prove Theorem VII.1.2.

3. a. Show that a symmetric n × n matrix A generates a hermitian form

   on n-dimensional space E^n through the formula $f(x,y) = Ax \cdot y$, where

   x, y are in E^n.

   b. Show that if the eigenvalues of the n × n matrix A are greater

   than 0, then the hermitian form $f(x,y) = Ax \cdot y$ is positive definite.

4. Show that in the Euclidean spaces E^2 and E^3 with the standard dis-

   tance formulas, that the angle ω between two vectors x and y is

   determined by $\cos \omega = (x,y)/\|x\|\|y\|$. Thus x and y are orthogonal

   if and only if $(x,y) = 0$. The definition given by VII.2.4 has the

   usual meaning in these special circumstances.

5. Prove that in Example VII.2.6, part 1, that the hermitian form   f
   is positive definite.

6. Show that, when   X   is a normed linear space and   M   is a closed
   subspace then   X/M   is a normed linear space under $\|\{x\}\| = \inf\limits_{m \in M} \|x+m\|_X$,
   for all   [x]   in   X/M.   Show further that if   X   is complete and   M
   is closed, then   X/M   is complete.

7. Given a pseudonorm $/\cdot/$,   show $/x+y/ \leq /x/ + /y/$ implies
   $\left| /x/ - /y/ \right| \leq /x-y/$ for all appropriate   x, y.

8. Give several additional examples of Hilbert spaces.

9. Let   H   be a Hilbert space, and   y   be an element in   H.   Show that
   $L_y(x) = (x,y)$ is a continuous linear functional defined for all   x   in
   H.

10. Let   L   be a continuous linear functional defined on a Hilbert space   H.
    Show that for all   x   in   H   there exist unique elements   y   and   z   in
    H   such that   z   is in the subspace   M = {z:L(z) = 0}, (y,z) = 0, and
    x = y + z.

11. Show that two continuous linear functionals with the same null space
    (the subspace which is transformed into   0) are linearly dependent.

12. In Euclidean n-dimensional space E^n,   give an example of an orthonormal
    set and its resulting Fourier expansion for arbitrary elements   x   in
    E^n.

13. Orthonormalize $1,t,\dots,t^n,\dots$ in $L^2(-1,1)$,   generated by the inner
    product $(x,y) = \int_{-1}^{1} \overline{y}x\,dt$.   The resulting orthonormal set consists of

the Legendre polynomials.

14. Show that $\{e_k\}_{k=1}^{\infty}$ is a complete orthonormal set if and only if

$$\sum_{k=1}^{\infty} |(x,e_k)|^2 = \|x\|^2 \quad \text{for all} \quad x \quad \text{in a Hilbert space} \quad H.$$

$$\sum_{k=1}^{\infty} |(x,e_k)|^2 = \|x\|^2$$

is called Parseval's equality.

References

1. N.I. Akhiezer and I.M. Glazman, "Theory of Linear Operators in Hilbert
    Space, vols. I and II, "Frederick Ungar, New York, 1961.

2. P.R. Halmos, "Finite Dimensional Vector Spaces," Van Nostrand, Princeton,
    1958.

3. P.R. Halmos, "Introduction to Hilbert Space and the Theory of Multiplicity,"
    Chelsea, New York, 1951.

4. M.A. Naimark, "Normed Rings," P. Noordhoff, Groningen, The Netherlands,
    1964.

5. F. Riesz and B. Sz.-Nagy, "Functional Analysis," Frederick Ungar, New
    York, 1955.

6. M.H. Stone, "Linear Transformations in Hilbert Space," American
    Mathematical Society, New York, 1966.

7. A.E. Taylor, "Introduction to Functional Analysis," John Wiley and Sons,
    New York, 1958.

VIII.   LINEAR OPERATORS ON A HILBERT SPACE

In Chapter II we have briefly considered linear operators. It is now desirable to explore in some detail the nature of bounded linear operators on a Hilbert space. We shall again encounter an adjoint operator, eigenvalues, and operator convergence, but in a different context. Our final goal in this chapter will be to write a large class of operators as integrals. When we finally encounter partial differential equations, these integral representations will help immeasurably in formulating solutions to them.

VIII.1. <u>Regular Operators on a Hilbert Space</u>. We recall that $L(H,H)$ is the Banach space of bounded linear operators from a Hilbert space H into H under the norm

$$\|A\| = \sup_{\substack{x \in H \\ x \neq \theta}} \|Ax\| / \|x\|$$

From Theorem III.1.4 we have that if A is in $L(H,H)$ and $|\lambda| > \|A\|$, then $(\lambda I - A)^{-1}$ exists, and $(\lambda I - A)^{-1}y = \sum_{n=0}^{\infty} (A^n / \lambda^{n+1})y$ for all y in H. Further $\|(\lambda I - A)^{-1}\| \leq (|\lambda| - \|A\|)^{-1}$. The following theorem is a corollary to these results.

VIII.1.1. <u>Theorem</u>. 1. Let H be a Hilbert space, and let A, B, A^{-1} be in $L(H,H)$, with $\|B\| < 1/\|A^{-1}\|$. Then $(A+B)^{-1}$ exists.

2. The set of all regular operators is an open set in $L(H,H)$.

Proof. 1. We recall that an operator is regular if $D_A = D_{A^{-1}} = H$. From
the remarks preceding this theorem, we see that if $|\lambda| > \|A\|$, then
$(\lambda I \pm A)$ is regular. Now in our particular case we let $\lambda = 1$ and replace
A by $A^{-1}B$. Since

$$\|A^{-1}B\| \leq \|A^{-1}\|\|B\| < 1,$$

we see that $(I+A^{-1}B)$ is regular. Since A is also regular, $A(I+A^{-1}B) = (A+B)$ is regular.

2. To show the second part, we need only observe that the set

$$\{X: X \in L(H,H), \quad \|X-A\| < 1/\|A^{-1}\|\}$$

is an open neighborhood in $L(H,H)$ when A is regular in $L(H,H)$.

VIII.1.2. Definition. Let A be an element in $L(H,H)$.

1. The spectrum of A, $\sigma(A)$, is the set of all complex numbers λ
    such that $(\lambda I - A)^{-1}$ does not exist.

2. The resolvent of A, $\rho(A)$, is the complement of $\sigma(A)$.
    $\rho(A) = K - \sigma(A)$.

3. If λ is in $\rho(A)$, then $(\lambda I - A)^{-1}$ is the resolvent operator.

We shall loosely use the term resolvent to denote both the resolvent set $\rho(A)$ and the resolvent operator $(\lambda I-A)^{-1}$. The meaning will always be clear from the context.

We now set about to refine and classify the spectrum of linear operators.

VIII.1.3. <u>Theorem</u>. <u>Let A be an element in $L(H,H)$. Then</u>

1. <u>$\sigma(A)$ is a compact set of complex numbers.</u>

2. <u>If λ is in $\sigma(A)$, then $|\lambda| \leq \|A\|$.</u>

3. <u>$\rho(A)$ is open.</u>

Proof. From the remarks made at the beginning of this section, if $|\lambda| > \|A\|$, then λ is in $\rho(A)$. The negation of this statement shows 2. Thus if λ is in $\sigma(A)$, then $|\lambda| \leq \|A\|$, and $\sigma(A)$ is bounded.

Now suppose λ is in $\rho(A)$, or, alternatively, that $(\lambda I-A^{-1})$ exists. Then by Theorem VIII.1. $([\lambda+\varepsilon]I-A)^{-1}$ also exists for ε sufficiently small. Thus $\rho(A)$ is open; $\sigma(A)$ is closed.

The two paragraphs preceding show that $\sigma(A)$ is closed and bounded, or compact.

VIII.1.4. <u>Examples</u>. 1. Let H be complex n-dimensional Euclidean space K^n, and let the transformation A be represented by the $n \times n$ matrices following:

$$A = U \begin{pmatrix} \lambda_1 & & 0 \\ & \ddots & \\ 0 & & \lambda_n \end{pmatrix} V,$$

where UV = VU = I.   Then

$$(\lambda I - A)^{-1} = U \begin{pmatrix} (\lambda-\lambda_1)^{-1} & & 0 \\ & \ddots & \\ 0 & & (\lambda-\lambda_n)^{-1} \end{pmatrix} \cdot V.$$

The spectrum of A, $\sigma(A) = \{\lambda_j\}_{j=1}^n$.

    2. Let H be $L^2(-1,1)$, and let the transformation A be defined by

$$Ax = m(t)x,$$

where m is continuous on $[-1,1]$. Then A is bounded. $\|A\| = \displaystyle\sup_{t \in [-1,1]} |m(t)|$,

since if

$$(\lambda I - A)x = f,$$

or

$$(\lambda - m(t))x(t) = f(t),$$

then

$$x(t) = f(t)/(\lambda - m(t)),$$

$$\|(\lambda I - A)^{-1}f\|^2 = \int_{-1}^{1} [|f(t)|^2/|\lambda - m(t)|^2]dt.$$

If $\quad \sup_{t \in [-1,1]} \left| \lambda - m(t) \right|^{-1} < M$, $\quad$ then

$$\left\| (\lambda I - A)^{-1} f \right\| \leq M^2 \| f \|^2,$$

and $(\lambda I - A)$ is regular. $\sigma(A) = \overline{\underset{t \in [-1,1]}{\text{range}} \quad m(t)}$.

VIII.2. <u>Bilinear Forms, the Adjoint Operator</u>. We have already briefly studied hermitian forms in Chapter VII, where they were used to generate an inner product. A slight weakening of the assumptions made there gives us the slight generalization we now need. The result is called a bilinear form

VIII.2.1. <u>Definition</u>. <u>Let X be a complex (real) linear space. A bilinear form on X is a function f whose domain is X × X and whose range is K(R) and also satisfies</u>

1. $f(x_1 + x_2, y) = f(x_1, y) + f(x_2, y)$, $\quad f(\alpha x, y) = \alpha f(x,y)$ <u>for all</u>

    $\underline{x_1, x_2, x, y}$ <u>in X and all α in K(R)</u>.

2. $f(x, y_1 + y_2) = f(x, y_1) + f(x, y_2)$, $\quad f(x, \alpha y) = \overline{\alpha} f(x,y)$ <u>for all</u>

    $\underline{x, y, y_1, y_2}$ <u>in X and all α in K(R)</u>.

Part 1 requires that f be linear in the first variable. Part 2 requires that f be conjugate linear in the second variable. In a real linear space X, f is linear in both variables. If $f(x,y) = \overline{f(y,x)}$ for all x, y in X, then f is hermitian.

VIII.2.2. Definition. Let X = H, a Hilbert space. If the bilinear form f satisfies

$$\|f\| = \sup\{|f(x,y)|/\|x\|\|y\|; \; x,y \in H, \; x \neq \theta, \; y \neq \theta\} < \infty,$$

then f is bounded.

Clearly f satisfies $|f(x,y)| \leqq \|f\|\|x\|\|y\|$ for all x, y in H.

Just as bounded linear functionals were characterized, so can bounded bilinear forms.

VIII.2.3. Theorem. Let f be a bounded bilinear form, defined on a Hilbert space H. Then there exists a bounded linear operator A in $L(H,H)$ such that

$$f(x,y) = (Ax,y)$$

for all x, y in H. Further $\|f\| = \|A\|$.

Proof. Consider the linear functional

$$\ell_x(y) = \overline{f(x,y)},$$

defined for all y in H, when x is an arbitrary but fixed element. Let $\|f\| = M.$ Then

$$\|\ell_x\| = \sup_{\substack{y \in H \\ y \neq \theta}} |f(x,y)|/\|y\| \leq M\|x\|.$$

So ℓ_x is bounded. Therefore there exists a unique element z_x in H such that for all y in H

$$\ell_x(y) = \overline{f(x,y)} = (y,z_x),$$

and

$$\|\ell_x\| = \|z_x\| \leq M\|x\|.$$

We define the operator A by letting $Ax = z_x$. Since $f(x,y) = (z_x,y)$ for all x, y in H, from part 1 of Definition VIII.2.1 we find

$$(Ax_1 + Ax_2, y) = (z_{x_1} + z_{x_2}, y) = f(x_1,y) + f(x_2,y),$$

$$= f(x_1 + x_2, y) = (z_{x_1 + x_2}, y),$$

$$= (A[x_1 + x_2], y),$$

and

$$(A(\alpha x), y) = (z_{\alpha x}, y) = f(\alpha x, y),$$

$$= \alpha f(x,y) = \alpha(z_x, y),$$

$$= \alpha(Ax, y) = (\alpha Ax, y),$$

for all x_1, x_2, x, y in H and α in $K(R)$. In the first case this is equivalent to

$$(Ax_1+Ax_2-A[x_1+x_2],y) = 0.$$

We let $y = Ax_1+Ax_2-A[x_1+x_2]$ to find

$$Ax_1+Ax_2 = A[x_1+x_2].$$

In an analogous manner, we find

$$A(\alpha x) = \alpha Ax.$$

Thus A is linear. Since $\|Ax\| = \|z_x\| \leq M\|x\|$ for all x in H, A is bounded.

Now $\|Ax\| \leq \|f\|\|x\|$, so $\|A\| \leq \|f\|$. On the other hand $f(x,y) = (Ax,y)$, so

$$|f(x,y)| \leq \|A\|\|x\|\|y\|.$$

Dividing and taking the maximum over all $x \neq \theta$, $y \neq \theta$,

$$\|f\| \leq \|A\|.$$

Hence they are equal.

VIII.2.4. Theorem. Let H be a Hilbert space. If A is in $L(H,H)$,
there exists a unique element A^* in $L(H,H)$ such that $\|A\| = \|A^*\|$ and

$$(Ax,y) = (x,A^*y)$$

for all x, y in H.

Proof. Let $f(x,y) = (x,Ay)$. Then for all x, y in H, f is a bounded
bilinear form. Therefore there exists a bounded linear transformation A^*
such that $f(x,y) = (A^*x,y)$. We therefore have

$$(x,Ay) = (A^*x,y),$$

and, taking conjugates,

$$(Ay,x) = (y,A^*x)$$

for all x, y in H. Now if $y = A^*x$,

$$\|A^*x\|^2 = (AA^*x,x),$$

$$\leq \|AA^*x\|\|x\|,$$

$$\leq \|A\|\|A^*x\|\|x\|.$$

Thus $\|A^*x\| \leq \|A\|\|x\|$, and $\|A^*\| \leq \|A\|$. Interchanging the roles of A, A^*

and x, y, we find $\|A\| \leq \|A^*\|$, so the two are equal.

VIII.2.5. Definition. Let A be a bounded linear operator in $L(H,H)$,
where H is a Hilbert space. Then the unique operator A^* in $L(H,H)$
which satisfies

$$(Ax,y) = (x,A^*y)$$

for all x, y in H is called the adjoint operator of A.

VIII.2.6. Theorem. Let A and B be operators in $L(H,H)$, where H is
a Hilbert space, and let α be a complex (real) number. Then

1. $(A^*)^* = A$.

2. $(A+B)^* = A^*+B^*$, $(\alpha A)^* = \bar{\alpha} A^*$.

3. $(AB)^* = B^*A^*$.

4. $\|A\| = \|A^*\|$.

5. $I^* = I$, $0^* = 0$. (The identity and zero operators.)

6. $\|A^*A\| = \|AA^*\| = \|A\|^2$.

7. $A^*A = 0$ if and only if $A = 0$.

8. If A is regular, then A^* is regular, and

$$(A^*)^{-1} = (A^{-1})^*.$$

Proof. 1. For all x, y in H,

$$(Ax,y) = (x,A^*y) = \overline{(A^*y,x)},$$

$$= \overline{(y,A^{**}x)} = (A^{**}x,y).$$

This implies $A = A^{**}$.

2,3,4 and 5 are trivial.

6. $\|A^*A\| \leq \|A^*\|\|A\| = \|A\|^2$. Conversely for all x in H,

$$\|Ax\|^2 = (Ax,Ax) = (A^*Ax,x),$$

$$\leq \|A^*A\|\|x\|^2.$$

So $\|A\|^2 \leq \|A^*A\|$, and they are equal.

7 is trivial.

8. From 3 we see that

$$(A^{-1})^*A^* = (AA^{-1})^* = I^*,$$

$$= I = (A^*)^{-1}A^*.$$

Since the inverse of A^* is unique, $(A^{-1})^* = (A^*)^{-1}$.

We conclude this section with the following remark: Let X be a complex Banach algebra which has an identity element I. Suppose that there exists an abstract operator * which transforms X into X such that for all A, B in X and all complex numbers α

1. $(A^*)^* = A$.

2. $(A+B)^* = A^* + B^*$, $(\alpha A)^* = \bar{\alpha}A^*$.

3. $(AB)^* = B^* A^*$.

4. $\|A^* A\| = \|A\|^2$.

5. $(I+A^* A)^{-1}$ is in X.

Then X is called a C^*-algebra. I.M. Gelfand and M.A. Naimark of the Soviet Union have proved that every C^*-algebra is isomorphic and isometric to a closed subalgebra of linear operators on a suitable Hilbert space.

VIII.3. <u>Self-Adjoint Operators</u>. There are certáin classes of operators about which considerably more can be said.

VIII.3.1. <u>Definition</u>. Let H be a Hilbert space, and let A be in $L(H,H)$.

1. If $A = A^*$, then A is self-adjoint.

2. If $A^* A = AA^*$, then A is normal.

3. If $A^* = A^{-1}$, then A is unitary.

These operators have strong analogies to various classes of numbers. The self-adjoint operators strongly resemble the real number system. Normal operators behave very similarly to the complex numbers. Unitary operators resemble the complex numbers on the unit circle (with modulus 1). In fact, such linear transformations on complex 1-dimensional space coincide with these classes of numbers.

VIII.3.2. <u>Theorem</u>. <u>Let H be a Hilbert space, and let f be a bounded hermitian form on H × H. Then there exists a unique self-adjoint operator A such that f(x,y) = (Ax,y) for all x, y in H.</u>

Proof. We know that an operator A exists such that $f(x,y) = (Ax,y)$. Since f is hermitian, $f(x,y) = \overline{f(y,x)}$ implies that $(Ax,y) = \overline{(Ay,x)} = (x,Ay)$.

VIII.3.3. <u>Examples</u>. 1. Let $H = K^2$, the Euclidean space of complex 2-dimensional vectors. For $x = (x_1,x_2)$ and $y = (y_1,y_2)$ in H, the operator A, as represented by its matrix (a_{ij}), generates a bilinear form

$$(Ax,y) = \sum_{i=1}^{2} (\sum_{j=1}^{2} a_{ij}x_j)\overline{y}_i$$

$$= \sum_{j=1}^{2} x_j \sum_{i=1}^{2} \overline{(a_{ij}y_i)}$$

$$= (x,A^*y).$$

Thus the matrix representation for A^* is $(\bar{a}_{ji})$. If $A = A^*$, then

$a_{ij} = \bar{a}_{ji}$.

2. Let $H = L^2(\sigma)$. That is, H consists of all σ-measurable functions x which satisfy $\|x\| = [\int |x(t)|^2 \sigma(dt)]^{1/2} < \infty$, where the domain of σ is a measurable set in real Euclidean n-dimensional space. Let $K(\cdot,\cdot)$ be σ-measurable in both variables and satisfy

$$\int \int |K(t,s)|^2 \sigma(ds)\sigma(dt) < \infty.$$

Define the operator A by

$$Ax(t) = \int K(t,s)x(s)\sigma(ds).$$

Then A is bounded and defined for all x in H, since

$$\|Ax\|^2 = \int |\int K(t,s)x(s)\sigma(ds)|^2 \sigma(dt),$$

$$\leq \int [\int |K(t,s)|^2 \sigma(ds)\int |x(s)|^2 \sigma(ds)]\sigma(dt)$$

from Schwarz's inequality. Hence

$$\|Ax\|^2 \leq [\int \int |K(t,s)|^2 \sigma(ds)\sigma(dt)]\|x\|^2.$$

Now if x, y are are in H,

$$(Ax,y) = \int [\int K(t,s)x(s)\sigma(ds)]\overline{y(t)}\sigma(dt),$$

$$= \int \int K(t,s)x(s)\overline{y}(t)\sigma(ds)\sigma(dt),$$

$$= \int \int K(t,s)x(s)\overline{y}(t)\sigma(dt)\sigma(ds),$$

$$= \int x(s)[\int \overline{K(t,s)}y(t)\sigma(dt)]^{*}\sigma(ds),$$

$$= (x,A^{*}y).$$

Thus

$$A^{*}y(t) = \int \overline{K(s,t)}y(s)\sigma(ds).$$

Such an operator as A is called a __Hilbert-Schmidt__ operator after the mathematicians who first studied it. A is self-adjoint if $K(t,s) = \overline{K(s,t)}$.

3. Again let $H = L^2(\sigma)$. Let the operator A be defined by

$$Ax(t) = m(t)x(t),$$

where m is essentially bounded $(|m(t)| \leq M$ for all t except for a set of σ-measure $0)$. Then

$$\|Ax\|^2 = \int |m(t)|^2|x(t)|^2\sigma(dt),$$

$$\leq M^2\|x\|^2.$$

So $\|A\| \leq M.$ Now

$$(Ax,y) = \int m(t)x(t)\overline{y}(t)\sigma(dt)$$

$$= \int x(t)\overline{\overline{m}(t)y(t)}\sigma(dt)$$

$$= (x,A^*y).$$

So $A^*y(t) = \overline{m}(t)y(t).$ Clearly Ax is defined for all x in $H.$ A is self-adjoint if $m = \overline{m}.$

Through the definition we already have the means for computing the norm of an operator $A.$ When the operator A is self-adjoint this can be considerably simplified.

VIII.3.4. <u>Theorem.</u> <u>Let A be a self-adjoint operator defined on a Hilbert space $H.$ Then A can be computed from any of the following expressions</u>:

1. $\|A\| = \sup_{x \neq \theta} \|Ax\|/\|x\|$ (<u>By definition</u>).

2. $\|A\| = \sup_{\|x\|=1} \|Ax\|.$

3. $\|A\| = \sup_{\|x\|=1} |(Ax,x)|.$

Proof. The first is by definition. The second, in fact, is valid for all operators. Since $1/\|x\|$ is real,

$$\|Ax\|/\|x\| = \|A(x/\|x\|)\|.$$

The expression in parenthesis has norm 1.

To show the third, let $\lambda = \sup\limits_{\|x\|=1} |(Ax,x)|$. Then when $\|x\| = 1$, by Schwarz's inequality

$$|(Ax,x)| \leq \|A\|\|x\|\|x\|,$$

$$= \|A\|.$$

So $\lambda \leq \|A\|$.

To show the reverse inequality, first note

$$|(Ax,x)| \leq \lambda\|x\|^2.$$

for all x in H. Now let y, z be in H, $\|y\| = 1$, $\|z\| = 1$. Since $(Ax,x) = (x,A^*x) = \overline{(A^*x,x)} = \overline{(Ax,x)}$, (Ax,x) is real for all x in H, and

$$(A[y+z],[y+z]) \leq \lambda\|y+z\|^2,$$

$$(A[y-z],[y-z]) \geq -\lambda\|y-z\|^2.$$

Expanding and subtracting these inequalities, we find

$$4\,\mathrm{Re}(Ay,z) = 2[(Ay,z) + \overline{(Ay,z)}],$$

$$= 2[(Ay,z) + (y,Az)],$$

$$= (A[y+z],[y+z]) - (A[y-z],[y-z]),$$

$$\leq \lambda[\|y+z\|^2 + \|y-z\|^2],$$

$$= 2\lambda[\|y\|^2 + \|z\|^2],$$

$$= 4\lambda.$$

Or $\mathrm{Re}(Ay,z) \leq \lambda$. Letting $z = Ay/\|Ay\|$, when $Ay \neq \theta$, we find

$$\mathrm{Re}(Ay,Ay/\|Ay\|) = \|Ay\| \leq \lambda.$$

Taking a maximum with respect to y, $\|A\| \leq \lambda$, and we are finished.

The following theorem is not only valid in a Hilbert space, but also in a Banach space. It is therefore stated and proved in that setting.

VIII.3.5. <u>Theorem</u>. <u>Let X be a Banach space and, let A be an operator in $L(X,X)$. Then A is regular if and only if</u>

1. <u>The range of A is dense in X.</u>

2. <u>There exists an $\alpha > 0$ such that</u>

$$\|Ax\| > \alpha\|x\|$$

<u>for all x in X, $x \neq \theta$.</u>

Proof. If A is regular, then the range of A is X. Thus 1 is satisfied.

Further, for all y in X

$$\|A^{-1}y\| \leq \|A^{-1}\|\|y\|.$$

If y = Ax, then

$$\|x\| \leq \|A^{-1}\|\|Ax\|,$$

and $\alpha = 1/\|A^{-1}\|$.

Conversely, suppose the range of A is dense in X, and there exists
an $\alpha > 0$ such that $\|Ax\| > \alpha\|x\|$ for all x in X, $x \neq \theta$. Suppose that
the sequence $Ax_n = y_n$, n = 1,..., has limit y. Then since

$$\|A(x_n - x_m)\| > \alpha\|x_n - x_m\|,$$

the sequence $\{x_n\}_{n=1}^{\infty}$ is also a Cauchy sequence and has a limit x. Since
A is continuous, Ax_n approaches Ax. Since Ax_n approaches y by
assumption, y = Ax. Thus the range of A is closed. Since the range is
dense in X, the range of A is therefore all of X.

Now if $Ax = \theta$, then

$$0 = \|\theta\| = \|Ax\| \geq \alpha\|x\| \geq 0.$$

Or $x = \theta$. Thus by Theorem II.2.16 A^{-1} exists, and by definition A is
regular.

VIII.4. Projections. One special type of self-adjoint operator which we have already encountered in Section VII.4, and which is especially important is the projection operator:

VIII.4.1. Definition. Let H be a Hilbert space. A projection operator P on H is a self-adjoint, idempotent operator. That is, an operator P which satisfies

    1. $P = P^*$,

    2. $P^2 = P$.

    This definition is closely related to Theorem VII.4.8.

VIII.4.2. Theorem. Let P be a projection on a Hilbert space H with range M. Then M is a closed linear manifold, and $P = P_M$, the projection of H onto M.

Proof. If y, y_1, y_2 are in the range of P and α is a complex (real) number, then there exists (at least one) x, x_1, x_2 such that $y = Px$, $y_1 = Px_1$, $y_2 = Px_2$. Thus

$$y_1 + y_2 = Px_1 + Px_2 = P(x_1 + x_2),$$

$$\alpha y = \alpha Px = P(\alpha x).$$

So $y_1 + y_2$ and y are also in M, the range of P. M is a linear manifold.

Next we note that each x in H has the representation

$$x = Px + (I-P)x.$$

Since $(Px, [I-P]x) = (x, P[I-P]x) = 0$, by the Pythagorean theorem,

$$\|x\|^2 = \|Px\|^2 + \|[I-P]x\|^2,$$

and we conclude $\|Px\| \leq \|x\|$, $\|P\| \leq 1$. Now let $\{y_n\}_{n=1}^{\infty}$ be a Cauchy sequence in M. Since $\{y_n\}_{n=1}^{\infty}$ is in H, $\lim_{n \to \infty} y_n = y_0$ exists in H. Since for each y_n, there exists an x_n such that $Px_n = y_n$, we find $Py_n = P^2 x_n = Px_n = y_n$. Thus we have

$$y_0 = y_n + (y_0 - y_n),$$

$$Py_0 = y_n + P(y_0 - y_n),$$

and

$$y_0 - Py_0 = (I-P)(y_0 - y_n).$$

Taking norms, we find

$$\|y_0 - Py_0\| = \|(I-P)(y_0 - y_n)\|,$$

$$\leq \|I-P\|\|y_0 - y_n\|,$$

$$\leq [\|I\| + \|P\|]\|y_0 - y_n\|,$$

$$\leq 2\|y_0 - y_n\|.$$

Since the left side is fixed, and the last expression on the right approaches 0, $y_0 = Py_0$, and y_0 is also in M. Thus M is closed. That $P = P_M$ is now obvious.

VIII.4.3. **Theorem.** Let P_1 and P_2 be projection operators on a Hilbert space H. Then

1. $P_1 + P_2$ is a projection if and only if $P_1 P_2 = P_2 P_1 = 0$.

2. $P_1 P_2$ is a projection if and only if $P_1 P_2 = P_2 P_1$.

3. $P_1 P_2 = P_2$ if and only if $(P_1 x, x) \geq (P_2 x, x)$ for all x in H. $((P_1 x, x) \geq (P_2 x, x)$ for all x in H will later be written $P_1 \geq P_2$.)

Proof. 1. If $P_1 P_2 = P_2 P_1 = 0$, then $(P_1 + P_2)^2 = P_1 + P_2$, $(P_1 + P_2)^* = P_1^* + P_2^* = P_1 + P_2$, and $P_1 + P_2$ is a projection.

Conversely, $(P_1 + P_2)^2 = P_1 + P_2$ implies that $P_1 P_2 + P_2 P_1 = 0$. Now if x is in H,

$$(P_1P_2x,x) = (P_1P_2{}^2x,x) = -(P_2P_1P_2x,x),$$

$$= -(P_1P_2x,P_2x) = -(P_1{}^2P_2x,P_2x),$$

$$= -\|P_1P_2x\|^2 \le 0.$$

But

$$(P_1P_2x,x) = -(P_2P_1x,x) = -(P_2P_1{}^2x,x),$$

$$= (P_1P_2P_1x,x) = (P_2P_1x,P_1x),$$

$$= (P_2{}^2P_1x,P_1x) = \|P_2P_1x\|^2,$$

$$\ge 0.$$

Thus $P_1P_2 = P_2P_1 = 0$.

2. Clearly $P_1P_2 = P_2P_1$ implies that P_1P_2 is a projection. Conversely if P_1P_2 is a projection, then

$$(P_1P_2) = (P_1P_2)^* = P_2{}^*P_1{}^* = P_2P_1.$$

3. If $P_1P_2 = P_2$, then $P_2P_1 = (P_1P_2)^* = (P_2)^* = P_2$. So $P_2P_1 = P_2$ also. Then

$$(P_1-P_2)^2 = P_1{}^2 - P_1P_2 - P_2P_1 + P_2{}^2$$

$$= P_1 - P_2 .$$

and $P_1 - P_2$ is also a projection. Since for all x in H

$$([P_1-P_2]x,x) = ([P_1-P_2]^2 x,x)$$

$$= \|[P_1-P_2]x\|^2$$

$$\geq 0,$$

we have $(P_1 x,x) \geq (P_2 x,x)$ for all x in H.

Conversely if $(P_1 x,x) \geq (P_2 x,x)$ for all x in H, then

$$([I-P_2]x,x) \geq ([I-P_1]x,x) .$$

Thus

$$\|(I-P_1)P_2 x\|^2 = ([I-P_1]^2 P_2 x, P_2 x),$$

$$= ([I-P_1]P_2 x, P_2 x),$$

$$\leq ([I-P_2]P_2 x, P_2 x),$$

$$= 0.$$

VIII.5. <u>Some Spectral Theorems</u>. In Section VII.1 the concept of the spectrum
of an operator was introduced. The purpose of this section is to refine that
concept, to classify the spectrum into two catagories. As we stated at the
beginning of the chapter our goal is to write certain operators as integrals.
The domain over which integration is performed is the spectrum of the operator
in question.

VIII.5.1. <u>Theorem (Spectral Mapping Theorem)</u>. <u>Let X be a Banach space,</u>
<u>let A be an element in $L(X,X)$, and let $p(\lambda) = \sum\limits_{j=0}^{n} c_j \lambda^j$ be a polynomial</u>
<u>with complex valued coefficients. We define the operator $p(A)$ by</u>

$$p(A) = \sum_{j=0}^{n} c_j A^j .$$

<u>Then</u>

   1. $\sigma(p(A)) = p(\sigma(A)) = \{p(\lambda):\lambda \in \sigma(A)\}.$

   2. <u>If A^{-1} exists, then</u>

$$\sigma(A^{-1}) = [\sigma(A)]^{-1} = \{1/\lambda:\lambda \in \sigma(A)\}.$$

   3. <u>If X = H, a Hilbert space, then</u>

$$\sigma(A^*) = \overline{\sigma}(A).$$

Proof. 1. For any fixed number λ_0,

$$p(\lambda)-p(\lambda_0) = (\lambda-\lambda_0)q(\lambda),$$

where $q(\lambda)$ is also a polynomial. If λ_0 is in $\sigma(A)$, then $A-\lambda_0 I$ is not regular. Since

$$p(A)-p(\lambda_0)I = (A-\lambda_0 I)q(A),$$

then $p(A)-p(\lambda_0)I$ is not regular, for if it were $[p(A)-p(\lambda_0)I]^{-1}q(A)$ would be the inverse of $A-\lambda_0 I$, which is impossible. Thus λ_0 in $\sigma(A)$ implies that $p(\lambda_0)$ is in $\sigma(p(A))$, or

$$p(\sigma(A)) \subset \sigma(p(A)).$$

Conversely, let λ_0 be in $\sigma(p(A))$. If

$$p(\lambda)-\lambda_0 = \alpha(\lambda-\lambda_1)\ldots(\lambda-\lambda_n),$$

then

$$p(A)-\lambda_0 I = \alpha(A-\lambda_1 I)\ldots(A-\lambda_n I).$$

Since $p(A)-\lambda_0 I$ is not regular, one of the terms $A-\lambda_j I$ is not regular, $j = 1,\ldots,n$. That value λ_j is in $\sigma(A)$. Letting $\lambda = \lambda_j$, we find $p(\lambda_j) = \lambda_0$, and

$$\sigma(p(A)) \subset p(\sigma(A)).$$

Thus

$$\sigma(p(A)) = p(\sigma(A)).$$

2. If A is regular, then 0 is not in $\sigma(A)$ so $[\sigma(A)]^{-1}$ is well defined. If λ is not in $\sigma(A)$ and $\lambda \neq 0$, then the equation

$$A^{-1} - (1/\lambda)I = [\lambda I - A](1/\lambda)A^{-1}$$

shows that $1/\lambda$ is not in $\sigma(A^{-1})$. If $1/\lambda$ is in $\sigma(A^{-1})$, then it implies λ is in $\sigma(A)$, or $\lambda = 0$. In other words,

$$\sigma(A^{-1}) \subset [\sigma(A)]^{-1}.$$

Conversely, if we apply this result to A^{-1} we find

$$[\sigma(A)]^{-1} \subset \sigma(A^{-1}).$$

Thus

$$[\sigma(A)]^{-1} = \sigma(A^{-1}).$$

3. If λ is not in $\sigma(A)$, then $A - \lambda I$ is regular. Thus $A^* - \bar{\lambda}I$ is regular and $\bar{\lambda}$ is not in $\sigma(A^*)$. The contrapositive of this statement shows

$$\sigma(A^*) \subset \overline{\sigma(A)}$$

Applying this to A^*,

$$\overline{\sigma(A)} \subset \sigma(A^*).$$

Thus

$$\sigma(A^*) = \overline{\sigma(A)}$$

VIII.5.2. **An Example.** An operator A, defined on a Hilbert space H, is nilpotent if there exists an $n > 0$ such that $A^n = 0$. Since

$$\{0\} = \sigma(0) = \sigma(A^n)$$

$$= [\sigma(A)]^n,$$

we find $\{0\} = \sigma(A)$.

VIII.5.3. <u>Definition.</u> <u>Let X be a Banach space, and let A be in $L(X,X)$.</u>
<u>Then</u>

1. <u>λ is an eigenvalue of A with associated eigenfunction x in X</u>
    <u>if $Ax = \lambda x$ and $x \neq \theta$. We write $\lambda \in \sigma_p(A)$.</u>

2. <u>λ is an approximate eigenvalue of A if there exists a sequence</u>

of elements $\{x_n\}_{n=1}^{\infty}$ such that $\|x_n\| = 1$ for all n, and

$$\lim_{n\to\infty} \|Ax_n - \lambda x_n\| = 0.$$

We write $\lambda \in \sigma_{ap}(A)$.

3. The set of all eigenvalues of A is called the point spectrum of $A: \sigma_p(A)$.

4. The set of all approximate eigenvalues of A is called the approximate point spectrum of $A: \sigma_{ap}(A)$.

There is more than one way of subdividing the spectrum of an operator. There is another, which divides the spectrum into point spectrum, continuous spectrum and residual spectrum. In this case the subsets are mutually exclusive, while in our division $\sigma_p \subset \sigma_{ap} \subset \sigma$, as we shall show. There is also a third classification into point spectrum and essential spectrum. Here the sets may overlap, but may not be included in each other. Each classification has its own particular advantages.

VIII.5.4. <u>Examples.</u> 1. Let the Hilbert space $H = L^2(0,1)$, and let the operator A be defined by $Ax(t) = m(t)x(t)$, where m has the following graph

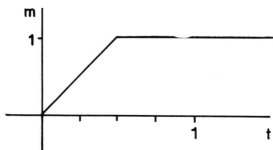

Then $\sigma(A) = \overline{\text{range } m(t)} = [0,1]$. The point spectrum $\sigma_p(A) = \{1\}$. Any function in $L^2(0,1)$ which is zero in $[0,1/2)$ is an eigenvector for A. The eigenvalue 1 has infinite multiplicity. $\sigma_{ap}(A) = [0,1]$.

2. Let $\{e_n\}_{n=1}^{\infty}$ be a complete orthonormal set in a Hilbert space H, and let A be defined by $Ae_j = e_{j+1}$, $j=1,\ldots,$ so that if

$$x = \sum_{j=1}^{\infty} (x,e_j)e_j$$

is an arbitrary element in H,

$$Ax = \sum_{j=1}^{\infty} (x,e_j)e_{j+1},$$

and

$$\|Ax\|^2 = \sum_{j=1}^{\infty} |(x,e_j)|^2 = \|x\|^2.$$

Since e_1 is missing from the expansion of Ax, the range of A is $\{e_1\}^{\perp}$. Thus the equation $Ax = e_1$ is unsolvable; A is not invertible, and 0 is in $\sigma(A)$.

Let us suppose for the moment that 0 is in $\sigma_{ap}(A)$. Then there exists a sequence $\{x_j\}_{j=1}^{\infty}$ with $\|x_j\| = 1$, $j=1,\ldots,$ such that $\|Ax_n - 0x_n\|$ approaches 0 as n approaches ∞. But $\|Ax_n - 0x_n\| = \|Ax_n\| = \|x_n\| = 1$. So 0 is not in $\sigma_{ap}(A)$. We see that in general $\sigma_{ap}(A) \neq \sigma(A)$.

VIII.5.5. **Theorem.** Let H be a Hilbert space and let A be an operator in $L(H,H)$. Then

$$\sigma_p(A) \subset \sigma_{ap}(A) \subset \sigma(A).$$

Proof. Clearly $\sigma_p(A) \subset \sigma_{ap}(A)$. To show the second inclusion, suppose λ is in $\sigma_{ap}(A)$ but not $\sigma(A)$. Then, since $A-\lambda I$ is invertible,

$$\|x\| = \|(A-\lambda I)^{-1}(A-\lambda I)x\|,$$

$$\leq \|(A-\lambda I)^{-1}\|\,\|(A-\lambda I)x\|.$$

But this shows that a sequence $\{x_j\}_{j=1}^{\infty}$, $\|x_j\| = 1$, $j=1,\ldots$, can not satisfy $\lim_{j\to\infty} \|(A-\lambda I)x_j\| = 0$, and we have a contradiction.

Examples VIII.5.4 may give a false impression. The following should make the situation clear.

VIII.5.6. **Theorem.** Let H be a Hilbert space, and A a self-adjoint opera in $L(H,H)$. Then

$$\sigma_{ap}(A) = \sigma(A).$$

Proof. We already know $\sigma_{ap}(A) \subset \sigma(A)$. Suppose λ is in $\sigma(A)$, but not in $\sigma_{ap}(A)$. Then there exists an $\varepsilon > 0$ such that

$$\|(A-\lambda I)x\| \geq \epsilon \|x\|$$

for all x in H. Since A is self-adjoint, and therefore normal, A-λI is normal, and

$$\|(A^*-\overline{\lambda}I)x\| = ([A^*-\overline{\lambda}I]x,[A^*-\overline{\lambda}I]x)^{1/2},$$

$$= ([A-\lambda I][A^*-\overline{\lambda}I]x,x)^{1/2},$$

$$= ([A^*-\overline{\lambda}I][A-\lambda I]x,x)^{1/2}$$

$$= ([A-\lambda I]x,[A-\lambda I]x)^{1/2},$$

$$= \|(A-\lambda I)x\|,$$

$$\geq \epsilon \|x\|.$$

Now since λ is in $\sigma(A)$, the range of A-λI is not all of H. Since, in view of the inequality

$$\|(A-\lambda I)x\| \geq \epsilon \|x\|,$$

the range of A-λI is a closed linear manifold, there exists an element y $\neq$ θ, orthogonal to the range of A-λI. Thus for all x in H,

$$([A-\lambda I]x,y) = (x,[A^*-\overline{\lambda}I]y) = 0.$$

If we let $x = [A^*-\bar{\lambda}I]y$, we find $[A^*-\bar{\lambda}I]y = \theta$, and thus $y = \theta$. We have arrived at a contradiction.

VIII.5.7. <u>Theorem</u>. <u>Let H be a Hilbert space, and let A be a self-adjoint operator in $L(H,H)$. Then $\sigma(A)$ is a compact set of real numbers.</u>

Proof. Suppose λ is in $\sigma(A) = \sigma_{ap}(A)$, and $\text{Im}(\lambda) \neq 0$. Then there exists a sequence $\{x_j\}_{j=1}^n$ such that $\|x_j\| = 1$, $j=1,\ldots,$ and $\lim_{j\to\infty} \|Ax_n - \lambda x_n\| = 0$. But then

$$0 < |\lambda-\bar{\lambda}|\|x_n\|^2,$$

$$= |([A-\lambda I]x_n,x_n) - (x_n,[A-\lambda I]x_n|,$$

$$\leq 2\|(A-\lambda I)x_n\|\|x_n\|$$

gives a contradiction, since the right side approaches 0, and $|\lambda-\bar{\lambda}|\|x_n\| = 2|\text{Im}(\lambda)| > 0$. Thus $\sigma(A)$ is real. It was already known to be closed and bounded.

VIII.5.8. <u>Theorem</u>. <u>Let H be a Hilbert space, and let A be a self-adjoint operator in $L(H,H)$. Then</u>

$$\|A\| = \sup\{|\lambda| : \lambda \in \sigma(A)\}.$$

<u>Thus $\sigma(A)$ is nonempty.</u>

Proof. If $|\lambda| > \|A\|$, then λ is in $\rho(A)$ by Theorem III.1.4, so if

λ is in $\sigma(A)$, $|\lambda| \leq \|A\|$, and $\sup\{|\lambda| : \lambda \in \sigma(A)\} \leq \|A\|$.

Conversely, let $|\lambda| = \|A\|$. Since $\|A\| = \sup_{\|x\|=1} \|Ax\|$, there exists a

sequence $\{x_j\}_{j=1}^{\infty}$ such that $\|x_j\| = 1$, $j=1,\ldots,$ and $\lim_{j \to \infty} \|Ax_j\| = \|A\|$. Then

$$\|A^2 x_j - \lambda^2 x_j\|^2 = \|A^2 x_j\|^2 - 2\lambda^2 \|Ax_j\|^2 + \lambda^4 \|x_j\|^2,$$

$$\leq \|A\|^2 \|Ax_j\|^2 - 2\lambda^2 \|Ax_j\|^2 + \lambda^4 \|x_j\|^2.$$

This last expression approaches $\lambda^2 \cdot \lambda^2 - 2\lambda^2 \cdot \lambda^2 + \lambda^4 = 0$, as j approaches

∞. Thus λ^2 is in $\sigma(A^2) = [\sigma(A)]^2$, and $\pm\lambda$ is in $\sigma(A)$. Therefore

$$\|A\| \leq \sup\{|\lambda| : \lambda \in \sigma(A)\}.$$

VIII.5.9. <u>Corollary</u>. <u>Let H be a Hilbert space, and let A be a self-</u>

<u>adjoint operator in $L(H,H)$. Let $p(\lambda)$ be a polynomial with real coefficients.</u>

<u>Then</u>

$$\|p(A)\| = \sup\{|p(\lambda)| : \lambda \in \sigma(A)\}.$$

The proof is left to the reader.

VIII.5.10. <u>Examples</u>. 1. Let H be n-dimensional Euclidean space E^n,

and let the operator A be defined by

$$Ax = \begin{pmatrix} \lambda_1 & & 0 \\ & \ddots & \\ 0 & & \lambda_n \end{pmatrix} x,$$

where $\{\lambda_j\}_{j=1}^n$ are real valued. Then $\|A\| = \sup\limits_{j=1,\ldots,n} |\lambda_j|$, and $\sigma(A) = \{\lambda_1, \ldots, \lambda_n\}$.

2. Let $H = L^2(0,1)$, and let the operator A be defined by

$$Ax(t) = m(t)x(t),$$

where m is real valued and continuous. Then $\|A\| = \sup\limits_{t \in [0,1]} |m(t)|$, and $\sigma(A) = \overline{[\text{range of } m(t)]}$.

VIII.6. <u>Operator Convergence</u>. The next two sections fulfill the major goal of this chapter: the writing of a bounded self-adjoint operator as an integral. In order to do so we shall need some additional ideas of convergence of operators, since the equality between the operator and its integral representation is not in the uniform sense, which provides the norm for $L(H,H)$. This section is devoted to providing those convergence techniques for the final step, which is the subject of the next section.

VIII.6.1. <u>Definition</u>. <u>Let H be a Hilbert space and let $\{A_j\}_{j=1}^\infty$ be a sequence of operators in $L(H,H)$. Then</u>

1. <u>The sequence $\{A_j\}_{j=1}^\infty$ converges UNIFORMLY to the operator A when</u>
    $$\lim_{j \to \infty} \|A_j - A\|_{op} = 0.$$

2. The sequence $\{A_j\}_{j=1}^{\infty}$ converges STRONGLY to the operator A if

$\lim_{j\to\infty} \|A_j x - Ax\| = 0$ for all x in H.

3. The sequence $\{A_j\}_{j=1}^{\infty}$ converges WEAKLY to the operator A if

$\lim_{j\to\infty} (A_j x - Ax, y) = 0$ for all x, y in H.

VIII.6.2. **Theorem.** In the space $L(H,H)$, uniform convergence implies strong convergence. Strong convergence implies weak convergence.

The proof is left to the reader.

We also need the concept of one operator being greater than another.

VIII.6.3. **Definition.** Let H be a Hilbert space and A, B be self-adjoint operators in $L(H,H)$. Then

1. If $(Ax,x) > (Bx,x)$ for all x in H, we say $A > B$.

2. If there exist constants m, M such that

$mI \leq A \leq MI$,

then m and M are lower and upper bounds for A.

3. If $A \geq 0$, then A is positive. If $A > 0$, then A is positive definite.

VIII.6.4. **Theorem.** Let H be a Hilbert space. Then every bounded monotonic sequence of self-adjoint operators $\{A_j\}_{j=1}^{\infty}$ in $L(H,H)$ converges strongly to a self-adjoint operator A in $L(H,H)$.

Proof. We may assume without loss of generality that $0 \leq A_1 \leq A_2 \leq \ldots \leq A_n$

$\ldots \leq I$, since we may multiply each by -1, if necessary, to achieve an

increasing sequence. We may then subtract the smallest from each, so each is

nonnegative. Finally, if necessary, we may divide by the upper bound. For

convenience we let $A_{ij} = A_j - A_i$ when $i < j$.

We note that if $A_0 \geq 0$, then $(A_0 x, x) \geq 0$ for all x in H. If

$x = y + \lambda (A_0 y, z) z$, where y, z are in H, and λ is real, then this

implies upon expanding

$$0 \leq (A_0 y, y) + 2\lambda |(A_0 y, z)|^2 + \lambda^2 |(A_0 y, z)|^2 (A_0 z, z).$$

The only way this can happen for all real λ is for

$$|(A_0 y, z)|^4 - |(A_0 y, z)|^2 (A_0 y, y)(A_0 z, z) \leq 0.$$

This implies

$$|(A_0 y, z)|^2 \leq (A_0 y, y)(A_0 z, z)$$

for all y, z in H. This is a generalization of Schwarz's inequality.

If $A_0 = I$, it is exactly. Replacing A_0 by A_{ij}, y by x, and z by

$A_{ij} x$, we have

$$\|A_{ij}x\|^4 = (A_{ij}x, A_{ij}x)^2,$$

$$\leq (A_{ij}x,x)(A_{ij}^2x, A_{ij}x),$$

$$\leq (A_{ij}x,x)\|A_{ij}\|^3\|x\|^2.$$

Since $\|A_{ij}\| \leq 1$, this implies

$$\|A_jx - A_ix\|^4 \leq [(A_jx,x) - (A_ix,x)]\|x\|^2$$

Now, since $\{(A_jx,x)\}_{j=1}^{\infty}$ is a bounded monotonically increasing sequence, it is a convergent sequence. This implies that if i, j are large enough, the term in the brackets will become arbitrarily small. This shows that $\{A_jx\}_{j=1}^{\infty}$ is a Cauchy sequence in H. We call the limit Ax. Since each of the A_j's is linear, bounded and self-adjoint, so is A.

VIII.6.5. <u>Theorem.</u> <u>Let H be a Hilbert space. For each positive self-adjoint operator in L(H,H) there exists a unique positive square root.</u>

Proof. Let the operator be denoted by A. By dividing A by $1/\|A\|$ if necessary, we may assume $\|A\| \leq 1$. We shall show that the equation

$$(I-Y)^2 = A$$

has a unique solution Y such that $0 \leq Y \leq I$. Then the square root in question will be I-Y. Letting A = I-B above, and solving for Y, we find

$$Y = (1/2)(B+Y^2)$$

is an equivalent equation. We shall solve this equation by successive

approximations.

We note that $0 \leq A \leq I$ implies $0 \leq B \leq I$, so $B \geq 0$. Further

$B^n \geq 0$, for if $n = 2m+1$ is odd, then

$$(B^n x, x) = (B[B^m x], [B^m x]) \geq 0$$

for all x in H. If $n = 2m$ is even, then

$$(B^n x, x) = (B^m x, B^m x) \geq 0.$$

This implies that all polynomials in B with real, nonnegative coefficients

are also positive.

We now let $Y_0 = 0$, $Y_1 = (1/2) B$ and, in general,

$$Y_{j+1} = (1/2)(B+Y_j^2).$$

By induction we find the following statements are valid for all $j \geq 0$.

1. Each Y_j is a polynomial in B with real, nonnegative coefficient

Therefore $Y_j \geq 0$.

2. $\|Y_{j+1}\| \le (1/2)[\|B\| + \|Y_j\|^2]$.   Thus $\|Y_j\| \le 1$.

3. $Y_{j+1} - Y_j = (1/2)(Y_j + Y_{j-1})(Y_j - Y_{j-1})$.   Thus each difference $Y_{j+1} - Y_j$ is

a polynomial in   B   with real, nonnegative coefficients, and

$$Y_{j+1} - Y_j \ge 0,$$

or

$$Y_{j+1} \ge Y_j.$$

So the sequence $\{Y_j\}_{j=1}^{\infty}$ is a bounded, monotonic sequence of self-adjoint

operators.   By Theorem VIII.6.4 there exists a unique strong limit   C.   Since

C   is the limit of $\{Y_j\}_{j=1}^{\infty}$,   C   is self-adjoint, $0 \le C \le I$,   and   C

commutes with   B   and   A.   The desired square root is $\sqrt{A} = I-C$.

To show $\sqrt{A}$ is unique, let us assume that   X   is an additional positive

square root.   Since $XA = X^3 = AX$,   X   commutes with   A,   all polynomials

in   A,   and thus also with $\sqrt{A}$.   Denote by   Z   and   W   the square roots of

$\sqrt{A}$ and   X,   respectively, obtained by the successive approximations.

If   x   is an arbitrary element in   H   and $y = (\sqrt{A}-X)x$,   then

$$\|Zy\|^2 + \|Wy\|^2 = (Z^2 y, y) + (W^2 y, y)$$

$$= ([\sqrt{A}+X]y, y)$$

$$= ([\sqrt{A}+X][\sqrt{A}-X]x, y)$$

$$= ([A-A]x,y)$$

$$= 0.$$

Thus Zy and Wy equal Θ. This implies $\sqrt{A}y = Z(Zy) = \Theta$, and $Xy = W(Wy) = \Theta$. Then

$$\| [\sqrt{A}-X]x \|^2 = ([\sqrt{A}-X]^2 x, x)$$

$$= ([\sqrt{A}-X]y, x)$$

$$= 0,$$

and $X = \sqrt{A}$ strongly.

VIII.6.6. Corollary. Let H be a Hilbert space, and let A, B be positiv commuting, self-adjoint operators in $L(H,H)$. Then their product AB is a positive self-adjoint operator.

Proof. Self-adjointness is trivial. If x is in H, then

$$(ABx,x) = (A\sqrt{B}\sqrt{B}x,x) = (A[\sqrt{B}x],[\sqrt{B}x]) \geqq 0.$$

VIII.6.7. Theorem. Let H be a Hilbert space and let A be a bounded-sel adjoint operator in $L(H,H)$ satisfying

$$mI \leq A \leq MI.$$

Let $p(\lambda) = \sum\limits_{j=0}^{n} c_j \lambda^j$ be a polynomial with real coefficients which satisfies

$p(\lambda) \geq 0$ when $m \leq \lambda \leq M$. Then $p(A) = \sum\limits_{j=0}^{n} c_j A^j \geq 0.$

Proof. We write $p(\lambda)$ in factored form

$$p(\lambda) = c\pi(\lambda-\alpha_j) \cdot \pi(\beta_j-\lambda) \cdot \pi([\lambda-\gamma_j]^2+\delta_j^2),$$

where $c \geq 0$, $\alpha_j \leq m$, $\beta_j \geq M$, and the other factors come from real zeros

between m and M or from complex zeros. Then

$$p(A) = c\pi(A-\alpha_j I)\pi(\beta_j I-A)\pi([A-\gamma_j I]^2+\delta_j^2 I).$$

Each of the individual terms is positive. Since they all commute, the

previous corollary shows $p(A) \geq 0.$

We have so far established that for every polynomial $p(\lambda) = \sum\limits_{j=0}^{n} c_j \lambda^j$

with real coefficients there exists a unique polynomial operator

$p(A) = \sum\limits_{j=0}^{n} c_j A^j$ such that $\|p(A)\| = \sup\{|p(\lambda)| : \lambda \in \sigma(A)\}$, and such that

if $p(\lambda) \geq 0$ when λ is in $[m,M]$, then $p(A)$ is a positive operator.

We would now like to extend this correspondence between function and operator

to those functions which are real valued and piecewise continuous on $[m,M]$.

We emphasize that throughout only those values of λ in the interval $[m,M]$

are of importance. In fact, it can be shown that only those values of λ in

$\sigma(A)$ are of importance.

VIII.6.8. Definition. Let S denote the class of nonnegative, piecewise
continuous, real valued functions, defined on the interval [m,M].

VIII.6.9. Theorem. For each element u in S there exists a monotonic,
decreasing sequence of polynomials p_n such that for all λ in [m,M]

$$\lim_{n\to\infty} p_n(\lambda) = u(\lambda).$$

Note that the convergence is not uniform for all values of λ. If this
were the case, u   would also be continuous.

Proof. Since the discontinuities of u are finite, by choosing a sequence
of slight modifications at these points, we may approximate u from above by
a sequence of continuous functions $\{q_j\}_{j=1}^{\infty}$ which converge to u at each
point λ in [m,M]. For example, the following graph shows how the sequences
$\{q_j\}_{j=1}^{n}$ might behave near a discontinuity of u.

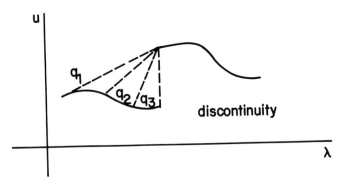

Then for each j, $q_j+(1/j)$ can be uniformly approximated to within $1/2^j$ by
a polynomial p_j, as proved by the Stone-Weierstrass approximation theorem.

Clearly $\{p_j\}_{j=1}^{\infty}$ has all the desired properties.

VIII.6.9. Theorem. Let H be a Hilbert space, and let A be a self-
adjoint operator in $L(H,H)$, bounded by m and M. Let $\{p_j(\lambda)\}_{j=1}^{\infty}$ be
an arbitrary sequence of monotonically decreasing polynomials which approach
the piecewise continuous function $u(\lambda)$ from above on the interval $[m,M]$.
Then the sequence of operators $\{p_j(A)\}_{j=1}^{\infty}$ converges strongly to a unique
limit $u(A)$ which is self-adjoint, positive, and in $L(H,H)$.

Proof. From Theorem VIII.6.7 we see that

$$0 \leq u(\lambda) \leq \ldots \leq p_j(\lambda) \leq \ldots \leq p_2(\lambda) \leq p_1(\lambda)$$

implies

$$0 \leq \ldots \leq p_j(A) \leq \ldots \leq p_2(A) \leq p_1(A).$$

Thus $\{p_j(A)\}_{j=1}^{\infty}$ is a bounded monotonic sequence of self-adjoint operators.
Thus it converges strongly to a limit, which we call $u(A)$.

   If two different polynomial sequences $\{p_j(\lambda)\}_{j=1}^{\infty}$ and $\{q_j(\lambda)\}_{j=1}^{\infty}$
converge monotonically to $u(\lambda)$ in $[m,M]$, there exist r and s
sufficiently large such that

$$p_s(\lambda) \leq q_r(\lambda) + 1/r,$$

$$q_s(\lambda) \leq p_r(\lambda) + 1/r,$$

for all λ in $[m,M]$. Thus

$$p_s(A) \leq q_r(A) + (1/r)I,$$

$$q_s(A) \leq p_r(A) + (1/r)I.$$

Letting s approach ∞ and then r approach ∞, we find

$$\lim_{s\to\infty} p_s(A) \leq \lim_{r\to\infty} q_r(A),$$

$$\lim_{s\to\infty} q_s(A) \leq \lim_{r\to\infty} p_r(A),$$

and the two limits are identical.

Thus for all functions u in S, there exists a unique self-adjoint operator $u(A)$, which is the strong limit of polynomial operators.

VIII.6.10. <u>Theorem.</u> <u>Let H be a Hilbert space, and let A be a bounded, self-adjoint operator in $L(H,H)$. Let u_1, u_2, u be in S and $\alpha \geq 0$. Then</u>

$$(u_1+u_2)(A) = u_1(A)+u_2(A)$$

$$(\alpha u)(A) = \alpha \cdot u(A)$$

<u>$S(A)$, the class of operators which correspond to functions in S, is additive and multiplicative by positive scalars.</u>

The proof depends upon the fact that if $\{p_j\}_{j=1}^{\infty}$ has u_1 as a limit from above, and $\{q_j\}_{j=1}^{\infty}$ has u_2 as a limit from above, $\{p_j+q_j\}_{j=1}^{\infty}$ approaches u_1+u_2 from above. The convergence properties are then inherited by the operators and their limits. The second part is similar. We invite the reader to fill in the details.

We need an additional extension.

VIII.6.11. <u>Definition.</u> <u>Let T denote the class of all piecewise continuous, real valued functions, defined on an interval [m,M].</u>

VIII.6.12. <u>Theorem.</u> <u>Each element w in T can be decomposed into the difference of two functions u, v in S.</u>

Proof. We let $u = \sup[w,0]$, $v = \sup[-w,0]$, then $w = u-v$. Since w is piecewise continuous, so are u and v. We might add that u and v are not unique. $u = \sup[w-a,a]$, $v = \sup[-w+a,a]$ will do just as well for any real number $a \geqq 0$.

VIII.6.13. <u>Theorem.</u> <u>Let H be a Hilbert space, and let A be a self-adjoint operator in $L(H,H)$ bounded by m and M. Then for each w in T, there exists a unique operator w(A) which is the difference of operators u(A) and v(A) in S(A). w(A) is the strong limit of polynomial operators $\{p_j(A)\}_{j=1}^{\infty}$, where $\{p_j(\lambda)\}_{j=1}^{\infty}$ approaches w(λ) for all λ in [m,M].</u>

Proof. Clearly $w(A) = u(A)-v(A)$ is defined. The operators $u(A)$ and $v(A)$ are not unique, but $w(\lambda) = u(\lambda)-v(\lambda)$ and $w(\lambda) = u_1(\lambda)-v_1(\lambda)$ imply

$u(\lambda)+v_1(\lambda) = u_1(\lambda)+v(\lambda)$. Since both sides of this equation are in S,
we find $u(A)+v_1(A) = u_1(A)+v(A)$. Thus $u(A)-v(A) = u_1(A)-v_1(A)$, and
$w(A)$ is unique.

VIII.6.14. Theorem. Let H be a Hilbert space, and let A be a bounded,
self-adjoint operator in $L(H,H)$. Then $T(A)$, the space of linear operators
which correspond to functions in T, is a real, normed, linear space under
the standard operator norm.

   The proof follows directly from Theorem VIII.6.10.

   Note that the correspondence between piecewise continuous functions and
linear operators can be extended to complex piecewise continuous functions.
The resulting space is a complex linear manifold in $L(H,H)$. We will not
need this extension.

VIII.7. The Spectral Resolution of a Self-Adjoint Operator. We are now in
a position to achieve the major goal of this chapter: the writing of a self-
adjoint operator as the uniform limit of sums of projections: As an integral.

VIII.7.1. Theorem. Let H be a Hilbert space, and let A be a self-adjoint
operator in $L(H,H)$, bounded by m and M. Let $e(\lambda)$ in S satisfy
$e(\lambda) = 0$ or 1, when λ is in $[m,M]$. Then the operator $e(A)$ corresponds
to $e(\lambda)$ is a projection.

Proof. Since $e(\lambda)$ is real valued, $e(A)$ is self-adjoint. Since
$e(\lambda)^2 = e(\lambda)$, $e(A)^2 = e(A)$. By Definition VIII.4.1, $e(A)$ is a projection.

VIII.7.2. Theorem. Let H be a Hilbert space, and let A be a self-adjoint operator in (H,H), bounded by m and M. Let

$$
e_\mu(\lambda) = \begin{cases} 1, & \lambda \leq \mu, \\ \\ 0, & \lambda > \mu, \end{cases}
$$

and let E(μ) be the corresponding projection. Then

1. $E(\mu)E(\nu) = E(\mu)$, when $\mu \leq \nu$. $E(\mu) \leq E(\nu)$, when $\mu \leq \nu$.

2. $E(\mu) = 0$, when $\mu < m$.

3. $E(\mu) = I$, when $\mu \geq M$.

4. $E(\mu)$ is continuous from above.

Proof. 1 follows from $e_\mu(\lambda)e_\nu(\lambda) = e_\mu(\lambda)$ when $\mu \leq \nu$. Since if $\mu < m$, $e_\mu(\lambda) = 0$ when λ is in $[m,M]$, the limit as shown in Theorem VIII.6.9 is clearly 0. Similarly, if $\mu \geq M$, $e_\mu(\lambda) = 1$ when λ is in $[m,M]$, and the limit is I. To see that $E(\mu)$ is continuous from the right, let $\{p_j(\lambda)\}_{j=1}^\infty$ approach $e_\mu(\lambda)$ from above, and at the same time satisfy $p_j(\lambda) \geq e_{\mu+1/j}(\lambda) \geq e_\mu(\lambda)$. Then $p_j(A) \geq E(\mu+1/j) \geq E(\mu)$. But as j approaches ∞, $p_j(A)$ approaches $E(\mu)$. Thus $\lim_{j\to\infty} E(\mu+1/j) = E(\mu)$.

VIII.7.3. Theorem(Spectral Resolution of a Self-Adjoint Operator). Let H be a Hilbert space, and let A be a self-adjoint operator in $L(H,H)$, bounded by m and M. Then there exists a collection of projection operators $\{E(\mu), \mu \in [m,M]\}$, strong limits of polynomials in A, satisfying

1. $E(\mu) \leq E(\nu)$, when $\mu \leq \nu$,

2. $E(\mu) = 0$, when $\mu < m$,

3. $E(\mu) = I$, when $\mu \geq M$,

4. $E(\mu)$ is continuous from above,

such that

$$A = \lim_{\sup|\mu_k - \mu_{k-1}| \to 0} \sum_k \lambda_k [E(\mu_k) - E(\mu_{k-1})] = \int_{m-}^{M} \lambda E(d\lambda),$$

where λ_k is in $[\mu_{k+1}, \mu_k]$. The integral exists in the Riemann-Stieltjes sense. The equality between A and its integral representation is in the uniform sense.

Proof. If $\mu \leq \nu$, then

$$\mu[e_\nu(\lambda) - e_\mu(\lambda)] \leq \lambda[e_\nu(\lambda) - e_\mu(\lambda)]$$

$$\leq \nu[e_\nu(\lambda) - e_\mu(\lambda)].$$

Thus

$$\mu[E(\nu) - E(\mu)] \leq A[E(\nu) - E(\mu)]$$

$$\leq \nu[E(\nu) - E(\mu)].$$

If we let $m \leq \mu_0 \leq \mu_1 \leq \ldots \leq \mu_{n-1} \leq \mu_n = M$, then letting $\mu = \mu_{k-1}$, $\nu = \mu_k$, $k=1,\ldots,n$, and summing, we find

$$\sum_k \mu_{k-1}[E(\mu_k)-E(\mu_{k-1})] \leq A \sum_k [E(\mu_k)-E(\mu_{k-1})$$

$$\leq \sum_k \mu_k[E(\mu_k)-E(\mu_{k-1})].$$

The middle term is $A[E(\mu_n)-E(\mu_0)] = A[I-0] = A$. Thus

$$\sum_k \mu_{k-1}[E(\mu_k)-E(\mu_{k-1})] \leq A \leq \sum_k \mu_k[E(\mu_k)-E(\mu_{k-1})].$$

Now let λ_k be an arbitrary point in $[\mu_{k-1},\mu_k]$, $k=1,\ldots,n$. Subtracting $\sum_k \lambda_k[E(\mu_k)-E(\mu_{k-1})]$ from each term in the last inequality,

$$\sum_k (\mu_{k-1}-\lambda_k)[E(\mu_k)-E(\mu_{k-1})] \leq A - \sum_k \lambda_k[E(\mu_k)-E(\mu_{k-1})]$$

$$\leq \sum_k (\mu_k-\lambda_k)[E(\mu_k)-E(\mu_{k-1})].$$

Now let $\varepsilon > 0$ be arbitrary and assume that $\sup[\mu_k-\mu_{k-1}] < \varepsilon$. Then $-\varepsilon < \mu_{k-1}-\lambda_k$, and $\mu_k-\lambda_k < \varepsilon$. Thus

$$-\varepsilon I \leq \sum_k (\mu_{k-1}-\lambda_k)[E(\mu_k)-E(\mu_{k-1})]$$

$$\leq A - \sum_k \lambda_k[E(\mu_k)-E(\mu_{k-1})]$$

$$\leq \sum_k (\mu_k-\lambda_k)[E(\mu_k)-E(\mu_{k-1})] \leq \varepsilon I.$$

As ε approaches 0, we see that

$$A = \lim_{\sup |\mu_k - \mu_{k-1}| \to 0} \sum_k \lambda_k [E(\mu_k) - E(\mu_{k-1})]^{\checkmark}$$

$$= \int_{m-}^{M} \lambda E(d\lambda),$$

where the last equality is by definition. Convergence is in the uniform

sense since operator inequalities, written in their inner product form

and maximized over H yield the operator norm for self-adjoint operators.

VIII.7.4.   Theorem.   Let H be a Hilbert space, and let A be a self-adjoint

operator in $L(H,H)$, bounded by m and M.   Then there exists a collection

of projection operators $\{E(\mu);\ \mu$ in $[m,M]\}$ satisfying

1. $E(\mu) \leq E(\nu)$, when $\mu \leq \nu$.

2. $E(\mu) = 0$, when $\mu < m$.

3. $E(\mu) = I$, when $\mu \geq M$,

such that

1. $I = \int_{m-}^{M} E(d\lambda)$.

2. $p(A) = \int_{m-}^{M} p(\lambda) E(d\lambda)$,   uniformly, for all polynomials p.

3. $w(A) = \int_{m-}^{M} w(\lambda) E(d\lambda)$,   strongly, for all functions w in T.

4. $\int_{m-}^{M} \alpha w(\lambda) E(d\lambda) = \alpha \int_{m-}^{M} w(\lambda) E(d\lambda),$

$\int_{m-}^{M} [w_1(\lambda) + w_2(\lambda)] E(d\lambda) = \int_{m-}^{M} w_1(\lambda) E(d\lambda) + \int_{m-}^{M} w_2(\lambda) E(d\lambda),$

$\int_{m-}^{M} w_1(\lambda) E(d\lambda) \int_{m-}^{M} w_2(\lambda) E(d\lambda) = \int_{m-}^{M} w_1(\lambda) w_2(\lambda) E(d\lambda),$

where the type of equality depends upon the type of functions w, w_1, and w_2.

Proof. Parts 1 and 2 follow immediately by the same technique used in Theorem VIII.7.3. Part 3 follows from part 2 by choosing proper approximating sequences, as do the first two statements of part 4. In the last statement, if w_1 and w_2 are polynomials

$$\int_{m-}^{M} w_1(\lambda) E(d\lambda) \int_{m-}^{M} w_2(\lambda) E(d\lambda) = w_1(A) w_2(A)$$

$$= \int_{m-}^{M} w_1(\lambda) w_2(\lambda) E(d\lambda).$$

The other cases follow by taking limits.

The first two statements of part 4, which show that the integral is a linear operator, are not surprising. However, the last statement, which shows that the integral is also multiplicative in a sense, is unusual. It is not true when applied to the ordinary Riemann and Lebesgue integrals.

VIII.7.5. Examples. 1. Let A be the operator on complex Euclidean space

K^2, represented by the matrix $\begin{pmatrix} 2 & 1 \\ 1 & 2 \end{pmatrix}$. The spectrum of A then consists

of two eigenvalues 1 and 3, with eigenvectors $\begin{pmatrix} 1/\sqrt{2} \\ -1/\sqrt{2} \end{pmatrix}$ and $\begin{pmatrix} 1/\sqrt{2} \\ 1/\sqrt{2} \end{pmatrix}$. Since

this pair is a complete orthonormal sequence in K^2, any vector $\begin{pmatrix} x_1 \\ x_2 \end{pmatrix}$ ca

be represented by its Fourier expansion,

$$\begin{pmatrix} x_1 \\ x_2 \end{pmatrix} = \{ (1/\sqrt{2}-1/\sqrt{2}) \begin{pmatrix} x_1 \\ x_2 \end{pmatrix} \} \begin{pmatrix} 1/\sqrt{2} \\ -1/\sqrt{2} \end{pmatrix} + \{ (1/\sqrt{2} \;\; 1/\sqrt{2}) \begin{pmatrix} x_1 \\ x_2 \end{pmatrix} \} \begin{pmatrix} 1/\sqrt{2} \\ 1/\sqrt{2} \end{pmatrix}.$$

We can reorder this series and write

$$\begin{pmatrix} x_1 \\ x_2 \end{pmatrix} = \begin{pmatrix} 1/\sqrt{2} \\ -1/\sqrt{2} \end{pmatrix} (1/\sqrt{2}-1/\sqrt{2}) \begin{pmatrix} x_1 \\ x_2 \end{pmatrix} + \begin{pmatrix} 1/\sqrt{2} \\ -1/\sqrt{2} \end{pmatrix} (1/\sqrt{2} \;\; 1/\sqrt{2}) \begin{pmatrix} x_1 \\ x_2 \end{pmatrix},$$

from which we conclude that

$$\begin{pmatrix} 1 & 0 \\ 0 & 1 \end{pmatrix} = \begin{pmatrix} 1/\sqrt{2} \\ -1/\sqrt{2} \end{pmatrix} (1/\sqrt{2} - 1/\sqrt{2}) + \begin{pmatrix} 1/\sqrt{2} \\ 1/\sqrt{2} \end{pmatrix} (1/\sqrt{2} \;\; 1/\sqrt{2}).$$

This is easily confirmed by performing the indicated multiplication.

Multiplying by $\begin{pmatrix} 2 & 1 \\ 1 & 2 \end{pmatrix}$, we find

$$\begin{pmatrix} 2 & 1 \\ 1 & 2 \end{pmatrix} = \begin{pmatrix} 2 & 1 \\ 1 & 2 \end{pmatrix} \begin{pmatrix} 1/\sqrt{2} \\ -1/\sqrt{2} \end{pmatrix} (1/\sqrt{2} - 1/\sqrt{2}) + \begin{pmatrix} 2 & 1 \\ 1 & 2 \end{pmatrix} \begin{pmatrix} 1/\sqrt{2} \\ 1/\sqrt{2} \end{pmatrix} (1/\sqrt{2} \;\; 1/\sqrt{2})$$

$$= 1 \begin{pmatrix} 1/\sqrt{2} \\ -1/\sqrt{2} \end{pmatrix} (1/\sqrt{2} - 1/\sqrt{2}) + 3 \begin{pmatrix} 1/\sqrt{2} \\ 1/\sqrt{2} \end{pmatrix} (1/\sqrt{2} \;\; 1/\sqrt{2})$$

$$= 1 \begin{pmatrix} 1/2 & - & 1/2 \\ -1/2 & - & 1/2 \end{pmatrix} + 3 \begin{pmatrix} 1/2 & 1/2 \\ 1/2 & 1/2 \end{pmatrix}$$

By considering various powers of $\begin{pmatrix} 2 & 1 \\ 1 & 2 \end{pmatrix}$ and then arbitrary sums, we find for any T function w,

$$w\begin{pmatrix} 2 & 1 \\ 1 & 2 \end{pmatrix} = w(1)\begin{pmatrix} 1/2 & -1/2 \\ -1/2 & 1/2 \end{pmatrix} + w(3)\begin{pmatrix} 1/2 & 1/2 \\ 1/2 & 1/2 \end{pmatrix}.$$

The family of projections $E(\lambda)$ satisfy

$$E(\lambda) = \begin{cases} \begin{pmatrix} 0 & 0 \\ 0 & 0 \end{pmatrix}, & \text{when } \lambda < 1, \\[2mm] \begin{pmatrix} 1/2 & -1/2 \\ -1/2 & 1/2 \end{pmatrix}, & \text{when } 1 \leq \lambda < 3, \\[2mm] \begin{pmatrix} 1 & 0 \\ 0 & 1 \end{pmatrix}, & \text{when } 3 \leq \lambda. \end{cases}$$

Thus

$$E(d\lambda) = \begin{cases} 0, & \text{when } \lambda < 1, \\[2mm] \begin{pmatrix} 1/2 & -1/2 \\ -1/2 & 1/2 \end{pmatrix}, & \text{when } \lambda = 1, \\[2mm] 0, & \text{when } 1 < \lambda < 3, \\[2mm] \begin{pmatrix} 1/2 & 1/2 \\ 1/2 & 1/2 \end{pmatrix}, & \text{when } \lambda = 3, \\[2mm] 0, & \text{when } 3 < \lambda. \end{cases}$$

The integral in this case is a finite series.

2. Let A be an operator on complex Euclidean space K^n which is self-adjoint, and whose spectrum is $\sigma(A) = \{\lambda_1 \leq \lambda_2 \leq \ldots \leq \lambda_n\}$ with corresponding normalized eigenfunctions $\{x_1, \ldots, x_n\}$. The eigenfunctions form a complete orthonormal set in K^n, so any element x in K^n can be represented by its Fourier expansion.

$$x = \sum_{j=1}^{n} [x_j^* x] x_j.$$

As in the previous example we reorder the terms to find

$$x = \sum_{j=1}^{n} x_j \cdot x_j^* \cdot x,$$

from which we conclude

$$I = \sum_{j=1}^{n} x_j x_j^*,$$

$$A = \sum_{j=1}^{n} (Ax_j) x_j^*,$$

$$= \sum_{j=1}^{n} \lambda_j (x_j x_j^*).$$

In addition for any function w in T,

$$w(A) = \sum_{j=1}^{n} w(\lambda_j)(x_j x_j^*).$$

The family of projections $E(\lambda)$ satisfy

$$E(\lambda) = \begin{cases} 0, & \text{when } \lambda < \lambda_1, \\ \sum_{i \le j} (x_j x_j^*), & \text{when } \lambda_j \le \lambda < \lambda_{j+1}, \\ I, & \text{when } \lambda_n \le \lambda. \end{cases}$$

Thus

$$
E(d\lambda) = \begin{cases} 0, & \text{when } \lambda \neq \lambda_j, \ j=1,\ldots,n, \\ \\ x_j x_j^*, & \text{when } \lambda = \lambda_j, \ j=1,\ldots,n. \end{cases}
$$

The integral here also is a finite series.

3. If $H = \ell^2$, and A is a self-adjoint operator in $L(H,H)$ with $\sigma(A) = \{\lambda_1 \leq \lambda_2 \leq \ldots \leq \lambda_n \leq \ldots\}$ with an associated complete orthonormal set of eigenvectors $\{x_1,\ldots,x_n,\ldots\}$, then the calculations are essentially unchanged from the previous example. Operators on this space, however, need not necessarily have eigenvalues.

4. Let $H = L^2(-1,1)$. For all x in H, let the operator A be defined by

$$
Ax(t) = tx(t).
$$

For all w in T and x, y in H,

$$
\int_{-1}^{1} w(\lambda)(E(d\lambda)x,y) = (w(A)x,y),
$$

$$
= \int_{-1}^{1} w(t)x(t)\overline{y}(t)dt,
$$

$$
= \int_{-1}^{1} w(t)d[\int_{-1}^{1} K_{(-1,t)}(s)x(s)\overline{y}(s)ds],
$$

$$
= \int_{-1}^{1} w(t)(E(dt)x,y),
$$

where

$$
K_\Delta(s) = \begin{cases} 1, & \text{when } s \text{ is in the set } \Delta, \\ \\ 0, & \text{when } s \text{ is not in the set } \Delta, \end{cases}
$$

and

$$
(E(\Delta)x,y) = \int_{-1}^{1} K_\Delta(s)x(s)\overline{y}(s)ds.
$$

Thus

$$
E(\Delta)x(t) = K_\Delta(t)x(t)
$$

for all x in H. The projections E(dλ) are generated by multiplication by functions whose functional values are 0 or 1.

VIII.8.   The Spectral Resolution of a Normal Operator. As a corollary to the results just obtained we can achieve similar results for normal and unitary operators. We examine normal operators in this section and unitary operators in the next.

Recall from Definition VIII.3.1 that an operator A is normal if $AA^* = A^*A$.

Each operator A, whether it is self-adjoint, normal, unitary, or not, possesses a real part X and an imaginary part Y, given by

$$X = 1/2[A+A^*],$$

$$Y = 1/2i[A-A^*].$$

It is an easy matter to check to see that both X and Y are self-adjoint.

VIII.8.1. Theorem. Let H be a Hilbert space, and let A be a normal operator in $L(H,H)$. Then the real and imaginary parts of A,

$$X = 1/2[A+A^*],$$

$$Y = 1/2i[A-A^*],$$

commute. That is, $XY = YX$.

Proof. Since A is normal $AA^* = A^*A$. Thus $XY = (1/4i)[A^2-A^{*2}] = YX$.

Now X and Y are self-adjoint, so there exist families of projections $E_X(d\lambda)$ and $E_Y(d\mu)$ such that

$$X = \int_{m-}^{M} \lambda E_X(d\lambda), \quad \text{uniformly,}$$

$$Y = \int_{n-}^{N} \mu E_Y(d\mu), \quad \text{uniformly,}$$

where m, M and n, N are bounds for X and Y, respectively. Further, since E_X and E_Y are strong limits of polynomials in X and Y, respectively

and X, Y commute, so do E_X, E_Y for all values of the parameters λ
and μ. This is crucial.

VIII.8.2. <u>Theorem (Spectral Resolution of a Normal Operator)</u>. <u>Let H be</u>
<u>a Hilbert space, and let A be a normal operator in $L(H,H)$. Then there</u>
<u>exists a collection of projection operators $\{E_z(s)$, s is a rectangle R</u>
<u>in $E^2\}$, strong limits of polynomials in A and A^*, satisfying</u>

1. <u>$E_z(s_1) \leq E_z(s_2)$, when $s_1 \subset s_2$</u>

2. <u>$E_z(s) = 0$ when s lies below and to the left of the rectangle R,</u>

3. <u>$E_z(s) = I$ when s lies above and to the right of the rectangle R,</u>

<u>such that</u>

$$A = \lim_{\sup |\Delta R_k| \to 0} \sum_k z_k E_z(\Delta R_k),$$

$$= \int\int_R z E_z(dR),$$

<u>where z_k is in ΔR_k, $\underset{k}{\cup} \Delta R_k = R$. The integral exists in the Riemann-</u>
<u>Stieltjes sense. Equality is in the uniform sense.</u>

Proof. From the remarks preceeding, we see

$$A = \int_{n-}^{M} \lambda E_X(d\lambda) + i \int_{n-}^{N} \mu E_Y(d\mu),$$

$$= \int_{m-}^{M} \lambda E_X(d\lambda) \int_{n-}^{N} E_Y(d\mu) + i \int_{m-}^{M} E_X(d\lambda) \int_{n-}^{N} \mu E_Y(d\mu),$$

$$= \int_{m-}^{M} \int_{n-}^{N} (\lambda+i\mu) E_X(d\lambda) E_Y(d\mu).$$

We let $z = \lambda+i\mu$, $E_Z(dR) = E_X(d\lambda) E_Y(d\mu)$. Since E_X and E_Y commute, $E_Z(dR)$ possesses the properties stated above. Integration takes place over the rectangle $R = [m-,M] \times [n-,N]$, and

$$A = \int_R \int zE_Z(dR), \quad \text{uniformly.}$$

Although we shall not formally state it as a theorem, there exists an extension similar to that of Theorem VIII.7.4, which shows that

$$A^* = \int_R \int \overline{z} E_Z(dR), \quad \text{uniformly.}$$

If $w = w(z,\overline{z})$ is a function of two variables in T, then

$$w(A,A^*) = \int_R \int w(z,\overline{z}) E_Z(dR),$$

where the kind of equality depends upon the type of function w.

VIII.8.3. <u>An Example.</u> Let A be the operator on complex Euclidean space K^2 represented by the matrix $\begin{pmatrix} a & b \\ -b & a \end{pmatrix}$, where a and b are real numbers.

Then   A   is normal, with adjoint operator represented by $\begin{pmatrix} a & -b \\ b & a \end{pmatrix}$.   The

operators   X   and   Y   are given by

$$X = \begin{pmatrix} a & 0 \\ 0 & a \end{pmatrix},$$

$$Y = \begin{pmatrix} 0 & -ib \\ ib & 0 \end{pmatrix}.$$

Thus the spectral resolutions of   X   and   Y   are, respectively,

$$X = a\begin{pmatrix} 1 & 0 \\ 0 & 1 \end{pmatrix},$$

$$Y = b\begin{pmatrix} 1/2 & -i/2 \\ i/2 & 1/2 \end{pmatrix} - b\begin{pmatrix} 1/2 & i/2 \\ -i/2 & 1/2 \end{pmatrix}.$$

Inserting these in the expression   A = X + iY,   we have

$$A = (a+ib)\begin{pmatrix} 1/2 & -i/2 \\ i/2 & 1/2 \end{pmatrix} + (a-ib)\begin{pmatrix} 1/2 & i/2 \\ -i/2 & 1/2 \end{pmatrix},$$

$$A^* = (a-ib)\begin{pmatrix} 1/2 & -i/2 \\ i/2 & 1/2 \end{pmatrix} + (a+ib)\begin{pmatrix} 1/2 & i/2 \\ -i/2 & 1/2 \end{pmatrix}.$$

VIII.9.   <u>The Spectral Resolution of a Unitary Operator.</u>   As the final subject
of this chapter we decompose a unitary operator in terms of a spectral
resolution.   We recall from Definition VIII.3.1 that an operator   A   is
unitary if $A^* = A^{-1}$.   The real motivating force behind the definition,
however, is better illustrated by the following theorem.

VIII.9.1. Theorem. Let H be a Hilbert space, and let A be a unitary operator in $L(H,H)$. Then

$$(x,y) = (Ax,Ay)$$

for all x, y in H. In particular,

$$\|x\| = \|Ax\|$$

for all x in H. A is an isometric operator.

Proof. This follows from

$$(Ax,Ay) = (A^*Ax,y),$$

$$= (A^{-1}Ax,y),$$

$$= (x,y).$$

VIII.9.2. Theorem (Spectral Resolution of a Unitary Operator). Let H be a Hilbert space, and let A be a unitary operator in $L(H,H)$. Then there exists a collection of projection operators $\{E(\theta), \ 0 \leq \theta \leq 2\pi\}$, strong limits of polynomials in A and A^*, satisfying

1. $E(\theta_1) \leq E(\theta_2)$, when $0 \leq \theta_1 \leq \theta_2 < 2\pi$,

2. $E(\theta) = 0$, when $\theta = 0$,

3. $\underline{E(\theta) = I, \text{ when } \theta = 2\pi,}$

4. $\underline{E(\theta) \text{ is continuous from above,}}$

such that

$$A = \lim_{\sup|\phi_k - \phi_{k-1}| \to 0} \sum_k e^{i\phi_k}[E(\theta_k) - E(\theta_{k-1})] ,$$

$$= \int_{0-}^{2\pi} e^{i\theta} E(d\theta) ,$$

where $\underline{\phi_k}$ is in $\underline{[\theta_{k-1}, \theta_k]}$. The integral exists in the Riemann-Stieltjes sense. Equality is in the uniform sense.

The proof involves a bit of manipluation with trigonmetric polynomials in which we will need the following lemma.

VIII.9.3. <u>Lemma.</u> <u>Every real and nonegative trigonometric polynomial</u>

$$p(\phi) = \sum_{j=-n}^{n} a_j e^{ij\phi} ,$$

<u>can be written as the absolute square of another trigonometric polynomial</u> <u>$q(\phi)$ whose coefficients may be complex.</u>

Proof. If $p(\phi)$ ever is zero, we add $\varepsilon > 0$ to $p(\phi)$, then at the end let ε approach 0. Thus we may assume $p(\phi) > 0$ for all values of ϕ. We also note that $a_j = \bar{a}_{-j}$, since if this were not so $p'(\phi)$ would have a nonvanishing imaginary part. Now, assuming $a_n \neq 0$, we find that the polynomial

$$P(z) = \sum_{j=-n}^{n} a_j z^{n+j}$$

satisfies $P(z) = z^{2n}\overline{P(1/\bar{z})}$ and is nonzero on the unit circle, since if $P(z) = 0$ at $z = e^{i\phi}$, then $p(\phi) = e^{-in\phi}P(e^{i\phi})$ would also be zero.

Let $\alpha_1,\ldots,\alpha_k$ be the zeros of $P(z)$ inside the unit circle (a multiple zero written as many times as necessary) and $\beta_1,\ldots,\beta_s$ be the zeros outside the unit circle. Then

$$P(z) = C \prod_{j=1}^{n} (z-\alpha_j) \prod_{j=1}^{s} (z-\beta_j),$$

where C is constant, or rewriting,

$$P(z) = C'\prod_{j=1}^{n} (z-\alpha_j) z^s \prod_{j=1}^{s} (1/z-1/\beta_j).$$

Substituting this in the formula derived earlier,

$$P(z) = z^{2n}\overline{C'\prod_{j=1}^{r}(1/\bar{z}-\alpha_j)(1/\bar{z})^s \prod_{j=1}^{s}(\bar{z}-1/\beta_j)},$$

$$= C'\prod_{j=1}^{r} (1/z-\overline{\alpha}_j) z^{2n-s} \prod_{j=1}^{s} (z-1/\overline{\beta}_j).$$

Comparing this with the original factored form, we see that, after a possible reordering, $r = s = n$, and $\alpha_j = 1/\overline{\beta}_j$, $j=1,\ldots,n$. Thus

$$P(z) = c' \prod_{j=1}^{n} (z-\alpha_j) z^n \prod_{j=1}^{n} (1/z-\bar{\alpha}_j),$$

and

$$p(\phi) = c' \prod_{j=1}^{n} (e^{i\phi}-\alpha_j) \prod_{j=1}^{n} (e^{-i\phi}-\bar{\alpha}_j),$$

$$= \left| \sqrt{c'} \left[\prod_{j=1}^{n} (e^{i\phi}-\alpha_j) \right] \right|^2.$$

Proof of Theorem VIII.9.2. The procedure is quite similar to that used in section VIII.7. To each trigonometric polynomial $p(\phi) = \sum_{j=-n}^{n} a_j e^{ij\phi}$, we let correspond the operator $p(A) = \sum_{j=-n}^{n} a_j A^j$. This correspondence is linear and multiplicative for all trigonometric polynomials $p(\phi)$. Moreover if $p(\phi) \geq 0$, then $p(\phi) = |q(\phi)|^2$. The operator corresponding to $p(\phi)$ is $p(A) = q(A)^* q(A)$. Since $(p(A)x,x) = \|q(A)x\|^2 \geq 0$ for all x in H, the correspondence preserves positivity also. Finally it is apparent that the operator corresponding to the complex conjugate of a trigonometric polynomial is the adjoint of the operator corresponding to the original.

By using Theorem VIII.6.4 this correspondence can again be extended to functions which are piecewise continuous in $[0,2\pi]$, just as was done in Theorem VIII.6.9.

In particular let the functions $e_\theta(\phi)$, $0 \leq \theta \leq 2\pi$, $0 \leq \theta \leq 2\pi$, be defined as follows:

$$e_0(\phi) = 0 \quad \text{for all} \quad \phi,$$

$$
e_\theta(\phi) = \begin{cases} 1, & \text{when } 0 < \phi \leq \theta, \\ \\ 0, & \text{when } \theta < \phi \leq 2\pi, \end{cases}
$$

$$
e_{2\pi}(\phi) = 1 \text{ for all } \phi.
$$

These functions may be approached by continuous functions from above.   Since

the Stone-Weierstrass theorem shows that the trigonometric polynomials are

dense in the (complex) space of continuous functions on $[0, 2\pi]$ we can choose

a decreasing sequence of trigonometric polynomials which converge to each

function $e_\theta(\phi)$ $0 \leq \theta \leq 2\pi$.   Then the corresponding operators converge

strongly to projection operators, which we denote by $E(\theta)$.   These obviously

have properties 1, 2 and 3, as stated in the theorem.

To show $E(\theta)$ is continuous from the right, let θ be fixed but less

than 2π.   Let $\{p_j(e^{i\phi})\}_{j=1}^\infty$ be a sequence of trigonometric polynomials

approaching $e_\theta(\phi)$ from above and also satisfying

$$
p_j(e^{i\phi}) \geq e_{\theta+1/j}(\phi) \geq e_\theta(\phi).
$$

Then

$$
p_j(A) \geq E(\theta+1/j) \geq E(\theta).
$$

But $p_j(A)$ approaches $E(\theta)$, as j approaches ∞.   Thus $\lim_{\psi \to \theta+} E(\psi) = E(\theta)$.

Finally let $[0,2\pi]$ be decomposed into subintervals by the points $\{\theta_j\}_{j=0}^n$,

$$0 = \theta_0 < \theta_1 < \ldots < \theta_{n-1} < \theta_n = 2\pi,$$

in such a manner that $\sup[\theta_j-\theta_{j-1}] < \varepsilon$, where ε has been previously chosen. In each interval $[\theta_{k-1},\theta_k]$ choose a point ϕ_k arbitrarily. Then if ϕ is in $[\theta_{k-1},\theta_k]$,

$$\left| e^{i\phi} - \sum_{k=1}^n e^{i\phi_k}[e_{\theta_k}(\phi)-e_{\theta_{k-1}}(\phi)] \right| = \left| e^{i\phi}-e^{i\phi_k} \right|,$$

$$= \left| \int_{\phi_k}^{\phi} e^{ix}dx \right|,$$

$$\leq \left| \int_{\phi_k}^{\phi} 1\ dx \right|,$$

$$= \left| \phi-\phi_k \right| < \varepsilon.$$

Similarly, if $\phi = 0$,

$$\left| e^{i0} - \sum_{k=1}^n e^{i\phi_k}[e_{\theta_k}(\phi)-e_{\theta_{k-1}}(\phi)] \right| < \left| \phi_1 \right| < \varepsilon.$$

Thus for all values of ϕ,

$$0 \leq \{e^{i\phi} - \sum_{k=1}^{n} e^{i\phi_k}[e_{\theta_k}(\phi) - e_{\theta_{k-1}}(\phi)]\}^* \cdot$$

$$\{e^{i\phi} - \sum_{k=1}^{n} e^{i\phi_k}[e_{\theta_k}(\phi) - e_{\theta_{k-1}}(\phi)]\} < \varepsilon^2.$$

This implies

$$0 \leq \{A - \sum_{k=1}^{n} e^{i\phi_k}[E(\theta_k) - E(\theta_{k-1})]\}^* \cdot$$

$$\{A - \sum_{k=1}^{n} e^{i\phi_k}[E(\theta_k) - E(\theta_{k-1})]\} < \varepsilon^2 I.$$

Thus

$$\|A - \sum_{k=1}^{n} e^{i\phi_k}[E(\theta_k) - E(\theta_{k-1})]\| < \varepsilon,$$

which completes the proof. So

$$A = \int_{0}^{2\pi} e^{i\theta} E(d\theta).$$

There also exists an extension similar to that of Theorem VIII.7.4,

which shows

$$A^* = \int_{0-}^{2\pi} e^{-i\theta} E(d\theta), \quad \text{uniformly.}$$

If w is a function of two variables in T, then

$$w(A,A^*) = \int_{0-}^{2\pi} w(e^{is}, e^{-is}) E(ds),$$

where the kind of equality depends upon the type of function w.

VIII.9.3. <u>An Example</u>. Let A be the operator on complex Euclidean space K^2 represented by the matrix

$$\begin{pmatrix} a & \sqrt{1-a^2} \\ -\sqrt{1-a^2} & a \end{pmatrix}$$

where $-1 \leq a \leq 1$. Then A is unitary, and from VIII.8.3 we find

$$A = (a+i\sqrt{1-a^2})\begin{pmatrix} 1/2 & -i/2 \\ i/2 & 1/2 \end{pmatrix} + (a-i\sqrt{1-a^2})\begin{pmatrix} 1/2 & i/2 \\ -i/2 & 1/2 \end{pmatrix}.$$

$$A^* = (a-i\sqrt{1-a^2})\begin{pmatrix} 1/2 & -i/2 \\ i/2 & 1/2 \end{pmatrix} + (a+\sqrt{1-a^2})\begin{pmatrix} 1/2 & i/2 \\ -i/2 & 1/2 \end{pmatrix}.$$

VIII.   Exercises

1. (a) Find the spectrum of the operator represented by the matrix $\begin{pmatrix} a & b \\ b & a \end{pmatrix}$ in $L(E^2, E^2)$, where E^2 is real Euclidean space.

    (b) Let the Hilbert space $H = L^2(-1,1)$, and let A in $L(H,H)$ be defined by $Ax(t) = tx(t)$ for all x in H. Find $\sigma(A)$.

2. On complex Euclidean space K^n, let the transformation A be represented by the matrix (a_{ij}). What is the representation of A^*?

3. Show that if A is self-adjoint on an arbitrary Hilbert space H,

    then

$$e^{iAt} = \sum_{j=0}^{\infty} (iAt)^j/j!$$

    is unitary for all real values of t.

4. Show that in an arbitrary Hilbert space H if A is self-adjoint,

    $\alpha > 0$, and $(Ax,x) > \alpha(x,x)$ for all x in H, then A is a regular

    operator.

5. If H is real Euclidean space E^2, what form do projection matrices

    have?

6. If $H = L^2(-1,1)$, show that $P_M x(t) = K_M(t)x(t)$, where

    $K_M(t) = \begin{cases} 1, & \text{when } t \in M \\ 0, & \text{when } t \notin M \end{cases}$ is a projection of x into the subspace

    $L^2(M)$, when M is a measurable set in $[-1,1]$.

7. Prove Corollary VIII.5.9.

8. Prove that in an arbitrary Hilbert space if an operator A in $L(H,H)$

    is nilpotent and self-adjoint then $A = 0$.

9. Show that in an arbitrary Hilbert space H, if the projection operator

    P satisfies $0 \neq P \neq I$, then $\sigma(P) = \{0,1\}$. Compute $(\lambda I - P)^{-1}$ when

    it exists.

10. Prove Theorem VIII.6.2.

11. Prove Theorem VIII.6.10.

12. Prove Theorem VIII.6.14.

13. In part 2 of Example VIII.7.5, show that the expressions $x_j x_j^*$ are

    $n \times n$ matrices and are projections.

14. Find the spectral resolution of the operator on K^3 generated by the

    matrix

    $$\begin{pmatrix} 210 & -84 & -60 \\ -84 & 473 & 24 \\ -60 & 24 & 331 \end{pmatrix}.$$

15. Find the spectral resolution of the operator on K^3 generated by the

    matrix

    $$\begin{pmatrix} 2 & 1 & 0 \\ -1 & 2 & 1 \\ 0 & -1 & 2 \end{pmatrix}.$$

16. Characterize all unitary matrices on the real Euclidean spaces E^2

    and E^3.

17. Starting from the spectral resolution of a normal operator, derive

    the spectral resolution of a unitary operator. Do so by showing that

    outside and inside the unit circle the varying of the projection

    measure leads to a contradiction.

References

1. N.I. Akhiezer and I.M. Glazman, "Theory of Linear Operators in Hilbert
      Space, vols. I and II," Frederick Ungar, New York, 1961.

2. P.R. Halmos, "Finite Dimensional Vector Spaces," Van Nostrand,
      Princeton, 1958.

3. P.R. Halmos, "Introduction to Hilbert Space and the Theory of Multiplicity,"
      Chelsea, New York, 1951.

4. M.A. Naimark, "Normed Rings," P. Noordhoff, Groningen, The Netherlands,
      1964.

5. F. Riesz and B. Sz.-Nagy, "Functional Analysis," Frederick Ungar, New
      York, 1955.

6. M.H. Stone, "Linear Transformations in Hilbert Space," American
      Mathematical Society, New York, 1966.

7. A.E. Taylor, "Introduction to Functional Analysis, John Wiley and Sons,
      New York, 1958.

IX.   COMPACT OPERATORS ON A HILBERT SPACE

In the previous chapter we decomposed self-adjoint, normal, and unitary operators into integrals over their spectra. We should like to apply these results, especially in the self-adjoint case, to ordinary and partial differential operators. Unfortunately differential operators on the standard L^2 spaces are not bounded, and thus do not satisfy the hypotheses made previously. To circumvent this difficulty in part we apply the results to the inverse operators, which are not only bounded, but in addition are of a very special form: They are compact.

IX.1. <u>Compact Operators</u>. We shall now define, in part characterize, and give an example of a compact operator. Recall from Definition VI.1.2 that a compact set is one in which every infinite set contains at least one limit point.

IX.1.1. <u>Definition</u>. Let H be a Hilbert space, and let X be a subset of H. X is conditionally or relatively compact if its closure (X and all its limit points) is compact.

IX.1.2. <u>Definition</u>. Let H be a Hilbert space, and let A be a linear operator with domain and range in H. A is compact if A transforms bounded sets in H into relatively compact sets in H.

Another term used for compact is completely continuous. Over the last few years, however the word compact has become increasingly more popular, largely because it is shorter.

IX.1.3. Theorem. Let A be a compact operator on a Hilbert space H.
Then A is bounded. That is, A is in $L(H,H)$.

Proof. Let x be an arbitrary element in D_A. Then

$$\|Ax\| = \|A(x/\|x\|)\| \cdot \|x\|.$$

Now the set of all elements in H with norm 1 is bounded. Since A is
compact, the image of such elements when transformed by A is relatively
compact, and is therefore bounded. Thus for all x in H, there exists
$M(=\|A\|)$ such that

$$\|Ax\| \leq M\|x\|.$$

IX.1.4. Examples. 1. Let A be any linear operator on a finite dimensional
space H. If X is a bounded set, then its image under A, $A(X)$, is bounded,
and, thus, because of the finite dimension, $\overline{A(X)}$ is compact.

In fact any bounded linear operator with finite dimensional range is
compact.

2. On the other hand, the identity operator on a Hilbert space is not
compact. For if $\{e_j\}_{j=1}^{\infty}$ is an infinite orthonormal set, then its image
is the same. Since for all $i \neq j$, $\|e_i - e_j\| = \sqrt{2}$, the set does not form a
Cauchy sequence and is not convergent.

IX.1.5. Theorem. 1. The collection of all compact operators on a Hilbert space H form a closed linear manifold in $L(H,H)$.

2. If A and B are in $L(H,H)$ and A is compact, then AB and BA are compact.

Proof. 1. Suppose that A, B are compact operators on H, and $\{x_j\}_{j=1}^{\infty}$ is a bounded set. Then $\{x_j\}_{j=1}^{\infty}$ contains a subsequence $\{x_j'\}_{j=1}^{\infty}$ such that $\{Ax_j'\}_{j=1}^{\infty}$ converges. Further $\{x_j'\}_{j=1}^{\infty}$ contains a subsequence $\{x_j''\}_{j=1}^{\infty}$ such that $\{Bx_j''\}_{j=1}^{\infty}$ converges. Thus $\{x_j\}_{j=1}^{\infty}$ contains a subsequence $\{x_j''\}_{j=1}^{\infty}$ such that $\{(A+B)x_j''\}_{j=1}^{\infty}$ converges. That αA is compact for all complex numbers α is trivial. Thus the set of compact operators is a linear manifold in $L(H,H)$.

To show the linear manifold is closed, let us assume that $\{x_j\}_{j=1}^{\infty}$ is bounded by M, and that $\{A_n\}_{n=1}^{\infty}$ is a sequence of compact operators with uniform limit A. We choose a subsequence $\{x_j'\}_{j=1}^{\infty}$ such that $\{A_1 x_j'\}_{j=1}^{\infty}$ converges. From this subsequence, we choose another $\{x_j''\}_{j=1}^{\infty}$ such that $\{A_2 x_j''\}_{j=1}^{\infty}$ converges. By induction, we choose a subsequence $\{x_j^{(n)}\}_{j=1}^{\infty}$ from $\{x_j^{(n-1)}\}_{j=1}^{\infty}$ such that $\{A_n x_j^{(n)}\}_{j=1}^{\infty}$ converges. Now let $y_j = x_j^{(j)}$, $j=1,\ldots$ Then $\{A_n y_j\}_{j=1}^{\infty}$ converges for all n. Considering $\{Ay_j\}_{j=1}^{\infty}$, we find

$$\|Ay_j - Ay_k\| \le \|Ay_j - A_n y_j\|$$

$$+ \|A_n y_j - A_n y_k\|$$

$$+ \|A_n y_k - Ay_k\|.$$

If n is sufficiently large to guarantee $\|A-A_k\| < \varepsilon/3M$, then

$$\|Ay_j-A_ky_j\| \leq \|A-A_n\|\|y_k\| \leq \varepsilon/3.$$

The second term satisfies

$$\|A_ny_j-A_ny_k\| < \varepsilon/3$$

when j and k are sufficiently large, since $\{A_ny_j\}_{j=1}^{\infty}$ converges for all
n. The third term, like the first, will be less than $\varepsilon/3$ if n is
sufficiently large. Thus if j and k are sufficiently large,

$$\|Ay_j-Ay_k\| < \varepsilon.$$

$\{Ay_j\}_{j=1}^{\infty}$ is a Cauchy sequence and is thus convergent.   A is compact, and the
linear manifold of compact operators is closed in $L(H,H)$.

2. If B is bounded then $\{Bx_j\}_{j=1}^{\infty}$ is bounded and $\{A(Bx_j)\}_{j=1}^{\infty}$ contains a convergent subsequence. AB is compact.

Similarly, $\{Ax_j\}_{j=1}^{\infty}$ contains a convergent subsequence.   Thus
$\{B(Ax_j)\}_{j=1}^{\infty}$ does also, since B is bounded, and BA is compact.

IX.1.6. Corollary. Let S be the linear manifold of operators in $L(H,H)$
with finite dimensional range. Then the closure of S (under the uniform
operator norm) in $L(H,H)$ is a closed linear manifold in the subspace of
compact operators on H.

Proof. It became apparent in Example IX.1.4 that such operators were compact

From Theorem IX.1.5 the limits of such operators are also compact. The

linearity of such a manifold is obvious.

IX.2. Some Special Examples. This section is devoted to two examples, one

in $L(\ell^2,\ell^2)$, the other in $L(L^2,L^2)$. An isometric isomorphism is produced

between them, which shows that when one is compact, so is the other. The

results obtained herein are fundamental to the following sections.

IX.2.1. Theorem. Let $H = \ell^2$, the Hilbert space of infinite sequences

$\{x:x=(x_1,\ldots,x_j,\ldots), \sum_{j=1}^{\infty} |x_j|^2 < \infty\}$ generated by the inner product

$(x,y) = \sum_{j=1}^{\infty} \bar{y}_j x_j$ and norm $\|x\| = [\sum_{j=1}^{\infty} |x_j|^2]^{1/2}$. Let the linear operator

A be defined by

$$Ax = A(x_1,\ldots,x_n,\ldots) =$$

$$(\sum_{k=1}^{\infty} a_{1k}x_k,\ldots, \sum_{k=1}^{\infty} a_{jk}x_k,\ldots),$$

where $\sum_{j=1}^{\infty} \sum_{k=1}^{\infty} |a_{jk}|^2 < \infty.$ Then A is compact.

Proof. Let $\varepsilon > 0$ be chosen arbitrarily. Then there exists a number p_ε

such that

$$\sum_{j=p_\varepsilon+1}^{\infty} \sum_{k=1}^{\infty} |a_{jk}|^2 < \varepsilon^2.$$

We define A_ε by

$$A_\varepsilon x = A_\varepsilon(x_1,\ldots,x_j,\ldots),$$

$$= (\sum_{k=1}^{\infty} a_{1k}x_k, \ldots, \sum_{k=1}^{\infty} a_{p_\varepsilon k}x_k, 0, 0, \ldots).$$

After the p_ε-th term all entries are 0. Then, since the range of each

operator A_ε is finite dimensional, each A_ε is compact. Now

$$\|A_\varepsilon x - Ax\|^2 = \sum_{j=p_\varepsilon+1}^{\infty} | \sum_{k=1}^{\infty} a_{jk}x_k |^2$$

$$\leq \sum_{j=p_\varepsilon+1}^{\infty} [\sum_{k=1}^{\infty} |a_{jk}|^2][\sum_{k=1}^{\infty} |x_k|^2]$$

$$\leq \varepsilon^2 \|x\|^2,$$

or $\|A_\varepsilon - A\| < \varepsilon$. Thus A_ε approaches A uniformly, and A is also compact.

Before we introduce the second operator, we need some preliminary infor-

mation.

IX.2.2. <u>Theorem</u>. Let $H_1 = L^2(\sigma)$, the Hilbert space of σ-measurable

functions $\{x:\int|x(t)|^2\sigma(dt) < \infty\}$ generated by the inner product $(x,y) = $

$\int \overline{y(t)}x(t)\sigma(dt)$ and norm $\|x\| = [\int|x(t)|^2\sigma(dt)]^{1/2}$. Similarly, let

$H_2 = L^2(\tau)$. Further let $\{x_j(t)\}_{j=1}^{\infty}$ and $\{y_k(s)\}_{k=1}^{\infty}$ be complete orthonormal

sets in H_1 and H_2, respectively. Then $\{x_j(t) \cdot y_k(s)\}_{j,k=1}^{\infty}$ is a complete

orthonormal set in $H = L^2(\sigma \cdot \tau)$, the Hilbert space of $\sigma \cdot \tau$-measurable functions $\{z: \int \int |z(t,s)|^2 \sigma(dt)\tau(ds) < \infty\}$, generated by the inner product

$(z,w) = \int \int \overline{w}(t,s)z(t,s)\sigma(dt)\tau(ds)$ and norm $\|z\| = [\int \int |z(t,s)|^2 \sigma(dt)\tau(ds)]$

Proof. Suppose z is in $L^2(\sigma \cdot \tau)$ and is orthogonal to $\{x_j(t)y_k(s)\}_{j,k=1}^{\infty}$.
For fixed j we have

$$\int \int \overline{x_j(t)y_k(s)}z(t,s)\sigma(dt)\tau(ds) = 0,$$

for k=1,... . Thus

$$\int \overline{y_k}(s)[\int \overline{x_j}(t)z(t,s)\sigma(dt)]\tau(ds) = 0,$$

for k=1,... . The expression in the brackets is orthogonal to $\{y_k(s)\}_{k=1}^{\infty}$
in $L^2(\tau)$. So

$$\int \overline{x_j}(t)z(t,s)\sigma(dt) = 0$$

except on a set E_j of τ-measure 0. Thus

$$\int \overline{x_j}(t)z(t,s)\sigma(dt) = 0,$$

j=1,..., except on a set $E = \bigcup_{j=1}^{\infty} E_j$, which still has τ-measure 0.
    Let E′ be the complement of E, and let

$$K_{E'}(s) = \begin{cases} 1, & \text{when } s \in E' \\ 0, & \text{when } s \notin E'. \end{cases}$$

Then

$$\int \bar{x}_j(t)K_{E'}(s)z(t,s)\sigma(dt) = 0,$$

$j=1,\ldots,$ for all s.

But this statement shows that $K_{E'}(s)z(t,s)$ is orthogonal to $\{x_j(t)\}_{j=1}^{\infty}$ in $L^2(\sigma)$. Thus $K_{E'}(s)z(t,s) = 0$ except for ·a set F of σ-measure 0. If F′ is the complement of F, and

$$K_{F'}(t) = \begin{cases} 1, & \text{when } t \in F' \\ 0, & \text{when } t \notin F', \end{cases}$$

then $K_{F'}(t)K_{E'}(s)z(t,s) = 0$ for all t,x. Since the complement of F′ × E′ is of $\sigma \cdot \tau$-measure 0; $z(t,s) = 0$ in $L^2(\sigma \cdot \tau)$.

Next follows a generalization of Theorem II.2.3, which derived the matrix representation for a linear operator in a finite dimensional space.

IX.2.3. <u>Theorem.</u> <u>Let H be a separable Hilbert space, and let A be an</u> <u>operator in $L(H,H)$. Let $\{x_j\}_{j=1}^{\infty}$ be a complete, orthonormal set in H.</u> <u>We define the infinite matrix $B = (a_{jk})$, where $a_{jk} = (Ax_k, x_j)$,</u> <u>j,k=1,..., to be the matrix representation of A. Then under the isometric</u> <u>isomorphism U between H and ℓ^2, as given by Theorem VII.7.3,</u>

$$B = UAU^{-1}.$$

<u>A is compact on H if and only if B is compact on ℓ^2.</u>

Proof. Recall that if x is in H, then $x = \sum\limits_{j=1}^{\infty} (x,x_j)x_j$, and

$Ux = ((x,x_1),\ldots,(x,x_j),\ldots)$. If $(a_1,\ldots,a_j,\ldots)$ is in ℓ^2, then

$U^{-1}(a_1,\ldots,a_j,\ldots) = \sum\limits_{j=1}^{\infty} a_j x_j$.

Now let $(a_1,\ldots,a_j,\ldots)$ be in ℓ^2. Then

$$U^{-1}(a_1,\ldots,a_j,\ldots) = \sum_{j=1}^{\infty} a_j x_j,$$

$$AU^{-1}(a_1,\ldots,a_j,\ldots) = \sum_{j=1}^{\infty} a_j Ax_j.$$

Now for arbitrary k,

$$Ax_k = \sum_{j=1}^{\infty} (Ax_k,x_j)x_j,$$

so

$$AU^{-1}(a_1,\ldots,a_j,\ldots) = \sum_{j=1}^{\infty} a_j \sum_{n=1}^{\infty} (Ax_j,x_n)x_n,$$

$$= \sum_{n=1}^{\infty} [\sum_{j=1}^{\infty} a_j(Ax_j,x_n)]x_n,$$

and

$$UAU^{-1}(a_1,\ldots,a_k,\ldots) = (\sum_{j=1}^{\infty} a_j(Ax_j,x_1),\ldots,\sum_{j=1}^{\infty} a_j(Ax_j,x_k),\ldots)$$

$$= (\sum_{j=1}^{\infty} a_j a_{1j},\ldots,\sum_{j=1}^{\infty} a_j a_{kj},\ldots),$$

$$= B(a_1,\ldots,a_k,\ldots).$$

Thus $B = UAU^{-1}$. The statement concerning compactness is obvious.

IX.2.4. Theorem. Let $H = L^2(\sigma)$, the Hilbert space of σ-measurable functions $\{x; \int |x(t)|^2\sigma(dt) < \infty$ generated by the inner product $(x,y) = \int \bar{y}(t)x(t)\sigma(dt)$ and norm $\|x\| = [\int |x(t)|^2\sigma(dt)]^{1/2}$. Let the Hilbert-Schmidt operator K be defined on $L^2(\sigma)$ by

$$Kx(t) = \int K(t,s)x(s)\sigma(ds),$$

where $K(\cdot,\cdot)$ is σ-measurable in both variables and satisfies

$$\int \int |K(t,s)|^2\sigma(ds)\sigma(dt) < \infty.$$

Then the operator K is compact.

Proof. Let $\{x_j\}_{j=1}^\infty$ be a complete orthonormal set in H. Then

$$(Kx_j, x_k) = \int \int K(t,s)x_j(s)\overline{x_k(t)}\sigma(ds)\sigma(dt).$$

Since by Theorem IX.2.2, $\{x_k(t)\bar{x}_j(s)\}_{j,k=1}^\infty$ is complete in $L^2(\sigma \cdot \sigma)$,

$$K(t,s) = \sum_{j=1}^\infty \sum_{k=1}^\infty [\int \int K(v,u)x_j(u)\bar{x}_k(v)\sigma(du)\sigma(dv)]x_k(t)\bar{x}_j(s),$$

$$= \sum_{j=1}^\infty \sum_{k=1}^\infty (Kx_j, x_k)x_k(t)\bar{x}_j(s).$$

Then Parseval's equality shows

$$\sum_{j=1}^{\infty} \sum_{k=1}^{\infty} |(Kx_j, x_k)|^2 = \int \int |K(t,s)|^2 \sigma(ds)\sigma(dt) < \infty.$$

By Theorems IX.2.1 and IX.2.3, K is compact.

IX.3.   The Spectrum of a Compact, Self-Adjoint Operator. As in the case of
every self-adjoint operator, the spectrum of a self-adjoint, compact operator
is real. But unlike operators in general, whose spectrum can contain an ent.
interval, a compact operator has a spectrum which can accumulate only at 0.

IX.3.1.   Theorem.   Let H be a Hilbert space, and let A be a compact,
self-adjoint operator in $L(H,H)$.   Then

$$\sigma(A)-\{0\} = \sigma_p(A)-\{0\}.$$

Proof. We already know $\sigma_p(A)-\{0\} \subset \sigma(A)-\{0\}$. To show the reverse inclusio
let λ be in $\sigma(A)-\{0\}$. By Theorem VIII.5.6, λ is in $\sigma_{ap}(A)-\{0\}$.
Therefore there exists a sequence of elements $\{x_j\}_{j=1}^{\infty}$, $\|x_j\| = 1$, $j=1,\ldots,$
such that

$$\lim_{j\to\infty} \|(A-\lambda I)x_j\| = 0.$$

Since A is compact, and $\{x_j\}_{j=1}^{\infty}$ is bounded, we can extract a subsequence
which we also call $\{x_j\}_{j=1}^{\infty}$, such that $\{Ax_j\}_{j=1}^{\infty}$ converges. We call its
limit y.   Then

$$\|y-\lambda x_j\| \leq \|y-Ax_j\| + \|Ax_j-\lambda x_j\|.$$

Since the right side approaches 0, we see that $\{x_j\}_{j=1}^{\infty}$ also converges to y/λ. Further

$$\left|\|y\|-|\lambda|\right| = \left|\|y\|-|\lambda|\|x_j\|\right|,$$

$$= \left|\|y\|-\|\lambda x_j\|\right|,$$

$$\leq \|y-\lambda x_j\|,$$

which approaches 0, as we have just seen. So $\|y\| = |\lambda|$, and $y \neq \theta$. Thus

$$Ay = A \lim_{j\to\infty} \lambda x_j = \lambda \lim_{j\to\infty} Ax_j = \lambda y,$$

and λ is in $\sigma_p(A)-\{0\}$.

IX.3.2. <u>Corollary.</u> Let H be a Hilbert space, and let A be a compact, self-adjoint operator in $L(H,H)$. Then either $\|A\|$ or $-\|A\|$ is an eigenvalue of A.

Proof. By Theorem VIII.5.8 we know either $\|A\|$ or $-\|A\|$ is in $\sigma(A)$. Since $\|A\| \neq 0$, and A is self-adjoint, either $\|A\|$ or $-\|A\|$ is in $\sigma_p(A)$.

IX.3.3. <u>Definition.</u> <u>Let H be a Hilbert space, let A be linear operator</u>
<u>on H, and let λ be an eigenvalue of A. The eigenmanifold $m(\lambda)$,</u>
<u>associated with the eigenvalue λ, is the set of all elements x in H</u>
<u>such that $Ax = \lambda x$.</u>

Clearly the manifold $m(\lambda)$ is a linear manifold. If A is bounded it
is also easy to show $m(\lambda)$ is closed.

The following is true for all self-adjoint operators, not just those
which are compact.

IX.3.4. <u>Theorem.</u> <u>Let A be a self-adjoint operator on a Hilbert space</u>
<u>H. Let $m(\lambda_1)$ and $m(\lambda_2)$ be two eigenmanifolds corresponding to different</u>
<u>eigenvalues λ_1 and λ_2. Then $m(\lambda_1)$ and $m(\lambda_2)$ are mutually orthogonal.</u>

Proof. Let x_1 be in $m(\lambda_1)$, and x_2 be in $m(\lambda_2)$. Then

$$(\lambda_1 - \lambda_2)(x_1, x_2) = \lambda_1(x_1, x_2) - \lambda_2(x_1, x_2),$$

$$= (\lambda_1 x_1, x_2) - (x_1, \lambda_2 x_2),$$

$$= (Ax_1, x_2) - (x_1, Ax_2),$$

$$= 0.$$

Since $\lambda_1 \neq \lambda_2$, $(x_1, x_2) = 0$.

IX.3.5. <u>Theorem</u>. <u>Let H be a Hilbert space, and let A be a compact, self-</u>
<u>adjoint operator in $L(H,H)$.   Then</u>

1. <u>If $\lambda \neq 0$ is an eigenvalue of A, $m(\lambda)$ has finite dimension.</u>

2. <u>$\sigma(A)$ has no nonzero limit points.</u>

Proof. 1. If $m(\lambda)$ has infinite dimension, then there exists an infinite

orthonormal set $\{x_j\}_{j=1}^{\infty}$ in $m(\lambda)$. Since $\{x_j\}_{j=1}^{\infty}$ is bounded, and A is

compact, $\{Ax_j\}_{j=1}^{\infty}$ has a convergent subsequence. But $\{Ax_j\}_{j=1}^{\infty} = \{\lambda x_j\}_{j=1}^{\infty}$,

which cannot ever converge.

2. Let $\{\lambda_j\}_{j=1}^{\infty}$ be eigenvalues of A which have limit $\lambda \neq 0$, and

let $\{x_j\}_{j=1}^{\infty}$ be the associated normalized eigenfunctions. Then

$\{(1/\lambda_j)x_j\}_{j=1}^{\infty}$ is bounded, and $\{A(1/\lambda_j)x_j\}_{j=1}^{\infty}$ contains a convergent sub-

sequence. But $\{A(1/\lambda_j)x_j\}_{j=1}^{\infty} = \{x_j\}_{j=1}^{\infty}$, and cannot possibly converge. We

have a contradiction.

In conclusion we state the following as a summation.

IX.3.6. <u>Theorem</u>. <u>Let H be a Hilbert space, and let A be a compact,</u>
<u>self-adjoint operator in $L(H,H)$.   Then</u>

1. <u>There exists at most a countable number of distinct nonzero eigen-</u>
<u>values $\{\lambda_j\}_{j=1}^{\infty}$, which, when ordered by absolute value, satisfy</u>

$$\|A\| = |\lambda_1| > |\lambda_2| > \ldots > |\lambda_n| > \ldots \geq 0.$$

The sequence either terminates or approaches 0.

2. Each of the eigenmanifolds $m(\lambda_j)$, j=1,..., $\lambda_j \neq 0$, is finite
dimensional.

IX.4.   The Spectral Resolution of a Compact, Self-Adjoint Operator.   We alrea
have a spectral decomposition of a self-adjoint operator, which we derived in
the previous chapter.   Compact operators, however, because of their special
properties, have a spectral resolution which is substantially simpler in
form.   Instead of an abstract integral, it is an infinite series.   In the
special case of finite dimensional spaces the series is even simpler: It is
finite.

IX.4.1.   Theorem (Spectral Resolution of a Compact Self-Adjoint Operator).

Let   H   be a Hilbert space, and let   A   be a compact, self-adjoint operator
in $L(H,H)$.   Let   A   have eigenvalues $\{\lambda_j\}_{j=1}^{\infty}$, ordered with decreasing
absolute values,   let $\{m(\lambda_j)\}_{j=1}^{\infty}$ be the corresponding eigenmanifolds, and
let $\{P_j\}_{j=1}^{\infty}$ be the projections from   H   onto these eigenmanifolds.   Then

1. $P_j P_k = 0$ when $j \neq k$. $P_j^2 = P_j$, j=1,... .

2. $AP_j = P_j A$, j=1,...,   uniformly.

3. $A = \sum_{j=1}^{\infty} \lambda_j P_j$ uniformly.

Proof.   1 is obvious from the results of the previous section.   To show 2,
let   x,   y   be in   H.   Then $AP_j x = \lambda_j P_j x$, $AP_j y = \lambda_j P_j y$,   since $P_j x$, $P_j y$

are both in $m(\lambda_j)$. Thus

$$(P_j Ax,y) = (x,AP_j y) = (x,\lambda_j P_j y),$$

$$= (\lambda_j P_j x,y) = (AP_j x,y).$$

Thus $([P_j A-AP_j]x,y) = 0$ for all x, y in H. If $y = [P_j A-AP_j]x$, we see $\|[P_j A-AP_j]x\| = 0$ for all x in H. Dividing by $\|x\|$ and maximizing, we have

$$\|P_j A-AP_j\| = 0.$$

To show 3, let $A_n = A - \sum_{j=1}^{n} \lambda_j P_j$. A_n is compact since it is the difference of compact operators. If $1 \leqq k \leqq n$, $A_n P_k = P_k A_n = 0$, so A_n transforms $m(\lambda_k)$ onto $\{\theta\}$. The range of A_n is orthogonal to $m(\lambda_j)$, $j=1,..,n$.

Now suppose $A_n x = \lambda x$. If $1 \leqq k \leqq n$, then

$$\theta = P_k A_n x = \lambda P_k x,$$

so x is orthogonal to $m(\lambda_k)$. Further

$$Ax = A_n x + \sum_{j=1}^{n} \lambda_j P_j x,$$

$$= A_n x$$

$$= \lambda x.$$

So λ is an eigenvalue of A, and $|\lambda| \leq |\lambda_j|$, j=1,...,n.

Finally, let $\varepsilon > 0$. We choose n sufficiently large, such that $|\lambda_j| < \varepsilon$, when $j \geq n$. Since

$$\|A_n\| = \sup\{|\lambda| : \lambda \in \sigma(A_n)\} < \varepsilon,$$

$$\lim_{n\to\infty} A_n = 0,$$

and

$$A = \sum_{j=1}^{\infty} P_j$$

uniformly.

IX.4.2. Corollary. Let H be a Hilbert space, and let A be a compact, self-adjoint operator in $L(H,H)$. Let A have eigenvalues $\{\lambda_j\}_{j=1}^{\infty}$ ordered with decreasing absolute values, let $\{m(\lambda_j)\}_{j=1}^{\infty}$ be the corresponding eigen-manifolds, and let $\{P_j\}_{j=1}^{\infty}$ be the projections from H onto the eigenmani-folds. Finally let $P_0 = I - \sum_{j=1}^{\infty} P_j$. Then

1. $p(A) = \sum_{j=1}^{\infty} p(\lambda_j)P_j + p(0)P_0$ uniformly for all polynomials p.

2. $I = \sum_{j=0}^{\infty} P_j$,

$$P_j P_k = 0, \text{ when } j \neq k. \quad P_j^2 = P_j, \quad j=0,\ldots.$$

Proof. It is sufficient to prove the first statement for the polynomials $p(\lambda) = \lambda^n$. If $p(\lambda) = c$, a constant, then clearly

$$p(A) = cI = \sum_{j=1}^{\infty} cP_j - c[I - \sum_{j=1}^{\infty} P_j].$$

So it is true for $n = 0$. We have already shown in the previous theorem that it holds for $n=1$. If it holds for arbitrary n, then

$$A^n = \sum_{j=1}^{\infty} \lambda_j^n P_j + 0.$$

Since $AP_j = \lambda_j P_j$ for $j=1,\ldots,$

$$A^{n+1} = \sum_{j=1}^{\infty} A\lambda_j^n P_j + A0,$$

$$= \sum_{j=1}^{\infty} \lambda_j^n AP_j + 0,$$

$$= \sum_{j=1}^{\infty} \lambda_j^{n+1} P_j + 0.$$

By induction it holds for all n.

The first part of the second statement follows from the definition of P_0. The others are trivial computations.

To conclude this section we give conditions under which P_0, the projection onto $m(0)$, will be 0. This will be particularly useful in the section following.

IX.4.3. Theorem. Let H be a Hilbert space, and let A be a compact, self-adjoint operator in $L(H,H)$. Let P_0 be the projection operator onto the subspace of H, orthogonal to all the eigenmanifolds $\{m(\lambda_j)\}_{j=1}^{\infty}$ where $\lambda_j \neq 0$, $j=1,\ldots$. Then $P_0 = 0$ if the range of A is dense in H, or if 0 is not an eigenvalue of A.

Proof. The range of A is the linear manifold spanned by all the eigenmanifolds $m(\lambda_j)$, $j=1,\ldots$. If this is dense in H, then for any $\varepsilon > 0$ and x in H, there exists an element y in $\bigcup_{j=1}^{\infty} m(\lambda_j)$ such that $\|x-y\| < \varepsilon$.

Now

$$\|P_0 x\| \leq \|P_0 y\| + \|P_0(x-y)\| < 0+\varepsilon = \varepsilon.$$

Since ε is arbitrary, $P_0 = 0$.

If $P_0 \neq 0$, then

$$P_0 H = \{x : x = P_0 y \text{ for some } y \text{ in } H\}$$

is nonempty. If x is in $P_0 H$, then

$$Ax = AP_0 y,$$

$$= \sum_{j=1}^{\infty} \lambda_j P_j P_0 y,$$

$$= 0x,$$

and 0 is an eigenvalue of A.

IX.5. <u>The Regular Sturm-Liouville Problem</u>. The preliminary work for the regular Sturm-Liouville problem was given in sections V.4. and V.5. We recall briefly that it concerned a differential operator A defined by

$$Ax = [(px')' + qx]/w$$

over an interval $I = [\alpha, \beta]$, where p is in $C^2[I]$, q, w are in $C[I]$, and $p > 0$, $w > 0$ on I. The domain of A consists of all functions x which satisfy

1. x is in $C^2[I]$.

2. x satisfies

$$\alpha_1 x(\alpha) + \alpha_2 x'(\alpha) = 0,$$

$$\beta_1 x(\beta) + \beta_2 x'(\beta) = 0,$$

where $\alpha_1, \alpha_2, \beta_1, \beta_2$ are real numbers which satisfy

$$\alpha_1^2 + \alpha_2^2 \neq 0, \quad \beta_1^2 + \beta_2^2 \neq 0.$$

When 0 is not an eigenvalue of A, the inverse of A is given by the integral operator K,

$$Kx(t) = \int_\alpha^\beta G(t,s)x(s)w(s)ds,$$

where G(t,s) is the Green's function defined in V.5.1. Finally we recall that (Theorem V.5.3) G(t,s) has the following properties:

1. G is symmetric. That is, G(t,s) = G(s,t) for all s,t in I.

2. G is real valued.

3. If x_1 and x_2 are (linearly independent) solutions of $(px')' + qx = 0$ which satisfy $x_1(\alpha) = \alpha_2$, $x_1'(\alpha) = -\alpha_1$, $x_2(\beta) = \beta_2$, $x_2'(\beta) = -\beta_1$, then

$$G(t,s) = \begin{cases} (1/c)x_1(t)x_2(s), & \alpha \leq t \leq s \leq \beta, \\ (1/c)x_1(s)x_2(t), & \alpha \leq s \leq t \leq \beta, \end{cases}$$

where $c = p(t)W[x_1,x_2](t) \neq 0$ is constant.

Let us now introduce the Hilbert space $L^2(w)$, which consists of all w-measurable functions x satisfying $\int_\alpha^\beta |x(t)|^2 w(t)dt < \infty$, and which is generated by the inner product

$$(x,y) = \int_\alpha^\beta \overline{y}(t)x(t)w(t)dt,$$

and norm

$$\|x\| = [\int_\alpha^\beta |x(t)|^2 w(t)dt]^{1/2}.$$

It is easy to see that the domain of A is also in $L^2(w)$. We now wish to use $L^2(w)$ as the setting for the regular Sturm-Liouville problem, primarily so that we may use the theory of operators we have so laboriously built up. Unfortunately A is not a bounded operator. We are fortunate, however, that its inverse K is, and it is to the operator K that we turn our attention. We assume that 0 is not an eigenvalue of A. Therefore $A^{-1} = K$ will exist.

IX.5.1. Theorem. Let $H = L^2(w)$, and let the operator K, defined for all x in H by

$$Kx(t) = \int_\alpha^\beta G(t,s)x(s)w(s)ds$$

exist. Then

1. K is compact and self-adjoint.

2. 0 is not in $\sigma_p(K)$.

3. $\underline{\sigma(K) = \{\lambda, \ 1/\lambda \ \text{is an eigenvalue of} \ A\} \cup \{0\}.}$

4. $\underline{\text{The eigenmanifolds of} \ K \ \text{have dimension 1.}}$

5. $\underline{\text{Each eigenfunction of} \ K \ \text{corresponding to the eigenvalue} \ \lambda \ \text{is}}$
$\underline{\text{also an eigenfunction of} \ A \ \text{corresponding to the eigenvalue} \ 1/\lambda.}$

Proof.   1.   By applying Theorem IX.2.4 with $\sigma(dt) = w(t)dt$, we see that
K is a Hilbert-Schmidt operator and is compact.   By Example VIII.3.3, K is
self-adjoint.

2.   Suppose 0 is in $\sigma_p(K)$.   Then there exists an element x in $L^2(w)$
$\|x\| \neq 0$,   such that

$$Kx = 0x = \theta.$$

Applying the operator A, we find

$$x = AKx = \theta,$$

which is a contradiction.

3.   Suppose x satisfies $Kx = \lambda x$, where $\|x\| \neq 0$, and $\lambda \neq 0$.   Then
$x = AKx = \lambda Ax$.   So $Ax = (1/\lambda)x$.

4.   If for any eigenvalue λ of K, $m(\lambda) \geqq 2$, then the same would
be true for A and $(1/\lambda)$.   According to Theorem V.5.6 this is impossible.

5. follows easily from 3.

IX.5.2. <u>Theorem</u>. <u>The eigenvectors of the regular Sturm-Liouville operator</u>
<u>A form a complete orthogonal set in $L^2(w)$.</u>

Proof. Since 0 is not in $\sigma_p(A)$, by Theorem IX.4.3 the projection operator
$P_0 = 0$ and $I = \sum\limits_{j=1}^{\infty} P_j.$ Thus for any x in $L^2(w)$

$$x = \sum_{j=1}^{\infty} P_j x,$$

$$= \sum_{j=1}^{\infty} (x, x_j) x_j,$$

where $\{x_j\}_{j=1}^{\infty}$ are the normalized eigenfunctions of K and A.

In conclusion let us state these results in terms of the differential

operator A.

IX.5.3. <u>Theorem (The Sturm-Liouville Eigenfunction Expansion)</u>. <u>Let A be</u>
<u>a formally self-adjoint differential operator with domain D_A in $L^2(w)$</u>
<u>consisting of all functions x satisfying</u>

1. <u>x is in $L^2(w)$,</u>

2. <u>$[(px')' + qw]/w$ is in $L^2(w)$,</u>

3. <u>$a_1 x(\alpha) + a_2 x'(\alpha) = 0,$ $a_1^2 + a_2^2 \neq 0,$ $\beta_1 x(\beta) + \beta_2 x'(\beta) = 0,$</u>
<u>$\beta_1^2 + \beta_2^2 \neq 0.$</u>

<u>For all x in D_A define A by</u>

$$Ax = [(px')' + qw]/w.$$

Assume that 0 is not an eigenvalue of A. Then A has a countable number of eigenvalues $\{\lambda_j\}_{j=1}^{\infty}$, which approach ∞. The eigenmanifolds all have dimension 1. The normalized eigenfunctions $\{x_j\}_{j=1}^{\infty}$ form a complete orthonormal set in $L^2(w)$. That is, for each x in $L^2(w)$,

$$x = \sum_{j=1}^{\infty} a_j x_j,$$

where

$$a_j = \int_{\alpha}^{\beta} \overline{x}_j(t)x(t)w(t)dt.$$

IX.5.4. **Examples.** 1. Let $H = L^2(0,1)$, the Hilbert space of functions generated by the inner product

$$(x,y) = \int_0^1 \overline{y}(t)x(t)dt$$

and norm

$$\|x\| = \int_0^1 |x(t)|^2 dt.$$

Let the operator A have the form $Ax = -x''$, and the boundary conditions be $x(0) = 0$, $x(1) = 0$. Then the eigenvalues are $\{-j^2\pi^2\}_{j=1}^{\infty}$. The normalized eigenfunctions are $\{\sqrt{2} \sin j\pi t\}_{j=1}^{\infty}$. For each x in $L^2(0,1)$,

$$x(t) = 2 \sum_{j=1}^{\infty} [\int_0^1 x(s)\sin j\pi s \, ds]\sin j\pi t.$$

2. Let $H = L^2(-1,1)$, with inner product and norm similar to example 1.
Let the operator A have the form $Ax = x'' + x$, with boundary conditions
$x'(-1) = 0$, $x'(1) = 0$. Then the eigenvalues are $\{1-j^2\pi^2\}_{j=0}^{\infty}$,
$\{1-(j+1/2)^2\pi^2\}_{j=1}^{\infty}$. The corresponding eigenfunctions are $x_0 = 1/\sqrt{2}$,
$\{\cos j\pi t\}_{j=1}^{\infty}$, and $\{\sin(j+1/2)\pi t\}_{j=1}^{\infty}$. The resulting eigenfunction expansion
for all x in $L^2(-1,1)$ is

$$x(t) = [\int_{-1}^{1} x(s)ds]/2 + \sum_{j=1}^{\infty} [\int_{-1}^{1} x(s)\cos j\pi s\ ds]\cos j\pi t$$

$$+ \sum_{j=1}^{\infty} [\int_{-1}^{1} x(s)\sin(j+1/2)\pi s\ ds]\sin(j+1/2)\pi t.$$

IX.5.5. An Additional Example (Fourier Series). The regular Sturm-Liouville
problem is concerned with a second order boundary value problem. There are
similar problems for every order, however. The book "Theory of Ordinary
Differential Equations," by Coddington and Levinson, in fact solves what is
called the n-th order Sturm-Liouville problem. The particular example
following is the simplest of these. It is concerned with a first order
operator, and generates what is known as ordinary Fourier Series.

Let $H = L^2(-\pi,\pi)$, the Hilbert space of functions generated by the
inner product

$$(x,y) = \int_{-\pi}^{\pi} \overline{y}(t)x(t)dt,$$

and norm

$$x = [\int_{-\pi}^{\pi} |x(t)|^2 dt]^{1/2}.$$

Let D_A denote the set of all functions x which satisfy

1. x is in $L^2(-\pi,\pi)$.

2. x' exists and is in $L^2(-\pi,\pi)$.

3. $x(\pi) = x(-\pi)$.

Define the operator A by

$$Ax = (1/i)x'$$

for all x in D_A. Then A has the following properties.

1. $\sigma(A) = \{0,\pm1,\ldots,\pm j,\ldots\}$. The eigenfunction associated with each integer j is $e^{ijt}/\sqrt{2}$.

2. For each λ, not an eigenvalue, $A-\lambda I$ has an inverse K_λ of the form

$$K_\lambda x(t) = \int_{-\pi}^{\pi} G(\lambda,t,s)x(s)ds,$$

where K_λ is compact and, if λ is real, is self-adjoint. The eigenfunction of K_λ are the eigenfunctions of $A-\lambda I$ and also of A. The eigenvalues of K_λ are reciprocals of those of $A-\lambda I$, which are $\{j-\lambda\}_{j=-\infty}^{\infty}$.

3. The eigenfunctions of A form a complete orthonormal set in $L^2(-\pi,\pi)$.
Thus if x is an arbitrary element in $L^2(-\pi,\pi)$,

$$x(t) = \frac{1}{2\pi} \sum_{n=-\infty}^{\infty} [\int_{-\pi}^{\pi} e^{-ijs}x(s)ds]e^{ijt}.$$

Parseval's equality states further that

$$\|x\|^2 = \frac{1}{2\pi} \sum_{j=-\infty}^{\infty} |\int_{-\pi}^{\pi} e^{-ijs}x(s)ds|^2.$$

The equality between the function x and the infinite series is in the
$L^2(-\pi,\pi)$ sense. That is,

$$\lim_{n\to\infty} \int_{-\pi}^{\pi} |x(t) - \frac{1}{2\pi} \sum_{j=-n}^{n} [\int_{-\pi}^{\pi} e^{-ijs}x(s)ds]e^{ijt}|^2 dt = 0.$$

There is a rather interesting theorem which is convenient to present
at this point. It is essentially the converse to Parseval's equality.

IX.5.6. <u>Theorem (The Riesz-Fischer Theorem)</u>. <u>Let</u> $\{c_j\}_{j=-\infty}^{\infty}$ <u>be a sequence</u>
<u>of complex numbers satisfying</u>

$$\sum_{j=-\infty}^{\infty} |c_j|^2 < \infty.$$

<u>Then there exists an element x in</u> $L^2(-\pi,\pi)$ <u>such that</u>

$$c_j = \int_{-\pi}^{\pi} [e^{-ijs}/\sqrt{2\pi}]x(s)ds,$$

the Fourier coefficients of x, and Parseval's equality holds:

$$\|x\|^2 = \sum_{j=-\infty}^{\infty} |c_j|^2.$$

Proof. In view of Theorem II.3.7 we see that the series $\sum_{j=-n}^{n} c_j e^{ijt}$ converges in $L^2(-\pi,\pi)$. The limit, denoted by x, has all the desired properties.

IX. Exercises

1. Give a complete proof of Theorem IX.4.2.

2. Give a detailed proof of the Riesz-Fischer Theorem (Theorem IX.5.6).

3. Show that if A is self-adjoint and compact, and A^{-1} exists, then

$$A^{-1}x = \sum_{j=1}^{\infty} (1/\lambda_j)(x,x_j)x_j,$$

where $\{\lambda_j\}_{j=1}^{\infty}$, $\{x_j\}_{j=1}^{\infty}$ are the eigenvalues and normalized eigenfunctions of A.

Show that x is in $D_{A^{-1}}$ if and only if

$$\sum_{j=1}^{\infty} (1/\lambda_j)^2 |(x,x_j)|^2 < \infty.$$

4. Show for the regular Sturm-Liouville problem that

$$Ax = \sum_{j=1}^{\infty} \lambda_j [\int_{\alpha}^{\beta} \bar{x}_j(s)x(s)w(s)ds]x_j(t)$$

for all   x   in D_A,   where $\{\lambda_j\}_{j=1}^{\infty}$ are the eigenvalues of   A   and
$\{x_j\}_{j=1}^{\infty}$ are the corresponding normalized eigenfunctions.

Show   x   is in D_A if and only if

$$\sum_{j=1}^{\infty} \lambda_j^2 \ |\int_{\alpha}^{\beta} \overline{x}_j(s)x(s)w(s)ds|^2 < \infty \ .$$

Rewrite the series expansion for   Ax   as an integral similar to that derived in Chapter VIII.

Finally show that

$$Kx = \sum_{j=1}^{\infty} (1/\lambda_j)[\int_{\alpha}^{\beta} \overline{x}_j(s)x(s)w(s)ds]x_j(t)$$

exists for all   x   in $L^2(w)$.

5.   Show in Example IX.5.5 that   x   is in D_A if and only if

$$\sum_{j=-\infty}^{\infty} j^2 \ |\int_{-\pi}^{\pi} e^{-ijs}x(s)ds|^2 < \infty,$$

and then

$$(1/i)x'(t) = \frac{1}{2\pi} \sum_{j=-\infty}^{\infty} j[\int_{-\pi}^{\pi} e^{-ijs}x(s)ds]e^{ijt}.$$

6.   Show that the set of functions $\{(1/2\pi)e^{i(ns+mt)}\}_{n,m=-\infty}^{\infty}$ form a complete orthonormal set in $L^2([-\pi,\pi]\times[-\pi,\pi])$,   the Hilbert space of functions of two variables generated by the inner product

$$(x,y) = \int_{-\pi}^{\pi} \int_{\pi}^{\pi} \overline{y}(t,s)x(t,s)ds \ dt,$$

and norm

$$\|x\| = [\int_{-\pi}^{\pi} \int_{\pi}^{\pi} |x(t,s)|^2 ds\ dt]^{1/2}.$$

Show that for all x in $L^2([-\pi,\pi]\times[-\pi,\pi])$,

$$x = \sum_{n=-\infty}^{\infty} \sum_{m=-\infty}^{\infty} c_{nm}[e^{i(ns+mt)}/2\pi],$$

where

$$c_{nm} = \int_{-\pi}^{\pi} \int_{-\pi}^{\pi} [e^{-i(ns+mt)}/2\pi]x(t,s)ds\ dt,$$

which converges in a manner similar to the 1-dimensional case.

References

1. E.A. Coddington and N. Levinson, "Theory of Ordinary Differential Equati⟨
       McGraw-Hill, New York, 1955.

2. N. Dunford and J.T. Schwartz, "Linear Operators, vol. 1," Interscience,
       New York, 1958.

3. S. Goldberg, "Unbounded Linear Operators," McGraw-Hill, New York, 1966.

4. F. Riesz and B. Sz.-Nagy, "Functional Analysis," Frederick Ungar, New
       York, 1955.

5. M.H. Stone, "Linear Transformations in Hilbert Space," American
       Mathematical Society, New York, 1966.

6. A.E. Taylor, "Introduction to Functional Analysis," John Wiley and
       Sons, New York, 1958.

X.   SPECIAL FUNCTIONS

This chapter is devoted primarily to the classical orthogonal polynomials, although a final section also discusses the Bessel functions of the
first kind.   We are for the most part interested in the Legendre, Laguerre,
and Hermite polynomials.   However, our first few results also apply to such
polynomials as those of Jacobi and Gegenbauer with only minor modifications.

These polynomials have a number of characteristics in common.   We
cover thse common properties in Section X.1.   Succeeding sections are devoted
to the Legendre, Laguerre and Hermite polynomials, and, finally, the Bessel
functions.

X.1.   <u>Orthogonal Polynomials</u>.   Let (α,β) be a fixed interval of finite or
infinite length.   The classical orthogonal polynomials may be defined in the
following manner by a generalized Rodrigues formula.

X.1.1.   <u>Definition</u>.   <u>Let   X(t)   be a polynomial in   t.   Let   w(t)   be in</u>
$C^{\infty}(\alpha,\beta)$ <u>satisfy   w(t) > 0   when   t   is in (α,β), $w(\alpha) = w(\beta) = 0$</u>.
<u>Suppose further that</u>

$$\frac{d^n}{dt^n}\,[X^n(t)w(t)]/w(t)$$

<u>is a polynomial of degree   n.   Finally let k_n be a nonzero constant</u>
<u>dependent upon   n.   We then define the polynomials $\{p_n\}_{n=0}^{\infty}$ by</u>

$$P_n(t) = \frac{d^n}{dt^n} [X^n(t)w(t)]/k_n w(t),$$

n=0,1,... .

X.1.2. Examples. The following chart gives some details for each of the orthogonal sets previously mentioned.

Polynomials	(α,β)	$w(t)$	$X(t)$	k_n
Legendre	$(-1,1)$	1	$(1-t^2)$	$(-1)^n 2^n n!$
Gegenbauer	$(-1,1)$	$(1-t^2)^{\mu-1/2}$, $\mu > -1/2$	$(1-t^2)$	$(-1)^n 2^n n!$
Jacobi	$(-1,1)$	$(1-t)^\mu (1+t)^\nu$, $\mu > -1, \nu > -1$	$(1-t^2)$	$(-1)^n 2^n n!$
Laguerre	$(0,\infty)$	e^{-t}	1	1
Hermite	$(-\infty,\infty)$	e^{-t^2}	t	$(-1)^n$

The conditions $\mu > -1/2$ and $\mu > -1, \nu > -1$ are imposed so that $w(t)$ is integrable over (α,β) in every case.

It can be shown that each set above satisfies a second order ordinary differential equation of the form

$$(wXp_n')' = \lambda_n w p_n$$

where λ_n is constant and depends upon n. Each satisfies a recurrence relation

$$P_{n+1} + (\frac{k_{n+1}}{k_n} t + B_n)P_n + C_n P_{n-1} = 0,$$

where k_n, k_{n+1} were previously defined, B_n and C_n are constants, and $C_n \neq 0$. Further, each possesses a simple generating function. That is, a function $H(s,t)$ satisfying

$$H(s,t) = \sum_{n=0}^{\infty} \alpha_n P_n(t) s^n.$$

Because of the computation involved it is easier to show these results separately in each instance. However, the following results are easily shown in a general setting.

X.1.3. <u>Theorem</u>. <u>The polynomials</u> $\{P_n\}_{n=0}^{\infty}$ <u>form an orthogonal set in</u> $L^2(w)$.

Proof. Let y be a polynomial in t of degree less than n. Then in $L^2(w)$

$$(y, P_n) = \frac{1}{k_n} \int_{\alpha}^{\beta} \frac{d^n}{dt^n} [x(t)^n w(t)] y(t) dt.$$

Integration by parts n times shows

$$(y, P_n) = \frac{1}{k_n} (-1)^n \int_{\alpha}^{\beta} x(t)^n w(t) \frac{d^n}{dt^n} y(t) dt,$$

$$= 0,$$

since $\dfrac{d^n y}{dt^n} = 0$. Thus for all n, P_n is orthogonal to $P_0, P_1, \ldots, P_{n-1}$, and the proof is complete.

X.1.4.   Corollary. For each fixed n, p_n is orthogonal to every polynomial of degree less than n.

X.1.5.   Theorem. For each n, the zeros of p_n are simple, real, and in (α,β).

Proof. Suppose p_n has odd powered zeros at $t_1,\ldots,t_m$. Let

$$P(t) = (t-t_1)\ldots(t-t_m).$$

Then

$$\int_\alpha^\beta p_n(t)P(t)w(t)dt > 0.$$

If m < n, however, P is orthogonal to p_n and we have a contradiction. Thus m = n and all the zeros are simple.

X.1.6.   Theorem. If (α,β) is a finite interval then the polynomials $\{p_n\}_{n=0}^\infty$ form a complete orthogonal set in $L^2(w)$.

Proof. We may assume without loss of generality that $(\alpha,\beta) = (-1,1)$. Suppose that y is orthogonal to $\{p_n\}_{n=0}^\infty$. Since t^n is a linear combination of $p_0,\ldots,p_n$,

$$\int_{-1}^1 t^n y(t)w(t)dt = 0$$

for all $n \geq 0$.

If w is bounded by M, then

$$\int_{-1}^{1} |yw|^2 dt \leq M \int_{-1}^{1} |y|^2 w \, dt,$$

so yw is in $L^2(-1,1)$. If $\{q_n(t)\}_{n=0}^{\infty}$ is a sequence of polynomials con-

verging uniformly to $e^{i\lambda t}$, as is possible according to the Stone-Weierstrass

Theorem, then

$$\int_{-1}^{1} e^{i\lambda t} y(t) w(t) dt = \lim_{n \to \infty} \int_{-1}^{1} q_n(t) y(t) w(t) dt,$$

$$= 0,$$

for all λ. In particular, the integral is zero if $\lambda = n\pi$. Thus the

Fourier coefficients of yw are all zero and $yw = 0$ in $L^2(-1,1)$. Thus

$$\int_{-1}^{1} |yw|^2 dt = 0.$$

Now

$$\int_{-1}^{1} |y|^2 w \, dt = \int_{-1}^{-1+\delta} |y|^2 w \, dt + \int_{-1+\delta}^{1-\delta} |y|^2 w \, dt + \int_{1-\delta}^{1} |y|^2 w \, dt.$$

If $\varepsilon > 0$ it is possible to choose δ sufficiently small such that

$$\int_{-1}^{-1+\delta} |y|^2 w \, dt < \varepsilon/2,$$

$$\int_{1-\delta}^{1} |y|^2 w \, dt < \epsilon/2.$$

If $m = \min_{t \in [-1+\delta, 1-\delta]} w(t)$, then

$$\int_{-1+\delta}^{1-\delta} |y|^2 w \, dt \leq 1/m \int_{-1+\delta}^{1-\delta} |yw|^2 dt,$$

$$\leq 1/m \int_{-1}^{1} |yw|^2 dt,$$

$$= 0.$$

Thus

$$\int_{-1}^{1} |y|^2 w \, dt < \epsilon,$$

and therefore equals 0.

If w is not bounded, we modify the proof slightly. Let $0 < \epsilon < 1/2$, and let $0 < \delta < \epsilon$. Let $q_n(t)$ uniformly approach the function $\phi(t,\lambda)$ which equals $e^{i\lambda t}$ on $[-1+\delta+\epsilon, 1-\delta-\epsilon]$ which varies linearly from 0 to $e^{i\lambda(-1+\delta+\epsilon)}$ over the interval $[-1+\epsilon, -1+\epsilon+\delta]$, and from $e^{i\lambda(1-\delta-\epsilon)}$ to 0 on $[1-\delta-\epsilon, 1-\epsilon]$. Then

$$\int_{-1+\epsilon}^{1-\epsilon} \phi(t,\lambda) y(t) w(t) dt = 0.$$

If we let δ approach 0, we find

$$\int_{-1+\varepsilon}^{1-\varepsilon} e^{i\lambda t} y(t) w(t) dt = 0$$

Thus $yw = 0$ in $L^2(-1+\varepsilon, 1-\varepsilon)$. Since

$$0 \leq m \int_{-1+\varepsilon}^{1-\varepsilon} |y|^2 w \, dt \leq \int_{-1+\varepsilon}^{1-\varepsilon} |yw|^2 dt = 0,$$

$y=0$ in $L^2(w)$ over $(-1+\varepsilon, 1-\varepsilon)$. Since ε is arbitrary, $y=0$ in $L^2(w)$. With the Laguerre and Hermite polynomials, the proof is more difficult. We will prove the same result holds for these two polynomial sets in Chapter XI.

X.2. The Legendre Polynomials. It is convenient to begin with a direct definition of the Legendre polynomials $\{P_n\}_{n=0}^{\infty}$, and to derive their other properties from it.

X.2.1. Definition. The Legendre polynomials P_n are given by

$$P_n(t) = \frac{1}{2^n} \sum_{j=0}^{m} (-1)^j \frac{(2n-2j)!}{(n-2j)!(n-j)!j!} t^{n-2j},$$

where $m = n/2$ when n is even, $m = (n-1)/2$ when n is odd, $n=0,1,\dots$.

It is a tedious but elementary computation to show that P_n satisfies the following differential equation.

X.2.2. Theorem. For each $n=0,1,\dots,$ P_n satisfies

$$((1-t^2)P_n')' + n(n+1)P_n = 0.$$

X.2.3. Theorem. If $n \neq m$, then

$$\int_{-1}^{1} P_n(t)P_m(t)dt = 0.$$

Proof. From the differential equations we find

$$(n-m)(n+m+1)P_n P_m = [n(n+1)-m(m+1)]P_n P_m,$$

$$= P_n((1-t^2)P_m')' - P_m((1-t^2)P_n')';$$

$$= [(1-t^2)(P_n P_m'-P_n'P_m)]'.$$

Thus, since $1-t^2 = 0$ at 1 and -1,

$$(n-m)(n+m+1)\int_{-1}^{1} P_n P_m dt = 0.$$

If $n \neq m$, then $(n-m)(n+m+1) \neq 0$. The integral must then be zero.

We note that any polynomial π of degree n can be expressed as

$$\pi(t) = \sum_{m=0}^{n} c_m P_m(t),$$

where

$$c_m = \int_{-1}^{1} \pi(t) P_m(t) dt / \int_{-1}^{1} P_m(t)^2 dt.$$

X.2.4.   <u>Theorem</u>. $(1-2st+s^2)^{-1/2}$ is a generating function for the Legendre

<u>polynomials</u>:

$$(1-2st+s^2)^{-1/2} = \sum_{j=0}^{\infty} P_j(t) s^j.$$

Proof.

$$(1-2st+s^2)^{-1/2} = \sum_{j=0}^{\infty} \binom{-1/2}{j} 1^{-1/2-j} (-2st+s^2)^j,$$

$$= \sum_{j=0}^{\infty} \sum_{k=0}^{j} \binom{-1/2}{j} \binom{j}{k} (-2)^{j-k} t^{j-k} s^{j+k}.$$

If $j+k = n$,

$$(1-2st+s^2)^{-1/2} = \sum_{j=0}^{\infty} \sum_{n=j}^{2j} \binom{-1/2}{j} \binom{j}{n-j} (-2)^{2j-n} t^{2j-n} s^n,$$

$$= \sum_{n=0}^{\infty} [\sum_{j=m}^{n} \binom{-1/2}{j} \binom{j}{n-j} (-2)^{2j-n} t^{2j-n}] s^n,$$

where $m = n/2$ if n is even, $m = (n-1)/2$ if n is odd.   If we replace

j by $n-k$,

$$(1-2st+s^2)^{-1/2} = \sum_{n=0}^{\infty} [\sum_{k=0}^{m} \binom{-1/2}{n-k} \binom{n-k}{k} (-2)^{n-2k} t^{n-2k}] s^n.$$

Now

$$\binom{-1/2}{n-k}\binom{n-k}{k}2^{n-2k}(-1)^{n-2k} = \frac{(-1/2)(-1/2-1)\ldots(-1/2-n+k+1)}{(n-k)!} \cdot \frac{(n-k)!}{k!(n-2k)!} 2^{n-2k}(-1)$$

$$= \frac{(-1)^k(2n-2k)!}{2^n(n-k)!k!(n-2k)!} \cdot$$

Inserting this into the previous result completes the proof.

It is an easy computation to show the generating function

$$H(s,t) = (1-2st+s^2)^{-1/2}$$

satisfies

$$(1-2st+s^2)\frac{\partial H}{\partial s} - (t-s)H = 0.$$

If the expansion $H(s,t) = \sum_{j=0}^{\infty} P_j(t)s^j$ is inserted into this equation, and coefficients of like powers of s are collected together, the following result is immediate.

X.2.5.   Theorem.   The Legendre polynomials satisfy the recurrence relations

$$P_0-1 = 0,$$

$$P_1-tP_0 = 0,$$

and, in general, for $n \geq 1$,

$$(n+1)P_{n+1} - (2n+1)tP_n + nP_{n-1} = 0.$$

Using the recurrence relation we can now compute the norm of P_n in $L^2(-1,1)$.

X.2.6.   Theorem.   For   n=0,1,...,

$$\int_{-1}^{1} P_n(t)^2 dt = 2/(2n+1).$$

Proof.   Since

$$nP_n = (2n-1)tP_{n-1} - (n-1)P_{n-2},$$

$$P_n^2 = \frac{2n-1}{n} tP_{n-1}P_n - \frac{n-1}{n} P_{n-2}P_n.$$

Thus

$$\int_{-1}^{1} P_n^2 dt = \frac{2n-1}{n} \int_{-1}^{1} tP_{n-1}P_n dt.$$

Again from the recurrence relation, we find

$$(n+1)P_{n+1}P_{n-1} + nP_{n-1}^2 = (2n+1)tP_nP_{n-1}.$$

So

$$(2n+1) \int_{-1}^{1} tP_nP_{n-1} dt = n \int_{-1}^{1} P_{n-1}^2 dt.$$

Thus

$$\int_{-1}^{1} P_n^2 dt = \frac{2n-1}{2n+1} \int_{-1}^{1} P_{n-1}^2 dt.$$

By induction,

$$\int_{-1}^{1} P_n^2 dt = 1/(2n+1) \int_{-1}^{1} 1^2 dt,$$

$$= 2/(2n+1).$$

X.2.7.   <u>Theorem</u>.   The Legendre polynomials are generated by the Rodrigues formula

$$P_n(t) = [(-1)^n/2^n n!]\frac{d^n}{dt^n} [(1-t^2)^n],$$

$n=0,1,\ldots$.

Proof.   This follows immediately upon expanding $(1-t^2)^n$ and performing the differentiation.

X.3.   <u>The Laguerre Polynomials</u>.   It is convenient to use the Rodrigues formula to generate the Laguerre polynomials:

X.3.1.   <u>Definition</u>.   The Laguerre polynomials are given by

$$L_n(t) = e^t \frac{d^n}{dt^n} [t^n e^{-t}],$$

n=0,1,... .

As indicated in section X.1, if $n \geq m$, the integral

$$I = \int_0^\infty L_n(t) L_m(t) e^{-t} dt,$$

$$= \int_0^\infty L_m(t) \frac{d^n}{dt^n} [t^n e^{-t}] dt,$$

$$= (-1)^m \int_0^\infty L_m^{(m)}(t) \frac{d^{n-m}}{dt^{n-m}} [t^n e^{-t}] dt.$$

after integration by parts m times. If $n > m$, an additional integration by parts shows $I = 0$. If $n = m$,

$$I = (-1)^n \int_0^\infty n! t^n e^{-t} dt,$$

$$= (n!)^2.$$

We summarize:

X.3.2. Theorem.

$$\int_0^\infty L_n(t) L_m(t) e^{-t} dt = \begin{cases} 0, & \text{if } n \neq m, \\ (n!)^2, & \text{if } n = m, \end{cases}$$

for all $n \geq 0$, $m \geq 0$.

X.3.3. Theorem. The Laguerre polynomials are given by

$$L_n(t) = \sum_{j=0}^{n} (-1)^{n-j} \binom{n}{j}^2 j! \, t^{n-j}.$$

Proof. This follows from Rodrigues formula by carrying out the differential

By computing $L_n{}'$ and $L_n{}''$ directly it is also easy to show the

following theorem is true.

X.3.4. Theorem. The Laguerre polynomials satisfy

$$(te^{-t} L_n{}')' + n e^{-t} L_n = 0,$$

$n=0,1,\ldots$.

X.3.5. Theorem. $e^{-(st/(1-s))}/(1-s)$ is a generating function for the

Laguerre polynomials:

$$e^{-(st/(1-s))}/(1-s) = \sum_{n=0}^{\infty} \frac{L_n(t)}{n!} s^n.$$

Proof.

$$e^{-(st/(1-s))}/(1-s) = \frac{1}{1-s} \sum_{j=0}^{\infty} (-1)^j \left(\frac{st}{1-s}\right)^j / j!,$$

$$= \sum_{j=0}^{\infty} (1/j!)(-1)^j s^j t^j (1-s)^{-j-1},$$

$$= \sum_{j=0}^{\infty} (1/j!)(-1)^j s^j t^j \sum_{k=0}^{\infty} \frac{(-j-1)\cdots(-j-k)}{k!} (-1)^k s^k,$$

$$= \sum_{j=0}^{\infty} \sum_{k=0}^{\infty} \frac{(-1)^{j+2k}(j+k)!}{k!(j!)^2} t^j s^{j+k}.$$

If we let j+k = n and eliminate k,

$$e^{-(st/(1-s))}/(1-s) = \sum_{n=0}^{\infty} \sum_{j=0}^{n} \frac{(-1)^j n!}{(j!)^2 (n-j)!} t^j s^n.$$

Replacing j by n-k,

$$e^{-(st/(1-s))}/(1-s) = \sum_{n=0}^{\infty} [\frac{1}{n!} \sum_{k=0}^{n} (-1)^{n-k} \binom{n}{k}^2 k! t^{n-k}] s^n,$$

$$= \sum_{n=0}^{\infty} \frac{L_n(t)}{n!} s^n.$$

Finally by the same technique used while dicussing the Legendre polynomials it is possible to show that the generating function H(s,t) satisfies

$$(1-2s+s^2) \frac{\partial H}{\partial s} = (1-s-t)H.$$

The recurrence relation follows in the usual manner:

X.3.6.   Theorem.   The Laguerre polynomials satisfy the recurrence relations

$$L_0 - 1 = 0,$$

$$L_1 - (1-t)L_0 = 0,$$

and, in general, for $n \geq 1$,

$$L_{n+1} - (2n+1-t)L_n + n^2 L_{n-1} = 0.$$

X.4. The Hermite Polynomials. As an illustration of a different approach,

we will use a generating function to define the Hermite polynomials.

X.4.1. Definition. The Hermite polynomials are given by

$$e^{2ts-s^2} = \sum_{n=0}^{\infty} \frac{H_n(t)}{n!} s^n.$$

From this formula the series expansion for H_n can be derived quickly.

X.4.2. Theorem. The Hermite polynomials satisfy

$$H_n(t) = \sum_{j=0}^{m} \frac{(-1)^j 2^{n-2j} n!}{(n-2j)! j!} t^{n-2j},$$

where $m = n/2$, if n is even, $m = (n-1)/2$, if n is odd, $n=0,1,\ldots$.

Proof.
$$e^{2st-s^2} = \sum_{i=0}^{\infty} \frac{(2st)^i}{i!} \sum_{j=0}^{\infty} \frac{(-s^2)^j}{j!},$$

$$= \sum_{i=0}^{\infty} \sum_{j=0}^{\infty} \frac{(-1)^j 2^i t^i s^{i+2j}}{i! j!}.$$

If we let $i+2j = n$ and eliminate i,

$$e^{2st-s^2} = \sum_{n=0}^{\infty} [\sum_{j=0}^{m} \frac{(-1)^j 2^{n-2j} t^{n-2j}}{(n-2j)! j!}] s^n,$$

where $m = n/2$, if n is even, $m = (n-1)/2$, if n is odd.

To establish a recurrence relation and the differential equation

satisfied by the Hermite polynomials, we need two lemmas.

X.4.3. Lemma. $H'_{n+1} = 2(n+1)H_n$, $n=0,1,\ldots$.

Proof. If we differentiate the generating function with respect to t,

$$\frac{\partial}{\partial t} (e^{2st-s^2}) = 2se^{2st-s^2}.$$

Thus, since $H_0 = 1$,

$$\sum_{n=0}^{\infty} \frac{H'_{n+1}}{(n+1)!} s^{n+1} = \sum_{n=0}^{\infty} \frac{2H_n}{n!} s^{n+1}.$$

We then equate coefficients of s^{n+1} on each side.

X.4.4. Lemma. $tH'_n - nH'_{n-1} = nH_n$, $n=1,\ldots$.

Proof. If we differentiate the generating function with respect to s,

$$\frac{\partial}{\partial s} (e^{2st-s^2}) = 2(t-s)e^{2st-s^2}.$$

If we compare this with the differentiation performed in the proof of the previous lemma,

$$(t-s)\frac{\partial}{\partial t}(e^{2st-s^2}) = s\frac{\partial}{\partial s} (e^{2st-s^2}).$$

Inserting the expansion of the generating function in this equation,

$$\sum_{n=1}^{\infty} \frac{tH'_n}{n!} s^n - \sum_{n=2}^{\infty} \frac{nH'_{n-1}}{n!} s^n = \sum_{n=1}^{\infty} \frac{nH_n}{n!} s^n.$$

The result then follows from equating coefficients.

X.4.5. __Theorem.__ __The Hermite polynomials satisfy the recurrence relations__

$$H_0 - 1 = 0,$$

$$H_1 - 2tH_0 = 0,$$

__and, in general, for__ $n \geq 1$,

$$H_n - 2tH_{n-1} + 2(n-1)H_{n-2} = 0.$$

Proof. From Lemma X.4.3, $H_n' = 2nH_{n-1}$, and $H_{n-1}' = 2(n-1)H_{n-2}$, $n=1,\ldots$.
Inserting these in Lemma X.4.4, we find

$$nH_n - 2ntH_{n-1} + 2n(n-1)H_{n-2} = 0,$$

$n=1,\ldots$. The first two statements follow from the series expansions for
H_0 and H_1.

X.4.6. __Theorem.__ __The Hermite polynomials satisfy__

$$(e^{-t^2} H_n')' + 2ne^{-t^2} H_n = 0,$$

$n=0,1,\ldots$.

Proof. From Lemma X.4.3, $H'_n = 2nH_{n-1}$. Thus $H''_n = 2_n H'_{n-1}$. Inserting this expression in the equation of Lemma X.4.4,

$$tH'_n - nH''_n/2n = nH_n,$$

or

$$H''_n - 2tH'_n + 2nH_n = 0,$$

for $n=1,\ldots$. This equation obviously holds for $n=0$. If we multiply by e^{-t^2} the proof is complete.

To show Rodrigues formula exists, we return to the generating function.

X.4.7. **Theorem.** **The Hermite polynomials are given by**

$$H_n(t) = (-1)^n e^{t^2} \frac{d^n}{dt^n} [e^{-t^2}],$$

$n=0,1,\ldots$.

Proof. Differentiating the generating function n times with respect to s shows

$$\frac{\partial^n}{\partial s^n} (e^{2st-s^2})\bigg|_{s=0} = H_n(t).$$

If we multiply by e^{-t^2},

$$e^{-t^2} H_n(t) = \frac{\partial^n}{\partial s^n} (e^{-(t-s)^2})\Big|_{s=0} .$$

If we let $w = t-s$,

$$e^{-t^2} H_n(t) = (-1)^n \frac{\partial^n}{\partial w^n} (e^{-w^2})\Big|_{w=t},$$

$$= (-1)^n \frac{\partial^n}{\partial t^n} (e^{-t^2}).$$

Using the Rodrigues formula, orthogonality and norms may be calculated as was done for the Laguerre polynomials.

X.4.8. Theorem.

$$\int_{-\infty}^{\infty} H_n(t) H_m(t) e^{-t^2} dt = \begin{cases} 0, & \text{if } n \neq m, \\ 2^n n! \sqrt{\pi}, & \text{if } n = m. \end{cases}$$

Proof. If $n \geq m$, let

$$I = \int_{-\infty}^{\infty} H_n(t) H_m(t) e^{-t^2} dt,$$

$$= (-1)^n \int_{-\infty}^{\infty} H_m(t) \frac{d^n}{dt^n} [e^{-t^2}] dt.$$

Integrating by parts m times,

$$I = (-1)^{n+m} \int_{-\infty}^{\infty} H_m^{(m)}(t) \frac{d^{n-m}}{dt^{n-m}} [e^{-t^2}] dt.$$

If $n > m$, an additional integration by parts shows $I = 0$. If $n = m$,

$$I = 2^n n! \int_{-\infty}^{\infty} e^{-t^2} dt.$$

This integral is evaluated in Lemma XI.2.10. Using that result,
$I = 2^n n! \sqrt{\pi}$.

We remark in closing that the Gegenbauer and Jacobi polynomials, as well as others have similar properties. We recommend the book "Special Functions" by E.D. Rainville to those who wish further details.

X.5. <u>Bessel Functions</u>. Bessel functions share a great many of the properties of exhibited by the orthogonal polynomials. Although they are not polynomials, they satisfy a second order linear differential equation, can be written as a series, satisfy a recurrence relation and have a generating function. In addition their norms can be computed, and there is an orthogonality relation. Just as with the orthogonal polynomials, our purpose here is to develop some of their properties for later application.

Bessel functions are usually encountered as the solution to the differential equation

$$s^2 \ddot{x} + s\dot{x} + (\lambda^2 s^2 - n^2)x = 0,$$

where $(\ )^{\cdot} = \dfrac{d}{ds}$, λ and n are fixed constants. By letting $\lambda s = t$, the equation is simplified to

$$t^2 x'' + tx' + (t^2 - n^2)x = 0,$$

where $( )' = \dfrac{d}{dt}$. We shall assume throughout that n is a nonnegative integer, although a great deal of what follows is valid even if it is not.

There are two linearly independent solutions to the differential equation above. The first is given by

$$J_n(t) = \sum_{j=0}^{\infty} \frac{(-1)^j}{j!\,(n+j)!} \left(\frac{t}{2}\right)^{n+2j}.$$

For different values of n these are called the Bessel functions of the first kind, and it is these we wish to consider. Any form of the second solution involves a logarithmic term, which behaves badly near $t = 0$. We will not be interested in these.

X.5.1. Theorem. Let $J_{-n}(t) = (-1)^n J_n(t)$, $n=0,1,\ldots$. The Bessel functions satisfy a recurrence relation

$$tJ_{n+1} - 2nJ_n + tJ_{n-1} = 0.$$

Proof.

$$[t^n J_n]' = t^n J_{n-1},$$

$$[t^{-n} J_n]' = -t^{-n} J_{n+1}.$$

Expanding,

$$tJ'_n + nJ_n = tJ_{n-1},$$

$$tJ'_n - nJ_n = -tJ_{n+1}.$$

If we subtract the second from the first, the result is immediate.

In order to derive some additional properties, an integral formula is convenient.

X.5.2.   Theorem.

$$J_0(t) = \frac{2}{\pi} \int_0^{\pi/2} \cos[t \sin \theta] d\theta.$$

Proof.   If

$$x(t) = \frac{2}{\pi} \int_0^{\pi/2} \cos[t \sin \theta] d\theta,$$

we compute

$$x'(t) = -\frac{2}{\pi} \int_0^{\pi/2} \cos[t \sin \theta] \sin \theta \, d\theta,$$

$$x''(t) = -\frac{2}{\pi} \int_0^{\pi/2} \cos[t \sin \theta] \sin^2 \theta \, d\theta.$$

If we integrate x' by parts,

$$x'(t) = \frac{2}{\pi} \left[\sin[t \sin \theta]\cos \theta\right]_{\theta=0}^{\pi/2}$$

$$- \frac{2}{\pi} \int_0^{\pi/2} \cos[t \sin \theta] t \cos^2 \theta \, d\theta,$$

$$= - \frac{2}{\pi} \int_0^{\pi/2} \cos[t \sin \theta] t \cos^2 \theta \, d\theta.$$

Thus

$$t^2 x'' + t x' + t^2 x =$$

$$\frac{2}{\pi} \int_0^{\pi/2} \cos[t \sin \theta]\{-t^2 \sin \theta - t^2 \cos^2 \theta + t^2\} d\theta,$$

$$= 0.$$

So x is a solution to Bessel's equation of order 0. Now

$$x(0) = \lim_{t \to 0} \frac{2}{\pi} \int_0^{\pi/2} \cos[t \sin \theta]d\theta,$$

$$= \frac{2}{\pi} \int_0^{\pi/2} \lim_{t \to 0} \cos[t \sin \theta]d\theta,$$

$$= 1.$$

Since $J_0(t)$ also approaches 1, as t approaches 0, while any other solution becomes infinite, we conclude that $x(t) = J_0(t)$.

X.5.3. Corollary.

$$J_0(t) = \frac{2}{\pi} \int_0^t \frac{\cos s}{(t^2-s^2)^{1/2}} \, ds.$$

Proof. Let $s = t \sin \theta.$

X.5.4. Theorem. J_0 has an infinite number of positive zeros. These zeros are all simple and have ∞ as their only limit.

Proof. Let $a_m = (m+1/2)\pi.$ Then at $a_m,$ $\cos s$ is zero, and $\cos s/(t^2-s^2)^{1/2}$ changes sign. Now

$$J_0(a_m) = \frac{2}{\pi} \int_0^{a_m} \frac{\cos s}{(a_m^2-s^2)^{1/2}} \, ds,$$

$$= \frac{2}{\pi} \int_0^{a_0} \frac{\cos s}{(a_m^2-s^2)^{1/2}} \, ds + \sum_{j=0}^{m-1} (-1)^{j+1}(\frac{2}{\pi}) \int_{a_j}^{a_{j+1}} \frac{|\cos s|}{(a_m^2-s^2)^{1/2}} \, ds.$$

If we let

$$A_0 = \frac{2}{\pi} \int_0^{a_0} \frac{\cos s}{(a_m^2-s^2)^{1/2}} \, ds,$$

$$A_j = \frac{2}{\pi} \int_{a_{j-1}}^{a_j} \frac{|\cos s|}{(a_m^2-s^2)^{1/2}} \, ds,$$

$j=1,\ldots,m,$ then it is easy to see that

$$A_j < A_{j+1}.$$

Thus if m is even,

$$J_0(a_m) = A_0 + (A_2 - A_1) + \ldots + (A_m - A_{m-1}) > 0.$$

If m is odd,

$$J_0(a_m) = -(A_1 - A_0) - (A_3 - A_2) - \ldots - (A_m - A_{m-1}) < 0.$$

Since J_0 is continuous, between each pair (a_m, a_{m+1}) there exists a point λ_m such that $J_0(\lambda_m) = 0$. Clearly these points have ∞ as their only limit. If any one were a multiple zero, then J_0' would also vanish. The uniqueness part of Theorem III.4.8 would then require J_0 to be identically zero, which it is not.

X.5.5. <u>Theorem.</u> <u>For all integers $n \geq 0$, J_n has an infinite number of positive zeros. These zeros are all simple and have ∞ as their only limit.</u>

<u>Proof.</u>. In the proof of Theorem X.5.1 we observed that

$$[t^{-n+1} J_{n-1}]' = t^{-n+1} J_n.$$

We proceed by induction. Assume that J_{n-1} possesses an infinite number of simple zeros accumulating only at ∞. Then $t^{-n+1} J_{n-1}$ also has this propert
Rolle's Theorem then guarantees that between each pair, the derivative of
$t^{-n+1} J_{n-1}$, $t^{-n+1} J_n$, must vanish. Since $t^{-n+1} \neq 0$, $J_n = 0$. The proof tha

these zeros are simple is the same as that for J_0.

The orthogonality relation following for the Bessel functions is of a different type from that obtained for the classical polynomials.

X.5.6. Theorem. Let λ and μ be distinct zeros of J_n. Then $J_n(\lambda s)$ and $J_n(\mu s)$ are orthogonal in $L^2(s)$ over $(0,1)$.

Proof. $J_n(\lambda s)$ satisfies

$$s^2 \ddot{y} + s\dot{y} + (\lambda^2 s^2 - n^2)y = 0,$$

while $J_n(\mu s)$ satisfies

$$s^2 \ddot{z} + s\dot{z} + (\mu^2 s^2 - n^2)z = .$$

If we multiply the first by z, the second by y, divide by s and subtract

$$[s(\dot{y}z - y\dot{z})]^{\cdot} + (\lambda^2 - \mu^2)syz = 0.$$

Integrating from 0 to 1, the first term vanishes. Since $\lambda \neq \mu$, $y = J_n(\lambda s)$, $z = J_n(\mu s)$,

$$\int_0^1 J_n(\lambda s)J_n(\mu s)s\,ds = 0.$$

The norms of these functions can also be computed from the differential equation. If we multiply the equation in y above by $2y^{\cdot}$, we find

$$2(sy^{\cdot})(sy^{\cdot})^{\cdot} + 2(\lambda^2 s^2 - n^2)yy^{\cdot} = 0$$

or

$$\frac{d}{ds}(sy^{\cdot})^2 + (\lambda^2 s^2 - n^2)\frac{d}{ds}(y^2) = 0.$$

Integrating from 0 to 1,

$$(sy^{\cdot})^2|_0^1 + \int_0^1 (\lambda^2 s^2 - n^2)\frac{d}{ds}(y^2)ds = 0.$$

If the second term is integrated by parts,

$$(sy^{\cdot})^2|_0^1 + [(\lambda^2 s^2 - n^2)y^2]|_0^1 - \int_0^1 2\lambda^2 sy^2 ds = 0,$$

or

$$J_n^{\cdot}(\lambda)^2 + n^2 J_n(0)^2 = 2\lambda^2 \int_0^1 J_n(\lambda s)^2 s\, ds.$$

If $n=0$, the second term vanishes because of n. If $n > 0$, then $J_n(0) = 0$. Thus we have proved the following:

X.5.7.   Theorem.   If λ is a zero of J_n, then

$$\int_0^1 J_n(\lambda s)^2 s \, ds = J_n^{\bullet}(\lambda)^2/2\lambda^2.$$

We remark that since

$$\frac{d}{dt}[t^{-n}J_n] = -t^{-n}J_{n+1},$$

an easy computation shows

$$J_n^{\bullet}(\lambda)^2 = \lambda^2 J_{n+1}(\lambda)^2.$$

X.5.8. Corollary. If λ is a zero of J_n, then

$$\int_0^1 J_n(\lambda s)^2 s \, ds = J_{n+1}(\lambda)^2/2.$$

In conclusion we develop a generating function for Bessel functions.

X.5.9. Theorem.

$$e^{(t/2)(s-1/s)} = \sum_{n=-\infty}^{\infty} J_n(t)s^n.$$

Proof. We compute

$$e^{st/2} = \sum_{j=0}^{\infty} (\frac{st}{2})^j/j!,$$

$$e^{-t/2s} = \sum_{k=0}^{\infty} (-1)^k (\frac{t}{2s})^k/k! \; .$$

Thus

$$e^{(t/2)(s-1/s)} = \sum_{j=0}^{\infty} [\sum_{k=0}^{\infty} \frac{(-1)^k t^{j+k}}{2^{j+k} j! k!} s^{j-k}.$$

If we let $j-k = n$, and eliminate j,

$$e^{(t/2)(s-1/s)} = \sum_{n=-\infty}^{\infty} [\sum_{k=0}^{\infty} \frac{(-1)^k}{k!(n+k)!} (\tfrac{t}{2})^{n+2k}] s^n,$$

$$= \sum_{n=-\infty}^{\infty} J_n(t) s^n.$$

X. Exercises

1. Let $\{p_n\}_{n=0}^{\infty}$ be a set of polynomials, orthogonal over (α,β). Let

$$s_n = \sum_{j=0}^{n} (y,p_j) p_j / \|p_j\|^2,$$

where y is continuous on (α,β). Show $y-s_n$ has at least $n+1$ zer

in (α,β).

2. Calculate $P_0,\ldots,P_6,L_0,\ldots,L_0,H_0,\ldots,H_6$.

3. Show

$$P_n(t) = \frac{(-1)^n}{2^n n!} \frac{d^n}{dt^n} [(1-t^2)^n].$$

4. Show by integrating on the right that

$$P_n(t) = \frac{2}{n!\sqrt{\pi}} \int_0^\infty e^{-s^2} s^n H_n(st)ds.$$

5. Show

$$e^{st} J_0(s\sqrt{1-t^2}) = \sum_{n=0}^\infty \frac{P_n(t)}{n!} s^n$$

6. Show

$$J_n(t) = \frac{1}{\pi} \int_0^\pi \cos(n\theta - t\sin\theta)d\theta,$$

n=0,1,... .

References

1. E.A. Coddington and N. Levinson, Theory of Ordinary Differential Equations," McGraw-Hill, New York, 1955.

2. D. Jackson, "Fourier Series and Orthogonal Polynomials," Mathematical Association of America, 1941.

3. A.L. Rabenstein, "Introduction to Ordinary Differential Equations," Academic Press, New York, 1966.

4. E.D. Rainville, "Special Functions," MacMillan, New York, 1960.

5. G. Szego, "Orthogonal Polynomials," American Mathematical Society, New York, 1939.

XI.   THE FOURIER INTEGRAL

This chapter consists of three essential parts. First comes a brief statement of the essential features of the Lebesgue integral, including the definition and **pertinent** major convergence theorems.

All this is essential for the next, which is a discussion of the Fourier integral or transform in the space $L^1(-\infty,\infty)$. We find that $L^1(-\infty,\infty)$ is not only a Banach space, but with convolution multiplication, is an algebra. The Fourier integral is used to show that although the algebra does not contain an identity element, it does contain what is called an approximate identity. Finally the Fourier integral is used to show that the Laguerre polynomials are dense in $L^2(e^{-t})$ over $[0,\infty)$, and the Hermite polynomials are dense in $L^2(e^{-t^2})$ over $(-\infty,\infty)$.

The third section considers the Fourier integral as an operator on $L^2(-\infty,\infty)$. It is shown to be a unitary operator with eigenvalues 1,-1,i,-i and eigenfunctions the Hermite functions.

XI.1.   The Lebesgue Integral. We have been using the Lebesgue integral informally since the very beginning. Because the theory of the Fourier integral depends upon it very heavily, however, it is now time to describe it in a bit more detail. We shall not for the most part prove our statements (Proofs can be found in any textbook on real analysis.) Rather we would like to give the reader some intuitive idea of what it is all about.

XI.1.1.   Definition. Let X be a set, and let A be a collection of subsets of X with the following properties.

1. If A and B are in A, then $A \cup B$ and A-B are in A.

2. If A_j, j=1,..., are in A, then $\bigcup_{j=1}^{\infty} A_j$ is in A.

3. $\bigcup_{A \in A} A$ is in A.

Then A is called a σ-algebra of subsets of X.

A σ-algebra has some other properties. Since $A \cap B = A-(A-B)$, $A \cap B$ is also in A. Further $\bigcap_{j=1}^{\infty} A_j = \bigcup_{A \in A} A-[\bigcup_{j=1}^{\infty} \{ \bigcup_{A \in A} A-A_j \}]$, so the countable intersection of elements in A is in A. Finally since the empty set $\phi = A-A$, ϕ is in A.

XI.1.2. <u>Definition</u>. Let $X = E^n$. The Borel sets are the sets in the smallest σ-algebra which contains all the open sets in E^n.

It is these sets we will primarily use in constructing the Lebesgue integral. We now formally define what is meant by a measure.

XI.1.3. <u>Definition</u>. Let A be a σ-algebra of subsets of E^n which contains the Borel sets. A function μ whose domain is A and range is the nonnegative real numbers is called a (countably additive) measure if

1. $0 \leq \mu(A) \leq \infty$ for all A in A.

2. $\mu(\phi) = 0$, where ϕ is the empty set.

3. $\mu(A \cup B) = \mu(A)+\mu(B)$ when A, B are in A and $A \cap B = \phi$.

4. $\mu(\bigcup_{j=1}^{\infty} A_j) = \sum_{j=1}^{\infty} \mu(A_j)$, <u>when A_j is in A, j=1,..., and</u>

$\{A_j\}_{j=1}^{\infty}$ <u>are pairwise disjoint</u>.

XI.1.3. <u>Examples</u>. 1. Let A consist of the Borel sets of the real line.

the measure μ be defined by $\mu[\alpha,\beta] = \beta-\alpha$ for all intervals $[\alpha,\beta]$. Pro

erly extended this is the <u>ordinary Lebesgue measure</u>. Throughout this chapt

this is the <u>only</u> measure we shall use.

2. Again, on the real line let $\mu[\alpha,\beta] = \beta^3-\alpha^3$ for all intervals [α

This measure yields a different (but closely related) integral.

3. Again on the real line, let $\mu[\alpha,\beta] = \{$The number of integers in

$[\alpha,\beta]\}$ for all intervals $[\alpha,\beta]$. This is a counting measure. For example

$$\sum_{j=1}^{n} j^3 = \int_{1-}^{n+} t^3 \mu(dt).$$

These last two measures are called Stieltjes measures.

In order to define the Lebesgue integral we need a slight generalizati

of the concept of measure.

XI.1.4. <u>Definition</u>. 1. <u>Let X be a set in E^n. The outer measure of X</u>

$$\mu_e(X) = \inf\{\mu(0), \quad 0 \text{ is an open set, } X \subset 0\}.$$

2. <u>The inner measure of a set X is</u>

$$\mu_i(X) = \sup\{\mu(C), \quad C \text{ is a closed set, } C \subset X\}.$$

3. If $\mu_e(X) = \mu_i(X)$, then X is said to be measurable. Its measure, $\mu(X)$, is their common value.

4. If $\mu(X) = 0$, then X has measure 0. If a statement is true except on a set of measure 0, then the statement is true almost everywhere (a.e.).

In addition we extend the concept of measurability to functions.

X1.1.5. Definition. Let x be a function with domain E^n and range in the real numbers. x is said to be measurable if any of the sets $\{t : t \in E^n, x(t) \geq c\}$, $\{t : t \in E^n, x(t) > c\}$, $\{t : t \in E^n, x(t) \leq c\}$, $\{t : t \in E^n, x(t) < c\}$ is measurable for all c.

In fact, if any of the four sets is always measurable, the others are also.

One can show that unions, intersections, differences of measurable sets are all measurable. Even limits of measurable sets are measurable under suitable hypotheses. In particular all Borel sets are measurable.

Similarly, sums, differences, products, quotients, and limits of measurable functions are measurable. In particular all continuous and piece-wise continuous functions are measurable. In fact the Baire functions (the smallest class of functions which contains the continuous functions and which is closed under limit processes) are all measurable.

We now define the one-dimensional Lebesgue integral. Higher dimensional integrals follow the same pattern of definition.

XI.1.6. Definition. Let x be a measurable function, defined on a finite
interval $[\alpha,\beta]$.

1. If $x \equiv 1$, let $\int_{\alpha}^{\beta} 1\mu(dt) = \mu[\alpha,\beta] = \beta-\alpha$.

2. If $x \equiv k$, let $\int_{\alpha}^{\beta} k\mu(dt) = k\mu[\alpha,\beta] = k(\beta-\alpha)$.

3. More generally, if m and M are lower and upper bounds for x.
    Choose $y_0,\ldots,y_n$ such that

$$m = y_0 < y_1 < \cdots < y_{n-1} < y_n = M.$$

Now $E_j = \{t; y_j \leq x(t) < y_{j+1}\}$ is a measurable set, $j=0,\ldots,n-1$. Thus

$$s_n = \sum_{j=0}^{n-1} y_j \mu(E_j)$$

and

$$S_n = \sum_{j=0}^{n-1} y_{j+1}\mu(E_j),$$

exist and are finite. When

$$\lim_{\sup[y_j-y_{j-1}] \to 0} s_n = \lim_{\sup[y_j-y_{j-1}] \to 0} S_n,$$

we define $\int_{\alpha}^{\beta} x(t)\mu(dt)$ to be their common value.

XI.1.7. <u>Theorem.</u> Let x be a bounded, measurable function, defined on a finite interval $[\alpha, \beta]$. Then $\int_{\alpha}^{\beta} x(t)\mu(dt)$ exists and is finite.

Proof. Let $\varepsilon > 0$, and choose the points $\{y_j\}_{j=0}^{\infty}$ to be sufficiently numerous and sufficiently close together such that $\sup[y_j - y_{j-1}] < \varepsilon$. Then

$$S_n - s_n = \sum_{j=0}^{n-1} [y_{j+1} - y_j]\mu(E_j),$$

$$< \varepsilon \sum_{j=0}^{n-1} \mu(E_j),$$

$$= \varepsilon(\beta - \alpha).$$

Thus as $\sup[y_j - y_{j-1}]$ approaches 0, $S_n - s_n$ approaches 0.

Now for two different subdivisions $\{y_j\}_{j=0}^{n}$ and $\{y'_j\}_{j=0}^{m}$, there exists a subdivision consisting of both sets, which we call $\{y''_j\}_{j=0}^{n+m}$. We let the corresponding sums be s, S, s', S' and s'', S''. It is an easy calculation to see $s \leqq s''$, $s' \leqq s''$, $S'' \leqq S$, $S'' \leqq S'$. This implies that as more points of subdivision are taken, no matter how, the lower integrals s form a monotone increasing sequence

$$\acute{s} \leqq s' \leqq s'' \ldots .$$

The upper integrals S form a monotone decreasing sequence

$$\ldots S'' \leqq S' \leqq S.$$

It is also evident that each upper integral is greater than or equal to each

lower integral. Thus

$$s \leq s' \leq s'' \leq \ldots \leq S'' \leq S' \leq S.$$

Since the difference between corresponding upper and lower integrals converge

to 0, each sequence must converge to the same limit, the value of the integ

XI.1.8. **Definition.** Each measurable function for which the Lebesgue integr

exists is said to be integrable.

XI.1.9. **Corollary.** Let x be a continuous function over a finite interval

$[\alpha,\beta]$. Then x is integrable over $[\alpha,\beta]$.

Proof. Since x is continuous over a finite closed interval, it is bounded

and measurable. Hence, the result follows from Theorem XII.1.7.

It can also be proved that for bounded measurable functions on a finite

interval, when the Cauchy-Riemann integral (the one taught in elementary cal

culus) exists, the Lebesgue integral does also, and they are equal. On the

other hand, the Lebesgue integral may exist when the Cauchy-Riemann integral

does not. For instance, on the interval $[0,1]$, let

$$x(t) = \begin{cases} 1, & \text{when } t \text{ is rational,} \\ 0, & \text{when } t \text{ is irrational.} \end{cases}$$

The upper Riemann sum is always 1. The lower Riemann sum is always 0. Th

they do not approach each other. The Riemann integral does not exist. However, since the Lebesgue measure of the rational numbers is 0 (See

Exercise XII.1.), the upper and lower Lebesgue sums are always 0, and thus

$\int_0^1 x(t)\mu(dt) = 0.$

We remark at this point that the Banach spaces L^p, $1 \leq p < \infty$, which

were previously found by completing an incomplete normed linear space through

the use of sequences, may be obtained directly by using the Lebesgue integral.

If D is a measurable set in E^n, then $L^p(D)$ consists of those measurable

functions x which satisfy

$$\|x\|_p = [\int_D |x(t_1,\ldots,t_n)|^p \mu(dt_1 x \ldots x dt_n)]^{1/p} < \infty.$$

Equality between two functions x and y occurs only when

$$[\int_D |x(t_1,\ldots,t_n) - y(t_1,\ldots,t_n)|^p \mu(dt_1 x \ldots x dt_n)]^{1/p} = 0.$$

Thus x and y may differ on a set of measure 0 and still be equal in

$L^p(D)$.

Next following are some extremely important theorems concerning the

Lebesgue integral.   Since the proofs are not particularly illustrative, and

can be found in standard real analysis books, they are omitted.

The first says essentially that changing the order of integration is

permitted.

XI.1.10.   Theorem (Fubini-Tonelli Theorem).   Let x(s,t) be measurable on

the rectangle $D = [\alpha,\beta] x [\gamma,\delta]$ in E^2.

1. If

$$\iint\limits_{D} |x(s,t)| \mu(ds \times dt) < \infty$$

($\mu(ds \times dt)$ indicates Lebesgue measure in two dimensions), then for almost

all s,

$$f(s) = \int\limits_{\gamma}^{\delta} x(s,t) \mu(dt)$$

exists, and

$$\int\limits_{\alpha}^{\beta} |f(s)| \mu(ds) < \infty \ .$$

Further

$$\iint\limits_{D} x(s,t) \mu(ds \times dt) = \int\limits_{\alpha}^{\beta} [\int\limits_{\gamma}^{\delta} x(s,t) \mu(dt)] \mu(ds),$$

$$= \int\limits_{\gamma}^{\delta} [\int\limits_{\alpha}^{\beta} x(s,t) \mu(ds)] \mu(dt).$$

2. If for almost all s,

$$g(s) = \int\limits_{\gamma}^{\delta} |x(s,t)| \mu(dt) < \infty,$$

and

$$\int\limits_{\alpha}^{\beta} |g(s)| \mu(ds) < \infty,$$

then

$$\int_D \int |x(s,t)| \mu(ds \times dt) < \infty.$$

While the previous theorem has a counterpart valid for the Cauchy-Riemann integral, the next two theorems, concerning limits of integrals, do not without severely restricting the functions involved.

XI.1.11. Theorem (Lebesgue Dominated Convergence Theorem). Let D be a measurable set in E^n. Let $\{x_j\}_{j=1}^{\infty}$ be a sequence of functions, defined on D and in $L^1(D)$, and suppose that $\lim_{j \to \infty} x_j = x$ for almost all points in D. Finally suppose that there exists a function y such that $|x_n| < y$ for all points in D and

$$\int_D |y(t_1,\ldots,t_n)| \mu(dt_1 \times \ldots \times dt_n) < \infty.$$

Then

$$\lim_{j \to \infty} \int_D x_j(t_1,\ldots,t_n) \mu(dt_1 \times \ldots \times dt_n) = \int_D x(t_1,\ldots,t_n) \mu(dt_1 \times \ldots \times dt_n) < \infty.$$

XI.1.12. Theorem (Lebesgue Monotone Convergence Theorem). Let D be a measurable set in E^n. Let $\{x_j\}_{j=1}^{\infty}$ be a sequence of functions, defined on D and in $L^1(D)$, satisfying

$$x_1 \leq x_2 \leq \ldots \leq x_n \leq \ldots .$$

Then if either $\lim\limits_{j \to \infty} \int_D x_j(t_1,\ldots,t_n)\mu(dt_1 \times \ldots \times dt_n)$ or $\int_D \lim\limits_{j \to \infty} x_j(t_1 \times \ldots \times t$ $\mu(dt_1 \times \ldots \times dt_n)$ exists, the other does also, and they are equal.

The next theorem is extremely useful when dealing with the L^p spaces.

XI.1.13. Theorem. Let D be measurable set in E^n. Then the sets of all functions which are infinitely differentiable or which are piecewise constan and which vanish outside some compact subset of D are dense in $L^p(D)$, $1 \leqq p < \infty$.

XI.1.14. Corollary. Let x be in $L^1(-\infty,\infty)$. Then

$$\lim_{h \to 0} \int_{-\infty}^{\infty} |x(t+h)-x(t)|\mu(dt) = 0.$$

Proof. Since the infinitely differentiable functions with compact suppert (which vanish outside a compact set) are dense in $L^1(-\infty,\infty)$ we may choose such a function y such that for arbitrary $\varepsilon > 0$,

$$\int_{-\infty}^{\infty} |x(t)-y(t)|\mu(dt) < \varepsilon/3.$$

Then

$$\int_{-\infty}^{\infty} |x(t+h)-y(t+h)|\mu(dt) < \varepsilon/3.$$

Further since y vanishes outside some compact set, there exists an n suc that $y \equiv 0$ if t and t+h are greater than n or less than $-n$. Thus

$$\int_{-\infty}^{\infty} |y(t+h)-y(t)|\mu(dt) = \int_{-n}^{n} |y(t+h)-y(t)|\mu(dt).$$

Finally on $[-n,n]$ y is uniformly continuous. Thus if h is sufficiently small, $|y(t+h)-y(t)| < \varepsilon/6n$. Thus $\int_{-n}^{n} |y(t+h)-h(t)|\mu(dt) < 2n\varepsilon/6n = \varepsilon/3$.

$$\int_{-\infty}^{\infty} |x(t+h)-x(t)|\mu(dt) \leqq \int_{-\infty}^{\infty} |x(t+h)-y(t+h)|\mu(dt)$$

$$+ \int_{-\infty}^{\infty} |x(t)-y(t)|\mu(dt)$$

$$+ \int_{-\infty}^{\infty} |y(t+h)-y(t)|\mu(dt),$$

$$< \varepsilon.$$

Finally we need the concept of differentiation of an integral.

XI.1.15. <u>Definition.</u> <u>Let x be defined on an interval $[\alpha,\beta]$. x is</u> <u>absolutely continuous on $[\alpha,\beta]$ if there exists a function y in $L^1[\alpha,\beta]$</u> <u>such that</u>

$$x(t) = \int_{\alpha}^{t} y(t)\mu(dt)+C,$$

<u>where C is a constant.</u>

XI.1.16. <u>Theorem.</u> <u>Let y be in $L^1[\alpha,\beta]$. Then</u>

$$x(t) = \int_{\alpha}^{t} y(t)\mu(dt)$$

is absolutely continuous on $[\alpha,\beta]$, x' exists almost everywhere and equals y almost everywhere.

In closing this section we would like to make two remarks. First, that L^p convergence does not imply pointwise convergence, not does pointwise convergence imply L^p convergence. For consider: Let x_n be defined by

$$
x_n(t) = \begin{cases}
0, & \text{when} \quad -\infty < t \leq 0 \\[2mm]
(4n^2 t)^{1/p}, & \text{when} \quad 0 \leq t \leq 1/2n \\[2mm]
[4n^2(1/n-t)]^{1/p}, & \text{when} \quad 1/2n \leq t \leq 1/n \\[2mm]
0, & \text{when} \quad 1/n \leq t.
\end{cases}
$$

Then x_n has the following graph:

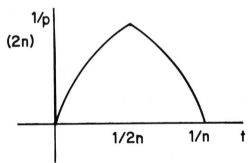

In $L^p(-\infty,\infty)$

$$
\|x_n\|_p = [\int_0^{1/n} |x_n(t)|^p \mu(dt)]^{1/p} = 1.
$$

But at every point t, $\lim_{n\to\infty} x_n(t) = 0$.

On the other hand, the sequence of functions as indicated by the graphs

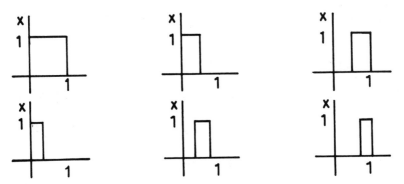

..., has $L^p(-\infty,\infty)$ limit 0, but does not converge pointwise for any point

in [0,1].

Secondly, we have been writing $\mu(dt)$ throughout this section to

emphasize the Lebesgue measure. We now discontinue this notation to adopt the

more standard dt in its place.

XI.2. The Fourier Integral in $L^1(-\infty,\infty)$. Now that we have the full power of

the Lebesgue integral at our disposal we begin the serious study of the

Fourier integral or transform. We choose to study it as an operator on

$L^1(-\infty,\infty)$ and $L^2(-\infty,\infty)$. The results in these two spaces are essentially

different, and accordingly are presented separately. We begin in $L^1(-\infty,\infty)$.

XI.2.1. Definition. Let x be in $L^1(-\infty,\infty)$. The Fourier transform of x

is

$$F(\omega) = (1/\sqrt{2\pi})\int_{-\infty}^{\infty} e^{-i\omega t}x(t)dt.$$

Since $\left| \int_{-\infty}^{\infty} e^{-i\omega t} x(t) dt \right| \leq \int_{-\infty}^{\infty} |x(t)| dt < \infty$, the integral is well defined.

In addition we also sometimes denote the Fourier transform by $F(x)(\omega)$.

XI.2.2. Theorem. If x is in $L^1(-\infty, \infty)$, then its Fourier transform F

is bounded, uniformly continuous and vanishes as ω approaches $\pm\infty$.

Proof. To show F is bounded we only need to observe that $|F| \leq \int_{-\infty}^{\infty} |x(t)| d$

To show uniform continuity, we compute

$$F(\omega+h) - F(\omega) = (1/\sqrt{2\pi}) \int_{-\infty}^{\infty} [e^{-i(\omega+h)t} - e^{-i\omega t}] x(t) dt,$$

$$= (\frac{2i}{\sqrt{2\pi}}) \int_{-\infty}^{\infty} [\frac{e^{-iht/2} - e^{iht/2}}{2i}] e^{-i\omega t} e^{-iht/2} x(t) dt,$$

$$= (\frac{-2i}{\sqrt{2\pi}}) \int_{-\infty}^{\infty} \sin(ht/2) e^{-i(\omega+h/2)t} x(t) dt.$$

Thus

$$\left| F(\omega+h) - F(\omega) \right| \leq \sqrt{\frac{2}{\pi}} \int_{-\infty}^{\infty} |\sin(ht/2)| \, |x(t)| dt.$$

Now choose $\varepsilon > 0$. If T is chosen large enough, we find $\sqrt{\frac{2}{\pi}} \int_{T}^{\infty} |x(t)| dt <$

and $\sqrt{\frac{2}{\pi}} \int_{-\infty}^{-T} |x(t)| dt < \varepsilon/3$. For t in $[-T, T]$, we can then choose h

sufficiently small so $|\sin(ht/2)| < \varepsilon/3[\sqrt{\frac{2}{\pi}} \int_{-T}^{T} |x(t)| dt + 1]$. Then

$\sqrt{\frac{2}{\pi}} \int_{-T}^{T} |\sin(ht/2)| \, |x(t)| dt < \varepsilon/3$ also, and $|F(\omega+h) - F(\omega)| < \varepsilon$.

Finally to show that F vanishes as ω approaches $\pm\infty$, we choose a function y which is continuously differentiable, vanishes outside some

compact interval, and satisfies

$$\int_{-\infty}^{\infty} |x(t)-y(t)|\,dt < \varepsilon/2,$$

where $\varepsilon > 0$ has been chosen arbitrarily. If the Fourier transform of y

is denoted by G, then

$$|F(\omega)-G(\omega)| < \varepsilon/2$$

also. Now, if y vanishes outside [-T,T], then

$$G(\omega) = (1/\sqrt{2\pi})\int_{-\infty}^{\infty} e^{-i\omega t}y(t)\,dt,$$

$$= (1/\sqrt{2\pi})\int_{-T}^{T} e^{-i\omega t}y(t)\,dt,$$

$$= (1/\sqrt{2\pi})[\frac{e^{-i\omega t}}{-i\omega} y(t)]_{-T}^{T} - (1/\sqrt{2\pi})\int_{-T}^{T} \frac{e^{-i\omega t}}{-i\omega} y'(t)\,dt,$$

$$= 0 + (1/\sqrt{2\pi})(1/i\omega)\int_{-T}^{T} e^{-i\omega t}y'(t)\,dt.$$

If ω is sufficiently large, this clearly is less than $\varepsilon/2$ in absolute

value. Then

$$|F(\omega)| \le |F(\omega)-G(\omega)| + |G(\omega)|,$$

XI.2.3. Corollary. If x is in $L^1(-\infty,\infty)$, then

$$\lim_{\omega\to\pm\infty} \int_{-\infty}^{\infty} \sin\omega t\ x(t)\ dt = 0$$

$$\lim_{\omega\to\pm\infty} \int_{-\infty}^{\infty} \cos\omega t\ x(t)\ dt = 0$$

While the product of two integrable functions may not be integrable, there is a form of multiplication of integrable functions which preserves integrability: the convolution.

XI.2.4. Definition. Let x and y be in $L^1(-\infty,\infty)$. Then the convolution of x and y is

$$x*y(t) = (1/\sqrt{2\pi})\int_{-\infty}^{\infty} x(s)y(t-s)ds.$$

XI.2.5. Theorem. Let x, y and z be in $L^1(-\infty,\infty)$. Then

1. x*y exists almost everywhere.

2. x*y is in $L^1(-\infty,\infty)$.

3. x*y = y*x.

4. (x*y)*z = x*(y*z).

5. If F(x) and F(y) denote the Fourier transforms of x and y, and F(x*y) that of their convolution, then

$$F(x*y) = F(x)\cdot F(y).$$

Proof.   To show 1 and 2, we consider

$$g(s) = |x(s)| \int_{-\infty}^{\infty} |y(t-s)| dt.$$

Then

$$\int_{-\infty}^{\infty} g(s) ds \leq \int_{-\infty}^{\infty} |x(s)| ds \int_{-\infty}^{\infty} |y(t)| dt.$$

So g is in $L^1(-\infty,\infty)$.   By Fubini's Theorem (XI.1.10) we see that

$$\int_{-\infty}^{\infty} |x(s)| \int_{-\infty}^{\infty} |y(t-s)| dt ds = \int_{-\infty}^{\infty} [\int_{-\infty}^{\infty} |x(s)| |y(t-s)| ds dt$$

$$\geq \int_{-\infty}^{\infty} | \int_{-\infty}^{\infty} x(s) y(t-s) ds | dt.$$

So x∗y is in $L^1(-\infty,\infty)$,   and must be finite almost everywhere.

To show 3 we merely make the change of variable   t−s = u   and eliminate

s.

$$x{*}y(t) = (1/\sqrt{2\pi}) \int_{-\infty}^{\infty} x(s) y(t-s) ds$$

$$= (1/\sqrt{2\pi}) \int_{-\infty}^{\infty} x(t-u) y(u) du$$

$$= y{*}x(t).$$

The 4th part is a similar but tedious computation.   To show part 5,

$$F(x*y)(\omega) = \frac{1}{2\pi} \int_{-\infty}^{\infty} e^{-i\omega t} [\int_{-\infty}^{\infty} x(t-s)y(s)ds] dt,$$

$$= \frac{1}{2\pi} \int_{-\infty}^{\infty} y(s) [\int_{-\infty}^{\infty} e^{-i\omega t} x(t-s) dt] ds,$$

$$= \frac{1}{2\pi} \int_{-\infty}^{\infty} e^{-i\omega s} y(s) [\int_{-\infty}^{\infty} e^{-i\omega(t-s)} x(t-s) dt] ds$$

Replacing t-s by u in the inner integral, we have

$$F(x*y)(\omega) = (1/\sqrt{2\pi}) \int_{-\infty}^{\infty} e^{-i\omega s} y(s) ds (1/\sqrt{2\pi}) \int_{-\infty}^{\infty} e^{-i\omega u} x(u) du,$$

$$= F(x)(\omega) \cdot F(y)(\omega).$$

XI.2.6. <u>Corollary</u>. $L^1(-\infty,\infty)$ under convolution multiplication forms a commutative Banach algebra.

A great many algebras (for instance the real and complex number systems or C[a,b], the Banach algebra of continuous functions) possess an identity element, which leaves all elements unchanged under multiplication. This is not the case in the Banach algebra $L^1(-\infty,\infty)$.

XI.2.7. <u>Theorem</u>. The Banach algebra generated by $L^1(-\infty,\infty)$ under convolut multiplication does not possess an identity element.

Proof. Suppose there exists an element j such that for all x in $L^1(-\infty$ x*j = x. Then, taking Fourier transforms,

$$F(x) \cdot F(j) = F(x).$$

Now we choose a representative for x. Let

$$x(t) = \begin{cases} 1, & \text{when } t \leq a, \\ \\ 0, & \text{when } t > a. \end{cases}$$

Then $F(x)(\omega) = -\sqrt{(2/\pi)}(\sin a\omega)/\omega$, which is nonzero except when $\omega = \pm \dfrac{n\pi}{a}$,

$n = 1, 2, \ldots$.

Thus by varying a we can find for any ω a function x such that
$F(x)(\omega) \neq 0$. Returning now to the equation involving the Fourier transforms,
for any value ω we choose x such that $F(x)(\omega) \neq 0$. Then

$$F(j)(\omega) = 1.$$

Since ω is arbitrary, $F(j) \equiv 1$. But this contradicts Theorem XI.2.2,
part 3.

Since there is no identity element, we look for the next best thing.

XI.2.8. **Definition.** A one parameter function $e_\alpha(t)$ is on approximate identity for $L^1(-\infty,\infty)$ if

1. $e_\alpha(t) \geqq 0$ for all α and t.

2. $\int_{-\infty}^{\infty} e_\alpha(t)dt = 1$ for all α.

3. for all $\varepsilon > 0$, $\lim\limits_{\alpha \to \infty} \int_{|t|>\varepsilon} e_\alpha(t)dt = 0$.

The so-called Weierstrass kernel $(\alpha/\sqrt{2\pi})e^{-\alpha^2 t^2/2}$ forms one example of an approximate identity.

XI.2.8. **Theorem.** Suppose that e_α is an approximate identity for $L^1(-\infty,$ Then for each x in $L^1(-\infty,\infty)$

$$\lim_{\alpha \to \infty} (\sqrt{2\pi}\, e_\alpha * x) = x$$

almost everywhere. That is,

$$\lim_{\alpha \to \infty} \int_{-\infty}^{\infty} e_\alpha(s)x(t-s)ds = x(t) \quad \text{a.e.}$$

Proof. Let $\varepsilon > 0$. Since $\lim\limits_{h \to 0} \int_{-\infty}^{\infty} |x(t+h)-x(t)|dt = 0$ by Corollary XI.1 there is a $\delta > 0$ such that if $|h| < \delta$, then $\int_{-\infty}^{\infty} |x(t+h)-x(t)|dt < \varepsilon/3$ Further $\int_{-\infty}^{\infty} |x(t+h)-x(t)|dt \doteqdot 2\|x\|$, where the norm is the $L^1(-\infty,\infty)$ norm $\|x\| = \int_{-\infty}^{\infty} |x(t)|dt$. Thus

$$\|\sqrt{2\pi}\ e_\alpha * x - x\| = \int_{-\infty}^{\infty} |\int_{-\infty}^{\infty} x(t-s)e_\alpha(s)ds - x(t)| dt,$$

$$= \int_{-\infty}^{\infty} |\int_{-\infty}^{\infty} [x(t-s)-x(t)]e_\alpha(s)ds| dt,$$

$$\leq \int_{-\infty}^{\infty} \int_{\infty}^{\infty} |x(t-s)-x(t)| e_\alpha(s)dsdt,$$

$$= \int_{-\infty}^{\infty} [\int_{-\infty}^{\infty} |x(t-s)-x(t)| dt]e_\alpha(s)ds.$$

Now we split this integral into three pieces.

$$\int_{-\infty}^{-\delta} [\int_{-\infty}^{\infty} |x(t-s)-x(t)| dt]e_\alpha(s)ds < 2\|x\| \int_{-\infty}^{-\delta} e_\alpha(s)ds,$$

$$\int_{\delta}^{\infty} [\int_{-\infty}^{\infty} |x(t-s)-x(t)| dt]e_\alpha(s)ds < 2\|x\| \int_{\delta}^{\infty} e_\alpha(s)ds,$$

$$\int_{-\delta}^{\delta} [\int_{-\infty}^{\infty} |x(t-s)-x(t)| dt]e_\alpha(s)ds < \int_{-\infty}^{\infty} |x(t-s)-x(t)| dt.$$

Now if δ is sufficiently small the last is less than $\varepsilon/3$. Because of third property satisfied by the approximate identity, if α is sufficiently large, each of the first two terms is less than $\varepsilon/3$. Thus if α is sufficiently large,

$$\|\sqrt{2\pi}\ e_\alpha * x - x\| < \varepsilon.$$

In passing we note that an approximate identity in $L^1(-\infty,\infty)$ is also one in $L^p(-\infty,\infty)$, $1 < p < \infty$, and that as α increases $e_\alpha(t)$ "converges" to what is known as the Dirac δ-function. We will encounter it again in the study of distributions.

We now turn our attention to finding the inverse of the Fourier transform. That is, we wish to solve the equation

$$F(x)(\omega) = F(\omega),$$

or

$$(1/\sqrt{2\pi})\int_{-\infty}^{\infty} e^{-i\omega t}x(t)dt = F(\omega),$$

for x.

XI.2.9. <u>Theorem</u>. <u>Let x be in $L^1(-\infty,\infty)$ with Fourier transform F(ω),</u> <u>and let</u>

$$e_\alpha(t) = (\alpha/\sqrt{2\pi})e^{-\alpha^2 t^2/2}$$

<u>be the Weierstrass kernel with Fourier transform $E_\alpha(\omega)$. Then</u>

$$\lim_{\alpha\to\infty} \|\int_{-\infty}^{\infty} E_\alpha(\omega)F(\omega)e^{i\omega t}d\omega - x(t)\| = 0.$$

To prove this theorem we need a result which depends upon the Cauchy integral theorem from the theory of functions of a complex variable.

XI.2.10. Lemma. If ω is real valued, then

$$(1/\sqrt{2\pi})\int_{-\infty}^{\infty} e^{-t^2/2} e^{-i\omega t} dt = e^{-\omega^2/2}.$$

In other words, the Fourier transform leaves $e^{-t^2/2}$ unchanged.

Proof. We consider $\int_{\Gamma} e^{-z^2/2} dz$ around the contour $I_1 = [-T,T]$,
$I_2 = [T,T+i\alpha]$, $I_3 = [T+i\alpha,-T+i\alpha]$, $I_4 = [-T+i\alpha,-T]$. Γ consists of I_1,
I_2, I_3 and I_4 described in a counter clockwise manner.

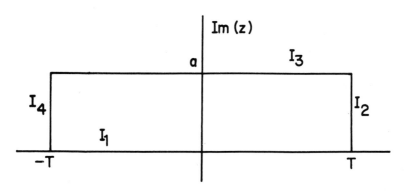

Using Cauchy's integral theorem we see that the value of the integral is 0.

Now

$$\left|\int_{I_2} e^{-z^2/2} dz\right| = \left|\int_0^{\alpha} e^{-(T+is)^2/2} i\,ds\right|,$$

$$\leq e^{-T^2/2} \int_0^{\alpha} e^{s^2/2} ds,$$

which approaches 0 as T approaches ∞. Similarly $\int_{I_4} e^{-z^2/2} dz$ approaches

0, as T approaches ∞. Thus

$$\int_{-\infty}^{\infty} e^{-t^2/2} dt = \int_{-\infty}^{\infty} e^{-(t+i\alpha)^2/2} dt.$$

Now let $I = \int_{-\infty}^{\infty} e^{-t^2/2} dt.$ Then

$$I^2 = \int_{-\infty}^{\infty} \int_{-\infty}^{\infty} e^{-t^2/2} e^{-s^2/2} ds\, dt,$$

$$= \int_{0}^{2\pi} \int_{0}^{\infty} e^{-R^2/2} R\, dR\, d\theta,$$

$$= 2\pi.$$

So $I = \sqrt{2\pi}.$ Finally,

$$1/\sqrt{2\pi} \int_{-\infty}^{\infty} e^{-i\omega t} e^{-t^2/2} dt = (1/\sqrt{2\pi}) \int_{-\infty}^{\infty} e^{-(t+i\omega)^2/2} e^{-\omega^2/2} dt,$$

$$= [(1/\sqrt{2\pi}) \int_{-\infty}^{\infty} e^{-t^2/2} dt] e^{-\omega^2/2},$$

$$= e^{-\omega^2/2}.$$

Proof of Theorem XI.2.9. We first need to compute the Fourier transform $E_\alpha(\omega)$. Using Lemma XI.2.10 we find

$$(1/\sqrt{2\pi}) \int_{-\infty}^{\infty} e^{-i\omega t} [(\alpha/\sqrt{2\pi}) e^{-\alpha^2 t^2/2}] dt =$$

$$(1/\sqrt{2\pi}) [(1/\sqrt{2\pi}) \int_{-\infty}^{\infty} e^{-i(\omega/\alpha)\alpha t} e^{-(\alpha t)^2/2} \alpha\, dt],$$

$$= (1/\sqrt{2\pi}) e^{-(\omega/\alpha)^2/2}.$$

A similar calculation shows

$$(1/\sqrt{2\pi})\int_{-\infty}^{\infty} e^{i\omega t}E_\alpha(\omega)\,d\omega = e_\alpha(t).$$

Thus

$$\int_{-\infty}^{\infty} E_\alpha(\omega)F(\omega)e^{i\omega t}\,d\omega = \int_{-\infty}^{\infty} E_\alpha(\omega)[(1/\sqrt{2\pi})\int_{-\infty}^{\infty} e^{-i\omega s}x(s)\,ds]e^{i\omega t}\,d\omega,$$

$$= \int_{-\infty}^{\infty} x(s)[(1/\sqrt{2\pi})\int_{-\infty}^{\infty} e^{i\omega(t-s)}E_\alpha(\omega)\,ds]\,ds,$$

$$= \int_{-\infty}^{\infty} x(s)e_\alpha(t-s)\,ds,$$

which approaches x in $L^1(-\infty,\infty)$ as α approaches ∞.

The question of whether two different elements can have the same Fourier transform is still unanswered.

XI.2.11. **Theorem.** **Suppose that** x **and** y **are in** $L^1(-\infty,\infty)$ **and that their Fourier transforms are equal. Then** $x = y$ **a.e.**

Proof. It is sufficient to show that if the Fourier transform $F(x)(\omega) = 0$, then $x=0$ a.e. If $F(x)(\omega) = 0$, then

$$\int_{-\infty}^{\infty} E_\alpha(\omega)F(x)(\omega)e^{i\omega t}\,d\omega = 0.$$

By Theorem XII.2.9, as α approaches ∞, this converges to x a.e., so $x=0$ a.e.

We get the following remarkable results as a corollary.

XI.2.12. Theorem. 1. Suppose that x is in $L^2(0,\infty)$ and is orthogonal to $\{e^{-t/2}L_j(t)\}_{j=0}^{\infty}$, the Laguerre functions. Then $x=0$ in $L^2(0,\infty)$.

2. Suppose that x is in $L^2(-\infty,\infty)$ and is orthogonal to $\{e^{-t^2/2}H_j(t)\}_{j=0}^{\infty}$, the Hermite functions. Then $x=0$ in $L^2(-\infty,\infty)$.

Proof. 1. If x is orthogonal to the Laguerre functions, then by taking appropriate linear combinations it is orthogonal to $\{e^{-t/2}t^n\}_{n=0}^{\infty}$. Multiplying each by $(-i\omega)^n/n!$ and summing, we find

$$\int_0^{\infty} e^{-t/2}e^{-i\omega t}x(t)dt = 0.$$

If we let

$$y(t) = \begin{cases} 0, & \text{when } t < 0, \\[2mm] e^{-t/2}x(t), & \text{when } t \geqq 0, \end{cases}$$

then y is in $L^1(-\infty,\infty)$, and satisfies $F(y)(\omega) = 0$. By Theorem XII.2.11, $y=0$ a.e., and $x=0$ in $L^2(0,\infty)$.

The second part is similarly proved.

XI.2.13. Corollary. 1. The Laguerre polynomials are dense in $L^2(e^{-t})$ over $[0,\infty)$.

2.   The Hermite polynomials are dense in $L^2(e^{-t^2})$ over $(-\infty,\infty)$.

If not only x, but its Fourier transform as well, is in $L^1(-\infty,\infty)$,
then the inversion formula of Theorem XI.2.9 can be considerably simplified.

XI.2.14.   Theorem.   Let x and its Fourier transform F(ω) both be in
$L^1(-\infty,\infty)$.   Then

$$x(t) = (1/\sqrt{2\pi})\int_{-\infty}^{\infty} e^{i\omega t}F(\omega)d\omega$$

a.e.

Proof.   From Theorem XI.2.9,

$$\lim_{\alpha\to\infty} \int_{-\infty}^{\infty} \left| \int_{-\infty}^{\infty} E_\alpha(\omega)F(\omega)e^{i\omega t}d\omega - x(t) \right| dt = 0.$$

Thus for any real T > 0,

$$\lim_{\alpha\to\infty} \int_{-T}^{T} \left| \int_{-\infty}^{\infty} E_\alpha(\omega)F(\omega)e^{i\omega t}d\omega - x(t) \right| dt = 0.$$

Now $E_\alpha(\omega) = (1/\sqrt{2\pi})e^{-(\omega/\alpha)^2/2}$. As α approaches ∞, E_α approaches
$(1/\sqrt{2\pi})$. Since F(ω) is in $L^1(-\infty,\infty)$, the integrand, the terms surrounded
by absolute value bars, is integrable over [-T,T], and, by using the
Lebesgue Dominated Convergence Theorem, we see we may pass to the limit
under the integral sign. So doing, we find

$$\int_{-T}^{T} \left| (1/\sqrt{2\pi})\int_{-\infty}^{\infty} e^{i\omega t}F(\omega)d\omega - x(t) \right| dt = 0.$$

Thus

$$x(t) = 1/\sqrt{2\pi} \int_{-\infty}^{\infty} e^{i\omega t} F(\omega) d\omega$$

a.e.

XI.2.15. <u>An Example (Fejer)</u>. Let $E_\alpha(\omega)$ be defined by

$$E_\alpha(\omega) = \begin{cases} (-1/\alpha\sqrt{2\pi})(\omega-\alpha), & \text{when } 0 \leq \omega \leq \alpha, \\ 0, & \text{when } |\omega| > \alpha, \\ (1/\alpha\sqrt{2\pi})(\omega+\alpha), & \text{when } -\alpha \leq \omega \leq 0. \end{cases}$$

$E_\alpha(\omega)$ then has the following graph:

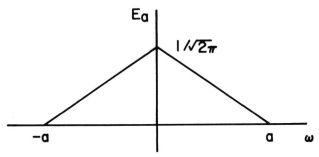

Since E_α is in $L^1(-\infty,\infty)$ we can directly compute e_α.

$$e_\alpha(t) = 2 \sin^2(\alpha t/2)/\pi \alpha t^2.$$

We note that $E_\alpha(\omega-\lambda)$ vanishes outside $(-\alpha+\lambda, \alpha+\lambda)$. By properly choosing α and λ we can find a function $e_{\alpha,\lambda}$ in $L^1(-\infty,\infty)$ whose Fourier transform vanishes outside any given interval.

XI.2.16. Theorem. The Banach algebra generated by $L^1(-\infty,\infty)$ under convolution multiplication has zero divisors. That is, there exist functions x
and y in $L^1(-\infty,\infty)$, neither identically 0, such that $x*y = 0$ a.e.

Proof. As is indicated possible in Example XI.2.15, let x and y in
$L^1(-\infty,\infty)$ have Fourier Transforms $F(x)$ and $F(y)$, which vanish outside
of disjoint intervals I_x and I_y respectively. Then

$$F(x*y) = F(x) \cdot F(y) = 0,$$

and $x*y = 0$ a.e.

XI.3. The Fourier Integral in $L^2(-\infty,\infty)$. The theory of the Fourier integral
in $L^2(-\infty,\infty)$ is essentially different from its counterpart in $L^1(-\infty,\infty)$.
Theoretically at least, there is no trouble in inverting the transform in
$L^2(-\infty,\infty)$, but in this setting, multiplication is not possible, as was done
in $L^1(-\infty,\infty)$, via the convolution. Thus we gain a little and lose a little.

We recall from the end of Chapter IX that ordinary Fourier series was
generated as the spectral resolution of an operator $(1/i)x'$ over a finite
interval. Starting from this result, we eventually find that the Fourier
integral is generated by a similar operator, but over the doubly infinite interval. The technique used is the same as that for the second order singular Sturm-
Liouville problem in Chapter XII.

Since we shall eventually show that the Fourier transform acts as a
unitary operator, we first develop the criterion we will need.

XI.3.1. Theorem. Let H be a Hilbert space, and let A be a linear operato

in $L(H,H)$. Then A is unitary if and only if

    1. $\|Ax\| = \|x\|$ for all x in H.

    2. The range of A is dense in H.

Proof. Recall that A is unitary if $A^* = A^{-1}$.

    Suppose that A is unitary. Then the domain of A^* is the same as the

domain of A^{-1}, which is the range of A. Since A^* is defined for all of

H, the range of A is H. Further,

$$\|Ax\|^2 = (A^*Ax,x) = \|x\|^2.$$

    Conversely, if assumptions 1 and 2 hold, then

$$(A^*Ax,x) = (x,x)$$

for all x in H. Using the polarization identities,

$$(A^*Ax,y) = (x,y)$$

for all x, y in H. Letting $y = A^*Ax-x$, we find

$$\|A^*Ax-x\| = 0,$$

or

$$A^*Ax = x$$

for all x in H. Applying A, we have

$$AA^*Ax = Ax.$$

Letting Ax = y, we have

$$AA^*y = y,$$

where the set of all y's, the range of A, is dense in H. For any
element z in H, we may choose a sequence $\{y_j\}_{j=1}^{\infty}$ in the range of
A such that $\lim_{j\to\infty} y_j = z.$ Then

$$\|AA^*z-z\| \leq \|[AA^*-I](z-y_j)\| + \|AA^*y_j-y_j\|,$$

$$\leq \|AA^*-I\|\|z-y_j\|,$$

which approaches 0 as j approaches ∞. So

$$AA^*z = z$$

for all z in H, and $A^* = A^{-1}.$

Finally before beginning the discussion of the $L^2(-\infty,\infty)$ Fourier transform we consider the relations between various L^1 and L^2 spaces.

XI.3.2.   Theorem.   1.   There exists an element x in $L^2(-\infty,\infty)$ which is not in $L^1(-\infty,\infty)$.

2.   There exists an element x in $L^1(-\infty,\infty)$ which is not in $L^2(-\infty,\infty)$.

3.   If (α,β) is a finite interval, then $L^2(\alpha,\beta) \subset L^1(\alpha,\beta)$.

Proof.   For the first let $x(t) = 1/(|t|+1)$ or $x(t) = \sin t/t$. For the second let

$$
x(t) = \begin{cases} t^{-2/3}, & \text{when } |t| \leq 1, \\\\ 0, & \text{when } |t| > 1, \end{cases}
$$

or $x(t) = e^{-t^2}/\sqrt{t}$. The third follows from Schwarz's inequality

$$
\int_\alpha^\beta 1 |x(t)| dt \leq (\int_\alpha^\beta 1^2 dt)^{1/2} (\int_\alpha^\beta |x(t)|^2 dt)^{1/2}.
$$

Any square integrable function is integrable.

The results concerning ordinary Fourier series in Chapter IX were set in the interval $[-\pi,\pi]$. Restated for an arbitrary interval $[-a,a]$ they have the following form.

XI.3.3.   <u>Theorem</u>.   <u>For any x in $L^2(-\infty,\infty)$,</u>

$$x(t) = \sum_{j=-\infty}^{\infty} (1/2\alpha)[\int_{-\alpha}^{\alpha} e^{-ij\pi s/\alpha} x(s)ds] e^{ij\pi t/\alpha},$$

$$\|x\|^2 = \sum_{j=-\infty}^{\infty} (1/2\alpha) |\int_{-\alpha}^{\alpha} e^{-ij\pi s/\alpha} x(s)ds|^2.$$

<u>The first is in the sense of $L^2(-\alpha,\alpha)$. The second is Parseval's equality.</u>

We now put these results into a new form.

XI.3.4.   <u>Theorem</u>.   <u>Let $\rho_\alpha(\omega)$ be a real valued, piecewise continuous</u>
<u>function, which increases by $1/2\alpha$ at $\omega = j\pi/\alpha$, j an integer, but is</u>
<u>otherwise constant, is continuous from the right at points of discontinuity,</u>
<u>and satisfies $\rho_\alpha(0) = 0$. Then for any x in $L^2(-\alpha,\alpha)$,</u>

$$x(t) = \int_{-\infty}^{\infty} [\int_{-\alpha}^{\alpha} e^{-i\omega s} x(s)ds] e^{i\omega t} d\rho_\alpha(\omega),$$

$$\|x\|^2 = \int_{-\infty}^{\infty} |\int_{-\alpha}^{\alpha} e^{-i\omega s} x(s)ds|^2 d\rho_\alpha(\omega).$$

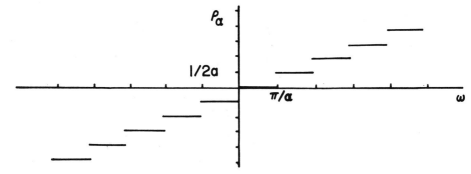

The proof is by inspection. These are old results in a new notation.

XI.3.5. <u>Theorem</u>. <u>As α approaches ∞, $\lim\limits_{\alpha \to \infty} \rho_\alpha(\omega)$ exists and</u>

$\lim\limits_{\alpha \to \infty} \rho_\alpha(\omega) = \omega/2\pi$.

Again this follows from inspection. (See Section XII.2.)

XI.3.6. <u>Theorem</u>. <u>Let x be continuously differentiable and vanish when</u>

<u>$|t|$ is sufficiently large. Further let</u>

$$F(\omega) = (1/\sqrt{2\pi}) \int_{-\infty}^{\infty} e^{-j\omega t} x(t) dt.$$

<u>Then</u>

$$\int_{-\infty}^{\infty} |x(t)|^2 dt = \int_{-\infty}^{\infty} |F(\omega)|^2 d\omega.$$

Proof. If α is sufficiently large, then

$$\|Lx\|^2 = \int_{-\infty}^{\infty} \left| \int_{-\alpha}^{\alpha} e^{-i\omega t} Lx(t) dt \right|^2 d\rho_\alpha(\omega),$$

where $Lx = (1/i)x'$. But

$$\int_{-\alpha}^{\alpha} e^{-i\omega t} Lx(t) dt = \int_{-\alpha}^{\alpha} \frac{-i\omega}{-i} e^{-i\omega t} x(t) dt,$$

$$= \omega \int_{-\alpha}^{\alpha} e^{-i\omega t} x(t) dt,$$

$$= \omega\sqrt{2\pi}\, F(\omega).$$

Thus

$$\|Lx\|^2 = 2\pi \int_{-\infty}^{\infty} |\omega|^2 |F(\omega)|^2 d\rho_\alpha(\omega).$$

Now let $\mu > 0$, and let $I = (-\infty,\infty)-(-\mu,\mu]$. Then

$$\|Lx\|^2 \geqq 2\pi\mu^2 \int_I |F(\omega)|^2 d\rho_\alpha(\omega),$$

or

$$2\pi \int_I |F(\omega)|^2 d\rho_\alpha(\omega) \leq (1/\mu^2)\|Lx\|^2.$$

Since

$$\int_{-\infty}^{\infty} |x(t)|^2 dt = 2\pi \int_{-\mu}^{\mu} |F(\omega)|^2 d\rho_\alpha(\omega)$$

$$+ 2\pi \int_I |F(\omega)|^2 d\rho_\alpha(\omega),$$

we find

$$\left| \int_{-\infty}^{\infty} |x(t)|^2 dt - 2\pi \int_{-\mu}^{\mu} |F(\omega)|^2 d\rho_\alpha(\omega) \right| < (1/\mu^2)\|Lx\|^2.$$

Letting α approach ∞, we find

$$\left| \int_{-\infty}^{\infty} |x(t)|^2 dt - \int_{-\mu}^{\mu} |F(\omega)|^2 d\omega \right| \leqq (1/\mu^2)\|Lx\|^2.$$

As μ approaches ∞, the result follows.

This is a rather weak form of the extended Parseval's equality. We continue to extend it.

XI.3.7. <u>Theorem</u>. <u>Let x be in</u> $L^2(-\infty,\infty)$ <u>and vanish for large</u> $|t|$. Furt

<u>let</u>

$$F(\omega) = (1/\sqrt{2\pi})\int_{-\infty}^{\infty} e^{-i\omega t}x(t)dt.$$

<u>Then</u>

$$\int_{-\infty}^{\infty} |x(t)|^2 dt = \int_{-\infty}^{\infty} |F(\omega)|^2 d\omega.$$

Proof. Since the set of all continuously differentiable functions which vanish outside a compact interval is dense in $L^2(-\infty,\infty)$, we can choose a se quence of such functions $\{x_j\}_{j=1}^{\infty}$ which are zero outside the same set as x and which converge to x in the sense of $L^2(-\infty,\infty)$. If

$$F_j(\omega) = (1/\sqrt{2\pi})\int_{-\infty}^{\infty} e^{-i\omega t}x_j(t)dt,$$

then for some N sufficiently large

$$|F(\omega)-F_j(\omega)| = |\int_{-\infty}^{\infty} e^{-i\omega t}[x(t)-x_j(t)]dt|,$$

$$\leq \int_{-N}^{N} 1|x(t)-x_j(t)|dt,$$

$$\leq (2N)^{1/2} (\int_{-N}^{N} |x(t)-x_j(t)|^2 dt)^{1/2}$$

$$\leq (2N)^{1/2} \|x-x_j\|.$$

Thus, since x_j converges in $L^2(-\infty,\infty)$ to x, $F_j(\omega)$ converges to $F(\omega)$. Finally employing Theorem XI.3.6 for each x_j, $j=1,\ldots$, we see that the left side converges to $\int_{-\infty}^{\infty} |x(t)|^2 dt$, while according to the Lebesgue Dominated Convergence Theorem XI.1.11, the right side approaches $\int_{-\infty}^{\infty} |F(\omega)|^2 d\omega$. Hence these two expressions are equal.

Finally we can extend the result to all x in $L^2(-\infty,\infty)$.

XI.3.8. Theorem (Parseval's Equality). Let x be in $L^2(-\infty,\infty)$, and let

$$F_\alpha(\omega) = (1/\sqrt{2\pi})\int_{-\alpha}^{\alpha} e^{-j\omega t} x(t) dt.$$

Then, as α approaches ∞, $F_\alpha(\omega)$ converges in $L^2(-\infty,\infty)$ to an element $F(\omega)$, and

$$\int_{-\infty}^{\infty} |x(t)|^2 dt = \int_{-\infty}^{\infty} |F(\omega)|^2 d\omega.$$

That is, $\|x\| = \|F\|$.

Proof. If $\beta > \alpha$, then

$$\int_{\alpha}^{\beta} |x(t)|^2 dt + \int_{-\beta}^{-\alpha} |x(t)|^2 dt = \int_{-\infty}^{\infty} |F_\beta(\omega)-F_\alpha(\omega)|^2 d\omega.$$

Since x is in $L^2(-\infty,\infty)$, as α, β approach ∞ the integrals on the left approach 0. This shows that as α approaches ∞, $\{F_\alpha\}$ is a Cauchy sequence in $L^2(-\infty,\infty)$. Thus $\lim_{\alpha\to\infty} F_\alpha$ exists in $L^2(-\infty,\infty)$, and we denote it by $F(\omega)$.

Now according to Theorem XI.3.7,

$$\int_{-\alpha}^{\alpha} |x(t)|^2 dt = \int_{-\infty}^{\infty} |F_\alpha(\omega)|^2 d\omega.$$

Again passing to the limit, we find

$$\int_{-\infty}^{\infty} |x(t)|^2 dt = \int_{-\infty}^{\infty} |F(\omega)|^2 d\omega.$$

Using this last theorem we are now in a position to derive the extension of ordinary Fourier series to the doubly infinite interval.

XI.3.9. <u>Theorem (The Fourier Integral)</u>. <u>Let</u> x <u>be in</u> $L^2(-\infty,\infty)$, <u>and let</u>

$$F(\omega) = \lim_{\alpha\to\infty} (1/\sqrt{2\pi}) \int_{-\alpha}^{\alpha} e^{-i\omega t} x(t)\, dt,$$

<u>where the limit is in the sense of</u> $L^2(-\infty,\infty)$. <u>Then</u>

$$x(t) = \lim_{\substack{\lambda\to\infty \\ \mu\to-\infty}} (1/\sqrt{2\pi}) \int_{\mu}^{\lambda} e^{i\omega t} F(\omega)\, d\omega,$$

<u>where here also the limit is in the sense of</u> $L^2(-\infty,\infty)$. <u>That is</u>,

$$\lim_{\substack{\lambda\to\infty \\ \mu\to-\infty}} \int_{-\infty}^{\infty} \left| x(t) - (1/\sqrt{2\pi}) \int_{\mu}^{\lambda} e^{i\omega t} F(\omega)\, d\omega \right|^2 dt = 0.$$

Proof. By using the polarization identity, Parseval's equality implies

$$\int_{-\infty}^{\infty} \overline{x}_2(t)x_1(t)dt = \int_{-\infty}^{\infty} \overline{F}_2(\omega)F_1(\omega)d\omega,$$

where x_1, x_2 are in $L^2(-\infty,\infty)$ and F_1, F_2 are their corresponding Fourier transforms as given by Theorem XI.3.8.

Let P be an arbitrary element in $L^2(-\infty,\infty)$ which vanishes when $|t| > \alpha$. Let $I = (\mu,\lambda]$, and let

$$x_I(t) = (1/\sqrt{2\pi})\int_I e^{i\omega t}F(\omega)d\omega,$$

where $F(\omega)$ is the Fourier transform of x. Then

$$\int_{-\infty}^{\infty} \overline{P}(t)x_I(t)dt = \int_{-\alpha}^{\alpha} \overline{P}(t)[\int_I e^{i\omega t}F(\omega)d\omega]dt,$$

$$= (1/\sqrt{2\pi})\int_I \int_{-\alpha}^{\alpha} \overline{P}(t)e^{i\omega t}F(\omega)dtd\omega,$$

$$= \int_I [1/\sqrt{2\pi}\int_{-\alpha}^{\alpha} e^{-i\omega t}P(t)dt]^* F(\omega)d\omega,$$

$$= \int_I \overline{Q}(\omega)F(\omega)d\omega,$$

where

$$Q(\omega) = (1/\sqrt{2\pi})\int_{-\infty}^{\infty} e^{-i\omega t}P(t)dt$$

is the Fourier transform of P.

Now in the inner product form of Parseval's equality (the first equation in the proof) let $x_1 = x$ and $x_2 = P$. Then

$$\int_{-\infty}^{\infty} \overline{P}(t) x(t) dt = \int_{-\infty}^{\infty} \overline{Q}(\omega) F(\omega) d\omega.$$

Subtracting from this the last derived equation, and letting $I^c = (-\infty, \infty) - I$ we have

$$\int_{-\infty}^{\infty} \overline{P}(t) [x(t) - x_I(t)] dt = \int_{I^c} \overline{Q}(\omega) F(\omega) d\omega.$$

In order to schieve the square of a norm on the left side, we let

$$P(t) = \begin{cases} x(t) - x_I(t), & \text{when} \quad 0 \le |t| \le a, \\ \\ 0, & \text{when} \quad |t| > a. \end{cases}$$

Then

$$\int_{-a}^{a} |x(t) - x_I(t)|^2 dt = \int_{I^c} \overline{Q}(\omega) F(\omega) d\omega,$$

where now Q, nonzero only on I^c, is the Fourier transform of $x - x_I$. Using Schwarz's inequality on the right,

$$\left(\int_{-a}^{a} |x(t) - x_I(t)|^2 dt \right)^2 \le \int_{I^c} |F(\omega)|^2 d\omega \int_{I^c} |Q(\omega)|^2 d\omega,$$

$$= \int_{I^C} |F(\omega)|^2 d\omega \int_{-\infty}^{\infty} |P(t)|^2 dt,$$

$$= \int_{I^C} |F(\omega)|^2 d\omega \int_{-\alpha}^{\alpha} |x(t)-x_I(t)|^2 dt.$$

Canceling $\int_{-\alpha}^{\alpha} |x(t)-x_I(t)|^2 dt$ on both sides, this yields

$$\int_{-\alpha}^{\alpha} |x(t)-x_I(t)|^2 dt \leq \int_{I^C} |F(\omega)|^2 d\omega.$$

The right side is independent of α, so, letting α approach ∞,

$$\int_{-\infty}^{\infty} |x(t)-x_I(t)|^2 dt \leq \int_{I^C} |F(\omega)|^2 d\omega.$$

Finally letting λ approach ∞ and μ approach $-\infty$, $\int_{I^C} |F(\omega)|^2 d\omega$ approaches 0, and

$$\lim_{\substack{\lambda \to \infty \\ \mu \to -\infty}} \int_{-\infty}^{\infty} \left| x(t)-(1/\sqrt{2\pi})\int_{\mu}^{\lambda} e^{i\omega t} F(\omega) d\omega \right|^2 dt = 0.$$

If we abuse notation slightly by agreeing that the infinite limits on the integrals are meant in the sense of $L^2(-\infty,\infty)$ convergence, then Theorems XI.3.8 and XI.3.9 can be written as

$$x(t) = (1/2\pi) \int_{-\infty}^{\infty} \left[\int_{-\infty}^{\infty} e^{-i\omega s} x(s) ds \right] e^{i\omega t} d\omega$$

and

$$\|x\|^2 = (1/2\pi) \int_{-\infty}^{\infty} |\int_{-\infty}^{\infty} e^{-i\omega s} x(s) ds|^2 d\omega.$$

In this form they preserve the appearance of the analagous formulas for ordinary Fourier series found in Example IX.5.5.

It is not too difficult to show that if D_L denotes the set of all functions x which satisfy

1. x is in $L^2(-\infty,\infty)$,

2. x' exists and is in $L^2(-\infty,\infty)$,

and if the self-adjoint operator L is defined by

$$Lx = (1/i)x'$$

for all x in D_L, then L has the following properties.

1. $\sigma(L) = (-\infty,\infty)$. Each point is in the approximate point spectrum, but not the point spectrum.

2. The spectral resolution of L is

$$Lx(t) = (1/2\pi) \int_{-\infty}^{\infty} \omega e^{i\omega t} [\int_{-\infty}^{\infty} e^{-i\omega s} x(s) ds] d\omega,$$

valid for all x in D_L.

The last item in this chapter is to show that the Fourier transform is a unitary operator on $L^2(-\infty,\infty)$.

XI.3.10. Theorem. Let the operator A be defined for all x in $L^2(-\infty,\infty)$

by

$$Ax(t) = \lim_{\alpha \to \infty} 1/\sqrt{2\pi} \int_{-\alpha}^{\alpha} e^{-its} x(s)ds,$$

where the limit is in the sense of $L^2(-\infty,\infty)$. Then A is a unitary operator

on $L^2(-\infty,\infty)$.

Proof. In view of Parseval's equality and Theorem XI.3.1 we must only show

that the range of A is dense in $L^2(-\infty,\infty)$. If the range of A were not

dense in $L^2(-\infty,\infty)$, there would exist an element y which is orthogonal to

the range. Thus

$$(F(x),y) = 0$$

for all x in $L^2(-\infty,\infty)$. Then, since

$$(F(x),y) = \int_{-\infty}^{\infty} \overline{y(\omega)} [(1/\sqrt{2\pi}) \int_{-\infty}^{\infty} e^{-i\omega t} x(t)dt]d\omega,$$

$$= \int_{-\infty}^{\infty} [(1/\sqrt{2\pi}) \int_{-\infty}^{\infty} e^{i\omega t} y(\omega)d\omega]^* x(t)dt,$$

the so called conjugate Fourier transform of y (with t replaced by -t)

is orthogonal to x. Since x is arbitrary, however, we can let x equal

the conjugate transform of y. This shows that y = θ.

It is now possible to derive the spectral resolution associated with the

Fourier transform. It is rather easy to show that the Hermite functions

$\{e^{-t^2/2}H_j(t)\}_{j=0}^{\infty}$ are all eigenfunctions of the Fourier transform with corresponding eigenvalues $\{(-i)^j\}_{j=0}^{\infty}$. That is, the eigenvalues are 1, i, -1 and -i. Since the Hermite functions are complete in $L^2(-\infty,\infty)$, for each x in $L^2(-\infty,\infty)$,

$$x(t) = \sum_{j=0}^{\infty} \frac{[\int_{-\infty}^{\infty} e^{-s^2/2}H_j(s)x(s)ds]}{2^j\ j!\ \sqrt{\pi}}\ e^{-t^2/2}H_j(t),$$

and

$$F(x)(t) = \sum_{j=0}^{\infty} (-i)^j \frac{[\int_{-\infty}^{\infty} e^{-s^2/2}H_j(s)x(s)ds]}{2^j\ j!\ \sqrt{\pi}}\ e^{-t^2/2}H_j(t).$$

XI. Exercises

1. Show that the Lebesgue measure of the rational numbers is 0.

2. Define the Cantor set in the following way. Begin with [0,1]. Then delete (1/3,2/3). From the remaining [0,1/3] and [2/3,1] delete the middle thirds (1/9,2/9) and (7/9,8/9). Continue deleting the middle third of each of the successive remaining intervals ad infinitum. The remainder is the Cantor set. Show that the Cantor set is not countable (in a 1-1 correspondence with the integers) and that the Lebesgue measure of the Cantor set is 0.

3. Let two measurable functions satisfy $x \equiv y$ if $x = y$ a.e. Show that this is an equivalence relation.

4. Give a counterexample to Theorem XI.1.13 when $p = \infty$.

5. Prove Theorem XI.2.5, part 4.

6. Show that the Weierstrass kernel is an approximate identity in $L^1(-\infty,\infty)$. Can you find others?

7. Compute $F(e^{-\alpha|t|})$. Use the result to show

$$(1/2\pi)\int_{-\infty}^{\infty} [\cos \omega t/(\omega^2-\alpha^2)]d\omega = e^{-\alpha|t|}.$$

8. Investigate the pointwise convergence of the Fourier integral. That is, when does

$$x(t) = (1/\sqrt{2\pi}) \int_{-\infty}^{\infty} e^{i\omega t}F(\omega)d\omega,$$

pointwise, where

$$F(\omega) = (1/\sqrt{2\pi}) \int_{-\infty}^{\infty} e^{-i\omega t}x(t)dt?$$

9. In the proof of Theorem XI.3.7 the Lebesgue dominated convergence theorem was quoted. What is an appropriate dominating function?

10. Examine in detail the examples of functions in $L^2(-\infty,\infty)$ but not in $L^1(-\infty,\infty)$, and those in $L^1(-\infty,\infty)$ but not in $L^2(-\infty,\infty)$. With regard to the Fourier transform, what is the significance of these examples?

11. Show that if x is in D_L (following Theorem XI.3.8), then

$$(1/i)x'(t) = (1/2\pi) \int_{-\infty}^{\infty} \omega e^{i\omega t}[\int_{-\infty}^{\infty} e^{-i\omega s}x(s)ds]d\omega$$

in the sense of $L^2(-\infty,\infty)$, and that x is in D_L if and only if

$$\|Lx\|^2 = \|(1/i)x'\|^2 = \int_{-\infty}^{\infty} \omega^2 |F(\omega)|^2 d\omega < \infty,$$

where, as usual, $F(\omega) = (1/\sqrt{2\pi}) \int_{-\infty}^{\infty} e^{-i\omega t} x(t) dt$.

12. Show that the Hermite functions $\{e^{-t^2/2} H_j(t)\}_{j=0}^{\infty}$ are eigenfunctions of the Fourier transform on $L^2(-\infty,\infty)$ with corresponding eigenvalues $\{(-i)^j\}_{j=0}^{\infty}$.

References

1. N. Dunford and J.T. Schwartz, "Linear Operators, vol. 1," Interscience, New York, 1958.

2. R.R. Goldberg, "Fourier Transforms," Cambridge University Press, Cambridge, 1961.

3. E. Hewitt and K. Stromberg, "Real and Abstract Analysis," Springer-Verlag, New York, 1965

4. E.J. McShane and T.A. Botts, "Real Analysis," Van Nostrand, New York, 1959

5. I.P. Natanson, "Theory of Functions of a Real Variable, vols. 1 and 2," Frederick Ungar, New York, 1955 and 1960.

6. H.L. Royden, "Real Analysis," MacMillan, New York, 1963.

7. E.C. Titchmarsh, "The Theory of Functions," Oxford University Press, Oxford, 1939.

XII.   THE SINGULAR STURM-LIOUVILLE PROBLEM

The Sturm-Liouville problem is singular when the results of Chapter IX are not applicable. This may happen when the coefficients p, q or w become infinite, when p or w approaches 0 at α or β, or when the interval $[\alpha, \beta]$ is infinite in length.

In order to make this more apparent, we make the following transformations: We replace the dependent variable x by y using

$$x = y/(pw)^{1/4}.$$

The independent variable t is replaced by s using

$$s = \int_{\gamma}^{t} (w/p)^{1/2} dt,$$

where γ is in (α, β). The differential equation

$$(px')'+qx = \lambda w x$$

is then transformed into

$$y''+Qy = \lambda y,$$

where

$$Q = \frac{q}{w} - \frac{1}{2}[wp]^{-1/4}([(wp)^{1/2}]'/[wp]^{1/4})'.$$

The interval $[\alpha,\beta]$ is transformed into the interval

$$[a,b] = [-\int_\alpha^\gamma (w/p)^{1/2} dt, \int_\gamma^\beta (w/p)^{1/2} dt].$$

Further, if two elements in the Hilbert space $L^2(w)$, y and z, are transformed into

$$f = (pw)^{1/4} y$$

and

$$g = (pw)^{1/4} z,$$

then the inner product in $L^2(w)$ becomes

$$\int_\alpha^\beta \overline{z}(t) y(t) w(t) dt = \int_a^b \overline{g}(s) f(s) (p(t) w(t))^{-1/2} w(t) (p(t)/w(t))^{1/2} ds,$$

$$= \int_a^b \overline{g}(s) f(s) ds,$$

which generates a Hilbert space with unit weight function.

In its transformed state the differential equation is said to be in Liouville normal form. The derivation of the transformation is a tedious chore and is left to the reader as an exercise. We note for the transformation to be valid, p and w must be in $C^2[\alpha,\beta]$ and nowhere zero in (α,β). This is more than was required before.

In terms of the transformed differential equation and interval, the Sturm-Liouville problem is singular if the transformed interval, now [a,b], is infinite in length, or failing that, if Q becomes infinite at its end points. In terms of p, q, w and [α,β] this may occur in a number of ways.

The theory in the singular case may be much more complicated than the results in Chapter IX. Eigenfunctions may fail to exist. The spectrum of the operator may become spread over portions or all of the real line. (The Fourier sine and cosine transforms illustrate this possibility.) Or, as in the regular case, both eigenfunctions and eigenvalues may exclusively exist (as the Legendre, Laguerre, and Hermite expansions will show).

We shall examine both possibilities. To do so, however, we shall need some additional tools. Hence the first two sections of this chapter are devoted to a property of circles under bilinear transformations in the complex plane and to the Helly convergence theorems. Following these, we begin the discussion of the singular Sturm-Liouville problem in earnest.

XII.1. <u>Circles under Bilinear Transformations</u>. The results in this section will be needed in Section XII.3 in which the real line in the complex plane is transformed by a bilinear transformation. We wish to show that the result is a circle. In what follows $z(=x+iy)$, p, q, w, a, b, c, d are complex valued, while ρ, r, k and the Greek letters θ and φ are real valued.

XII.1.1. <u>Definition</u>. We mean by the word Circle (capitalized) in the complex plane an ordinary circle or a straight line. A straight line has infinite radius and center at ∞.

This definition is made necessary by certain ambiguity which will arise shortly.

XII.1.2. Definition. Let $|z-z_0| = \rho$ be a circle with center z_0 and radius ρ. Two points p and q are inverse with respect to the circle if they are collinear with the center z_0, on the same side of it, and if the product of their distances from z_0 is equal to ρ^2.

Therefore, p and q are inverse points if

$$p = z_0 + re^{i\phi},$$

and

$$q = z_0 + \frac{\rho^2}{r} e^{i\phi}.$$

XII.1.3. Definition. Two points p and q are inverse with respect to a straight line L if they are equidistant from L, lie on opposite sides of it, and the line through p and q is perpendicular to L.

The following theorem unites the two concepts of inverse points.

XII.1.4. Theorem. Two points p and q are inverse with respect to a Circle C if and only if

$$\left|\frac{z-p}{z-q}\right| = k$$

for all points z on C. If k ≠ 1, C is an ordinary circle. If k = 1,

C is a straight line.

Proof. If p and q are inverse with respect to an ordinary circle with

center z_0 and radius ρ, then for each point z on the circle

$$z-z_0 = \rho e^{i\theta},$$

and

$$\left|\frac{z-p}{z-q}\right| = \frac{\left|\rho e^{i\theta} - re^{i\phi}\right|}{\left|\rho e^{i\theta} - \frac{\rho^2}{r}e^{i\phi}\right|},$$

$$= \frac{r}{\rho}\left|\frac{\rho e^{i\theta} - re^{i\phi}}{re^{i\theta} - \rho e^{i\phi}}\right|,$$

$$= \frac{r}{\rho}.$$

If p and q are inverse with respect to a straight line, then for

each point z on the line

$$\left|\frac{z-p}{z-q}\right| = 1.$$

Conversely, if

$$\left|\frac{z-p}{z-q}\right| = k,$$

then

$$|z-p|^2 = k^2|z-q|^2,$$

which is equivalent to

$$|z|^2 - 2\ \mathrm{Re}(\bar{p}z) + |p|^2 = k^2[|z|^2 - 2\ \mathrm{Re}(\bar{q}z) + |q|^2].$$

When $k \neq 1$,

$$|z|^2 - \frac{2\ \mathrm{Re}[(\bar{p}-k^2\bar{q})z]}{1-k^2} + \frac{|p|^2 - k^2|q|^2}{1-k^2} = 0,$$

and

$$\left|z - \frac{p-k^2q}{1-k^2}\right|^2 = \frac{|p-k^2q|^2}{(1-k^2)^2} - \frac{|p|^2-k^2|q|^2}{1-k^2},$$

$$= \frac{k^2|p-q|^2}{|1-k^2|^2}.$$

Thus if

$$z_0 = (p-k^2q)/(1-k^2),$$

$$\rho = k|p-q|/|1-k^2|,$$

then $|z-z_0| = \rho$. Further

$$p-z_0 = k^2(q-p)/(1-k^2),$$

and

$$q-z_0 = (q-p)/(1-k^2),$$

so $|p-z_0||q-z_0| = \rho^2$, and p and q are inverse points.

If $k = 1$, z is equidistant from both p and q. The locus of all such points is a straight line.

XII.1.5. <u>Definition</u>. <u>Let a,b,c,d be complex numbers satisfying $ad-bc \neq 0$.</u>
<u>The transformation</u>

$$w = (az+b)/(cz+d)$$

<u>of the complex plane into itself is called a bilinear transformation.</u>

If $ad-bc = 0$, of course, w is constant. When $ad-bc \neq 0$, the transformation is 1 to 1. The point $z = -d/c$ is transformed into ∞. $z = \infty$ corresponds to $w = a/c$. Other points are in a 1 to 1 correspondence through the inverse transformation

$$z = -(dw-b)/(cw-a),$$

which is also bilinear.

XII.1.6. Theorem. A bilinear transformation transforms Circles into Circle

and inverse points into inverse points.

Proof. We let any Circle in question be represented by the equation

$$\left|\frac{z-p}{z-q}\right| = k,$$

where p and q are inverse points, and the bilinear transformation in

question be

$$w = (az+b)/(cz+d).$$

Then

$$z = -(dw-b)/(cw-a).$$

Inserting this into the equation for the Circle, we find

$$\left|\frac{-(dw-b)/(cw-a)-p}{-(dw-b)/(cw-a)-q}\right| = k,$$

or

$$\left|\frac{dw-b+p(cw-a)}{dw-b+q(cw-a)}\right| = k.$$

If we divide the numerator by cp+d and the denominator by cq+d, this becomes

$$\frac{\left|w-\dfrac{ap+b}{cp+d}\right|}{\left|w-\dfrac{aq+b}{cq+d}\right|} = k\left|\frac{cq+d}{cp+d}\right| \; ,$$

from which the results easily follow.

XII.2. **Helly's Convergence Theorems.** The technique used in the discussion
of the singular Sturm-Liouville problem is to first consider the problem over
a subinterval, where the problem is regular, and then to permit the interval
to swell, approaching the original interval. As this occurs, certain limit-
ing processes take place. In order to make them precise, we need certain
convergence theorems attributed to Helly.

Our goal is to discuss the convergence of integrals of the form
$\int_a^b f(t)dF_n(t)$ as n approaches ∞. We first introduce the concept of a weak
(pointwise) limit.

XII.2.1. **Definition.** Let F, F_n, n=1,..., be piecewise continuous functions
over the interval [a,b]. Then the sequence $\{F_n\}_{n=1}^{\infty}$ converges weakly to F
if $\lim_{n\to\infty} F_n(t) = F(t)$ at all points of continuity t of F. We write

$$\mathrm{Lim}_{n\to\infty} F_n = F.$$

We point out that this is not the same as pointwise convergence.

XII.2.2. <u>Lemma</u>. Let $\{F_n\}_{n=1}^{\infty}$ be a sequence of nondecreasing functions defined on the real line R. Let D be dense in R, and suppose that $F_n(t)$
converges to F(t) for all t in D. Then

$$\lim_{n\to\infty} F_n = F$$

over R.

Proof. Let t be arbitrary and let t', t" be in D and satisfy
t' ≦ t ≲ t". Then

$$F_n(t') \lessapprox F_n(t) \lessapprox F_n(t").$$

Letting n approach ∞, we find

$$F(t') \leqq \lim\inf F_n(t) \lessapprox \lim\sup F(t) \lessapprox F(t").$$

Then since D is dense in R, we can let t' and t⁻ approach t. In
so doing we have

$$F(t-) \leqq \lim\inf F_n(t) \leqq \lim\sup F_n(t) \leqq F(t+).$$

If t is a point of continuity of F, all four quantities coincide.

XII.2.3. <u>Theorem (Helly's First Theorem)</u>. <u>Every sequence of nondecreasing</u>
<u>functions</u> $\{F_n\}_{n=1}^{\infty}$, <u>which is uniformly bounded on any compact subset of the</u>

real line R, contains a subsequence $\{F_k\}_{k=1}^{\infty}$ which converges weakly to a
bounded nondecreasing function F.

Proof. Let $r_1, r_2, \ldots r_n$ be a sequence of rational numbers which is dense in
R, and consider $\{F_n(r_1)\}_{n=1}^{\infty}$. Since this set is bounded, it contains a convergent subsequence, which we denote by $\{F_{1n}(r_1)\}_{n=1}^{\infty}$.

At r_2, the sequence $\{F_{1n}(r_2)\}_{n=1}^{\infty}$ is bounded, so an additional convergent subsequence $\{F_{2n}(r_2)\}_{n=1}^{\infty}$ can be extracted. We continue this process
$r_3, r_4, \ldots$, each time extracting a convergent subsequence.

Now consider $\{F_{nn}\}_{n=1}^{\infty}$. This set consists of nondecreasing functions
and converges to a limit ϕ at each rational point $r_k, k=1, \ldots$. That is,

$$\lim_{n \to \infty} F_{nn}(r_k) = \phi(r_k),$$

$k=1, \ldots$. Now define F by

$$F(t) = \inf_{r_k \geq t} \phi(r_k).$$

By Lemma XII.2.2,

$$\lim_{n \to \infty} F_{nn} = F.$$

We are now in a position to prove Helly's second convergence theorem
and an extension. The extension will be crucial to the derivation of a
spectral resolution for the singular Sturm-Liouville problem.

XII.2.4. <u>Theorem (Helly's Second Theorem)</u>. Let f be continuous on a finite

interval [a,b]. Let $\{F_n\}_{n=1}^{\infty}$ be a sequence of nondecreasing functions

which are uniformly bounded on compact subsets of [a,b] and which converge

weakly to F.   Then

$$\lim_{n\to\infty} \int_a^b f(t)dF_n(t) = \int_a^b f(t)dF(t).$$

Proof. Since f is continuous, for any $\varepsilon > 0$, we can construct a partition

of [a,b],

$$a = t_0 < t_1 < \ldots < t_{N-1} < t_N = b,$$

such that $|f(t)-f(t_j)| < \varepsilon$ when $t_j \leq t \leq t_{j+1}$, $j=0,\ldots,N-1$.

Next we note that for each point of discontinuity of F, c, there

corresponds an interval $(F(c-),F(c+))$ which is disjoint from the remainder

of the range of F, since F is nondecreasing, and which contains a

rational point. Since the rational numbers are countable, so are these

intervals, and so are the points of discontinuity of F. By avoiding this

countable set we may require that the partition points are points of

continuity of F.

Thus $|F_n(t_j)-F(t_j)|$ can be made arbitrarily small if n is sufficiently

large.

Let $M = \sup_{t \in [a,b]} |f(t)|$. Then if k is sufficiently large,

$$|F_k(t_j)-F(t_j)| < \varepsilon/MN.$$

Further, if $f_\varepsilon(t) = f(t_j)$, $t_j \leq t < t_{j+1}$, for $j = 0, \ldots, N-1$, then

$$|f(t) - f_\varepsilon(t)| < \varepsilon$$

for all t in $[a,b]$. Thus

$$\left| \int_a^b f(t) dF(t) - \int_a^b f(t) dF_n(t) \right| \leq$$

$$\left| \int_a^b f(t) dF(t) - \int_a^b f_\varepsilon(t) dF(t) \right|$$

$$+ \left| \int_a^b f_\varepsilon(t) dF(t) - \int_a^b f_\varepsilon(t) dF_n(t) \right|$$

$$+ \left| \int_a^b f_\varepsilon(t) dF_n(t) - \int_a^b f(t) dF_n(t) \right|.$$

On the right side of this inequality the first term is less than $\varepsilon \int_a^b dF(t) \leq c_1 \varepsilon$. The third is less than $\varepsilon \int_a^b dF_n(t)$, which is likewise less than $c_2 \varepsilon$. The second term is

$$\left| \sum_{j=0}^{N-1} f(t_j)[F(t_{j+1}) - F(t_j)] - \sum_{j=0}^{N-1} f(t_j)[F_n(t_{j+1}) - F_n(t_j)] \right|$$

$$\leq N[M\varepsilon/MN + M\varepsilon/MN] = 2\varepsilon.$$

Thus the left side is less than $(2 + c_1 + c_2)\varepsilon$ when n is sufficiently large.

XII.2.5.   Theorem (Helly's Second Theorem Extended).   Let f be continuous and bounded on $(-\infty, \infty)$. Let $\{F_n\}_{n=1}^{\infty}$ be a sequence of nondecreasing function which are uniformly bounded on compact subsets of $(-\infty, \infty)$ and which converge weakly to F. Suppose further that $\lim_{n\to\infty} F_n(-\infty) = F(-\infty)$ and $\lim_{n\to\infty} F_n(\infty) = F(\infty)$ are finite. Then

$$\lim_{n\to\infty} \int_{-\infty}^{\infty} f(t)dF_n(t) = \int_{-\infty}^{\infty} f(t)dF(t).$$

Proof. Let $-\infty < a < b < \infty$, and

$$J_1 = \left| \int_{-\infty}^{a} f(t)dF_n(t) - \int_{-\infty}^{a} f(t)dF(t) \right|,$$

$$J_2 = \left| \int_{a}^{b} f(t)dF_n(t) - \int_{a}^{b} f(t)dF(t) \right|,$$

$$J_3 = \left| \int_{b}^{\infty} f(t)dF_n(t) - \int_{b}^{\infty} f(t)dF(t) \right|.$$

The difference between $\int_{-\infty}^{\infty} f(t)dF_n(t)$ and $\int_{-\infty}^{\infty} f(t)dF(t)$ is less than $J_1+J_2+J_3$. Since f is bounded, we can choose a sufficiently close to $-\infty$, and b sufficiently close to ∞ to guarantee that $J_1 < \varepsilon$ and $J_3 < \varepsilon$ for some preassigned $\varepsilon > 0$. By Theorem XII.2.4, $J_2 < c\varepsilon$ if n is sufficiently large. Thus if n is sufficiently large

$$\left| \int_{-\infty}^{\infty} f(t)dF_n(t) - \int_{-\infty}^{\infty} f(t)dF(t) \right| < (2+c)\,\varepsilon.$$

XII.3. Limit Points and Limit Circles. Let us assume that the singular Sturm-Liouville problem has been put into Liouville normal form, so the differential operator in question has the form

$$Ly = y'' + Qy,$$

while the interval [a,b] may be finite or infinite. The Hilbert space in question is $L^2(a,b)$, which is generated by the inner product

$$(x,y) = \int_a^b \overline{y}(t)x(t)dt,$$

and norm

$$\|x\| = \int_a^b |x(t)|^2 dt.$$

In the discussion of the regular Sturm-Liouville problem everything under consideration was automatically well defined. Every solution of $Lx = \lambda x$ was in the Hilbert space. Further, when the Green's function existed, integrals involving it were always well defined.

This is not necessarily so when the problem is singular. Our immediate problem is to discover which solutions, if any, are in $L^2(a,b)$. Finding these, and when they exist, will give us the tools to continue.

Let us recall Green's formula from Theorem V.2.5. If t_1 and t_2 are points in (a,b) and x and y are in $C^2[t_1,t_2]$, then

$$\int_{t_1}^{t_2} [Lx \cdot \overline{y} - x \cdot \overline{Ly}]dt = [x,\overline{y}](t_2) - [x,\overline{y}](t_1),$$

where

$$[x,\overline{y}] = x'\overline{y} - x\overline{y}',$$

$$= -W[x,\overline{y}].$$

If x and y are solutions to $Lx = \lambda x$ and $Ly = \lambda y$, then $W[x,y]$ is constant.

We now choose an arbitrary point s in [a,b], and for the moment consider the interval [s,b]. After achieving results over the subspace $L^2(s,b)$ we shall turn our attention to its complement $L^2(a,s)$. Finally we shall put them together in the Cartesian product space

$$L^2(a,b) = L^2(a,s) \times L^2(s,b).$$

XII.3.1. Theorem. Let $\lambda = \mu+i\nu$ be a complex number. For all λ with nonzero imaginary part ν there exists a solution χ of $Lx = \lambda x$ which is in $L^2(s,b)$.

Proof. Let x_0 and y_0 be the solutions of $Lx = \lambda x$ which satisfy

$$x_0(s) = \cos \theta, \qquad x_0'(s) = - \sin \theta,$$

$$y_0(s) = - \sin \theta, \qquad y_0'(s) = - \cos \theta,$$

where $0 \leqq \theta < \pi$. These solutions are independent, and $W[x_0,y_0] = -1$. Let lie between 0 and π, and let

$$\chi = x_0 + My_0$$

be that solution of $Lx = \lambda x$ which satisfies

$$\cos \phi \, \chi(b') - \sin \phi \, \chi'(b') = 0,$$

where b' is in (s,b). M is therefore determined, and is given by

$$M = -\frac{\cot \phi \, x_0(b') - x_0'(b')}{\cot \phi \, y_0(b') - y_0'(b')} \; .$$

Let $\cot \phi = z$. Then as ϕ varies between 0 and π, z varies over the entire real line in the complex plane. Further, according to Theorem XII.1.6, as z varies over the real line, its image under

$$M = -\frac{x_0(b')z - x_0'(b')}{y_0(b')z - y_0'(b')}$$

is a Circle $C_{b'}$. We see, therefore, that χ satisfies a real boundary condition at b' if and only if M is on $C_{b'}$.

Now $C_{b'}$ is the image of the real axis in the z-plane. It corresponds to the equation $\text{Im}(z) = 0$. Using the inverse transformation, this is equivalent to

$$(\overline{x}_0(b') + \overline{My}_0(b'))(x_0'(b') + My_0'(b'))$$

$$- (x_0(b') + My_0(b'))(\overline{x}_0'(b') + \overline{My}_0'(b')) = 0,$$

which is a circle with center

$$\widetilde{M}_{b'} = -W[x_0,\overline{y}_0](b')/W[y_0,\overline{y}_0](b')$$

and radius

$$r_{b'} = 1/|W[y_0,\overline{y}_0](b')|.$$

If we divide the circle equation above by $-W[y,\overline{y}]$, the coefficient of $|M|^2$, it can be written with the coefficient of $|M|^2$ as 1 in the form

$$W[\chi,\overline{\chi}](b')/W[y_0,\overline{y}_0](b') = 0,$$

where $\chi = x_0 + My_0$. If M is replaced by the point at the center $\widetilde{M}_{b'}$, the left side becomes $-|W[x_0,y_0](b')|^2$, which is negative. We therefore conclude that, in general, M is in the interior of $C_{b'}$ if and only if

$$W[\chi,\overline{\chi}](b')/W[y_0,\overline{y}_0](b') < 0.$$

Now Green's formula yields

$$-W[y_0,y_0](b') + W[y_0,y_0](s) = 2i\nu \int_s^{b'} |y_0(t)|^2 dt.$$

Since $W[y_0, \bar{y}_0](s) = 0$,

$$W[y_0, \bar{y}_0](b') = -2i\nu \int_s^{b'} |y_0(t)|^2 dt,$$

and

$$r_{b'} = 1/2 |\nu| \int_s^{b'} |y_0(t)|^2 dt.$$

Further, Green's formula shows

$$-W[\chi, \bar{\chi}](b') + W[\chi, \bar{\chi}](s) = 2i\nu \int_s^{b'} |\chi(t)|^2 dt.$$

Since $W[\chi, \bar{\chi}](s) = 2i \, \text{Im}(M)$, we see that M is on $C_{b'}$ if and only if

$$\int_s^{b'} |\chi(t)|^2 dt = \text{Im}(M)/\nu;$$

M is in the interior of C_b if and only if

$$\frac{W[\chi\bar{\chi}](b')}{w[\bar{y}\bar{y}](b')} = \frac{\text{Im}(M) - \nu \int_s^{b'} |\chi(t)|^2 dt}{-\nu \int_s^{b'} |y_0(t)|^2 dt} < 0,$$

or

$$\int_s^{b'} |\chi(t)|^2 dt < \text{Im}(M)/\nu.$$

Finally, if $b' < b'' < b$, and M is in $C_{b''}$, then the inequality

$$\int\limits_{s}^{b'} |X(t)|^2 dt < \int\limits_{s}^{b''} |X(t)|^2 dt < \text{Im}(M)/\nu$$

implies that M is also in $C_{b''}$. Thus $C_{b''} \subset C_{b'}$. As b' increases the circles $C_{b'}$ are nested, each lying inside all those preceding it.

Therefore, as b' approaches b, the circles $C_{b'}$ contract, con-verging either to a limit circle, C_b, or to a limit point, M_b. If the limit is a circle C_b, then $r_b > 0$, which implies

$$\int\limits_{s}^{b} |y_0(t)|^2 dt < \infty,$$

or y_0 is in $L^2(s,b)$. If M is in C_b, then

$$\int\limits_{s}^{b} |X(t)|^2 dt < \text{Im}(M)/\nu,$$

so X is also in $L^2(s,b)$. Since $L^2(s,b)$ is a linear space, this implies that every solution of $Lx = \lambda x$ is in $L^2(s,b)$.

If the limit is a point, then $r_b = 0$. This implies

$$\int\limits_{s}^{b} |y_0(t)|^2 dt = \infty,$$

and y_0 is not in $L^2(s,b)$. Since, when $M = M_b$, $X = x_0 + My_0$ is in $L^2(s,b)$, we conclude that x_0 is not in $L^2(s,b)$, and that only multiples of X are in $L^2(s,b)$.

It is natural to wonder if the number of solutions in $L^2(s,b)$ varies as λ varies. This is, in a sense, not the case.

XII.3.2. Theorem. If for some λ_0 the equation $Lx = \lambda_0 x$ possesses two solutions in $L^2(s,b)$, then the equation $Lx = \lambda x$ possesses two solutions in $L^2(s,b)$ for all values of λ.

Proof. Let z satisfy $Lz = \lambda z$, and let x_0 and y_0 be solutions of $Lx = \lambda_0 x$ which lie in $L^2(s,b)$. Then, since $Lz = \lambda_0 z + (\lambda - \lambda_0)z$, variation of parameters yields

$$z(t) = c_1 x_0(t) + c_2 y_0(t) - (\lambda - \lambda_0) \int_s^t \frac{[x_0(t)y_0(\tau) - x_0(\tau)y_0(t)]}{W[x_0,y_0]} z(\tau) dz .$$

Using Schwarz's inequality,

$$\left| \int_s^t \frac{[x_0(t)y_0(\tau) - x_0(\tau)y_0(t)]}{W[x_0,y_0]} z(\tau) d\tau \right| \leq K[|x_0(t)| + |y_0(t)|][\int_s^t |z(\tau)|^2 d\tau]^{1/2}$$

where K is chosen such that

$$[\int_s^t x_0(\tau)^2 d\tau]^{1/2} < K \, W[x_0,y_0] ,$$

and

$$[\int_s^t |y_0(\tau)|^2 d\tau]^{1/2} < K|W[x_0,y_0]| .$$

Then

$$[\int_s^t |z(t)|^2 dt]^{1/2} \leq K|W[x_0,y_0]|[|c_1| + |c_2|]$$

$$+ 2|\lambda - \lambda_0|K^2|W[x_0,y_0]|^2[\int_s^t |z(t)|^2 dt]^{1/2}.$$

If s_0 is chosen such that

$$|\lambda-\lambda_0|K^2 |W[x_0,y_0]|^2 < 1/4,$$

then

$$[\int_{s_0}^{t} |z(t)|^2 dt]^{1/2} < 2K|W[x_0,y_0]|[|c_1| + |c_2|].$$

We then let t approach b.

We remark that the integral $\int_{s}^{s_0} |z(t)|^2 dt$ is not important, since it will automatically be finite. Only the integral near the end point b is critical.

Finally, because of the geometry involved in the proof of Theorem XII.3 the following definition is natural.

XII.3.3. <u>Definition</u>. <u>If for some λ the Sturm-Liouville equation $Lx = \lambda x$</u> <u>possesses two solutions in $L^2(s,b)$, then the singular Sturm-Liouville</u> <u>problem is in the limit circle case at b. If for some λ the Sturm-Liouvi</u> <u>equation $Lx = x$ possesses no solution or only one solution in $L^2(s,b)$,</u> <u>then the singular Sturm-Liouville problem is in the limit point case at b.</u>

We shall use these names to denote the two possible cases in the future.

We remark that Theorems XII.3.1 and XII.3.2 are also applicable to the interval [a,s] with only minor changes. The results are similar.

XII.4.   The Limit Point Case.   This section is devoted to the case where the
left end point   a   is regular, while the right end   b   is in the limit point
case.

We let   s = a,   let   b′   be in (a,b), and for the moment consider the
regular Sturm-Liouville problem on the interval   [a,b′]:

$$Lx = x'' + Qx,$$

$$\cos \theta \, x(a) - \sin \theta \, x'(a) = 0,$$

$$\cos \phi \, x(b') - \sin \phi \, x(b') = 0,$$

where $0 \leqq \theta < \pi$, and $0 \leqq \phi < \pi$.   If x_0 and y_0 represent those   solutions
satisfying the boundary conditions at   a   as stated in the proof of Theorem XII.
then since there exists a sequence of eigenvalues $\{\lambda_{b'n}\}_{n=1}^{\infty}$;   since y_0
satisfies the first boundary condition , the corresponding normalized eigen-
functions have the form $r_{b'n} y_0(t, \lambda_{b'n})$,   where $r_{b'n}$ is the normalizing factor.
Parseval's equality (Chapter VII, problem 14) then has the form

$$\int_a^{b'} |f(t)|^2 dt = \sum_{n=1}^{\infty} |r_{b'n}|^2 |\int_a^{b'} f(t) y_0(t, \lambda_{b'n}) dt|^2.$$

This can be put in a somewhat different form:

XII.4.1.   Definition.   We denote the n-th generalized Fourier coefficient of
an element   f   in $L^2(a,b')$ by

$$g(\lambda) = \int_a^{b'} f(t) y_0(t,\lambda) dt$$

XII.4.2. Definition. The spectral measure for the regular Sturm-Liouville
problem on (a,b') is a monotone nondecreasing function (of λ), $\rho_{b'}$, in-
creasing $|r_{b'n}|^2$ at $\lambda_{b'n}$, which satisfies in addition

$$\rho_{b'}(\lambda+0) = \rho_{b'}(\lambda),$$

$$\rho_{b'}(0) = 0.$$

With this notation, Parseval's equality becomes

$$\int_a^{b'} |f(t)|^2 dt = \int_{-\infty}^{\infty} |g(\lambda)|^2 d\rho_{b'}(\lambda).$$

One of our major goals is to evaluate this formula as b' approaches b.

XII.4.3. Theorem. There exists a monotone nondecreasing function ρ, defined on $(-\infty,\infty)$, such that

$$\rho(\lambda) - \rho(\mu) = \lim_{b' \to b} [\rho_{b'}(\lambda) - \rho_{b'}(\mu)].$$

Proof. Let $M_{b'}$ be the point on $C_{b'}$ generating the regular Sturm-Liouvill
problem, and let

$$\chi_{b'} = x_0 + M_{b'} y_0$$

satisfy $LX_{b'} = \lambda_0 X_{b'}$.

Applying Parseval's equality to $X_{b'}$, we find

$$\int_a^{b'} |X_{b'}(t)|^2 dt = \sum_{n=1}^{\infty} |r_{b'n}|^2 |\int_a^{b'} X_{b'}(t)y_0(t,\lambda_{b'n})dt|^2.$$

Since $LX_{b'} = \lambda_0 X_{b'}$ and $Ly_0(t,\lambda_{b'n}) = \lambda_{b'n} y_0(t,\lambda_{b'n})$, Green's formula yields

$$(\lambda_0 - \lambda_{b'n})\int_a^{b'} X_{b'}(t)y_0(t,\lambda_{b'n})dt$$

$$= [X_{b'},y_{b'n}](b') - [X_{b'},y_{b'n}](a),$$

where we have denoted $y_0(t,\lambda_{b'n})$ by $y_{b'n}$. At b' both satisfy the same boundary condition, so $[X_{b'},y_{b'n}](b') = 0$. At $a, [X_{b'},y_{b'n}] = 1$, so

$$\int_a^{b'} X_{b'}(t)y_0(t,\lambda_{b'n})dt = 1/(\lambda_{b'n}-\lambda_0).$$

Substituting this into Parseval's equality above,

$$\int_a^{b'} |X_{b'}(t)|^2 dt = \int_{-\infty}^{\infty} \frac{d\rho_{b'}(\lambda)}{|\lambda-\lambda_0|^2}$$

Now, since $M_{b'}$ is on $C_{b'}$,

$$\int_a^{b'} |X_{b'}(t)|^2 dt = \text{Im}(M_{b'})/\text{Im}(\lambda_0).$$

We let $\lambda_0 = i$. Since the circles $C_{b'}$ contract as b' increases, there exists a constant K such that

$$\int_{-\infty}^{\infty} \frac{d\rho_b,(\lambda)}{\lambda^2 + 1} < K,$$

or, if $\mu > 0$, such that

$$\int_{-\mu}^{\mu} d\rho_b,(\lambda) < K[1+\mu^2].$$

Since $\rho_b,(0) = 0$, this further guarantees that

$$\rho_b,(\lambda) < K[1+\lambda^2],$$

and $\rho_b,$ is uniformly bounded over compact intervals. By Theorem XII.2.3, Helly's First Theorem, there exists a subsequence of $\rho_b,(\lambda)$ which converges weakly to a bounded nondecreasing function $\rho(\lambda)$. It is clear that $\rho(\lambda)$ also satisfies

$$\rho(\lambda) < K[1+\lambda^2].$$

XII.4.4. Theorem. If f is in $L^2(a,b)$, then there exists a function g in $L^2(\rho)$, the Hilbert space defined over the interval $(-\infty,\infty)$ generated by the inner product

$$(g,h) = \int_{-\infty}^{\infty} g(\lambda)\overline{h}(\lambda)d\rho(\lambda)$$

and norm

$$\|g\| = [\int_{-\infty}^{\infty} |g(\lambda)|^2 d\rho(\lambda)]^{1/2},$$

such that

$$\lim_{b'\to b} \int_{-\infty}^{\infty} \left| g(\lambda) - \int_{a}^{b'} f(t)y_0(t,\lambda)dt \right|^2 d\rho(\lambda) = 0,$$

and (Parseval's equality)

$$\int_{a}^{b} |f(t)|^2 dt = \int_{-\infty}^{\infty} |g(\lambda)|^2 d\rho(\lambda).$$

Proof. We need to borrow a fact from the theory of Lebesgue integration. The functions which are twice continuously differentiable and which vanish near a and b are dense in $L^2(a,b)$. That is, for any f in $L^2(a,b)$, there exists a sequence of elements $\{f_n\}_{n=1}^{\infty}$, twice continuously differentiable and vanishing near a and b such that $\|f_n-f\|$ approaches 0 as n approaches ∞. We refer the reader to Theorem XI.1.13, although we do not prove it there either.

Let $C_0^2(a,b)$ stand for those functions which are twice continuously differentiable and vanish near a and b. If f is in $C_0^2(a,b)$. Then Parseval's equality applied to Lf yields

$$\int_{a}^{b} |Lf(t)|^2 dt = \int_{-\infty}^{\infty} \left| \int_{a}^{b} Lf(t)y_0(t,\lambda)dt \right|^2 d\rho_{b'}(\lambda).$$

Green's formula applied to f and y_0 shows

$$\int_{a}^{b} Lf(t)y_0(t,\lambda)dt = \int_{a}^{b} f(t)Ly_0(t,\lambda)dt,$$

$$= \lambda \int_{a}^{b} f(t)y_0(t,\lambda)dt,$$

$$= \lambda g(\lambda),$$

where

$$g(\lambda) = \int_a^b f(t) y_0(t,\lambda) dt.$$

Thus

$$\int_a^b |Lf(t)|^2 dt = \int_{-\infty}^{\infty} \lambda^2 |g(\lambda)|^2 d\rho_b,(\lambda).$$

Now, if $A > 0$,

$$\int_{-\infty}^{-A} |g(\lambda)|^2 d\rho_b,(\lambda) + \int_A^{\infty} |g(\lambda)|^2 d\rho_b,(\lambda)$$

$$\leq (1/A^2)[\int_{-\infty}^{-A} \lambda^2 |g(\lambda)|^2 d\rho_b,(\lambda) + \int_A^{\infty} \lambda^2 |g(\lambda)|^2 d\rho_b,(\lambda)],$$

$$\leq (1/A^2) \int_{-\infty}^{\infty} \lambda^2 |g(\lambda)|^2 d\rho_b,(\lambda),$$

$$= (1/A^2) \int_a^b |Lf(t)|^2 dt.$$

Applying this to Parseval's equality for f,

$$\int_a^b |f(t)|^2 dt = \int_{-\infty}^{-A} |g(\lambda)|^2 d\rho_b,(\lambda)$$

$$+ \int_{-A}^{A} |g(\lambda)|^2 d\rho_b,(\lambda)$$

$$+ \int_A^{\infty} |g(\lambda)|^2 d\rho_b,(\lambda),$$

and

$$\left| \int_a^b |f(t)|^2 dt - \int_{-A}^A |g(\lambda)|^2 d\rho_{b'}(\lambda) \right| \leq (1/A^2) \int_a^b |Lf(t)|^2 dt.$$

If we let b′ approach b, Helly's second theorem shows

$$\left| \int_a^b |f(t)|^2 dt - \int_{-A}^A |g(\lambda)|^2 d\rho(\lambda) \right| \leq (1/A^2) \int_a^b |Lf(t)|^2 dt.$$

Letting A approach ∞, we find

$$\int_a^b |f(t)|^2 dt = \int_{-\infty}^\infty |g(\lambda)|^2 d\rho(\lambda).$$

Now if f vanishes near b but is otherwise arbitrary in $L^2(a,b)$, we choose a sequence of elements $\{f_n\}_{n=1}^\infty$ in $C_0^2(a,b)$ such that

$$\lim_{n\to\infty} \int_a^b |f_n(t) - f_m(t)|^2 dt = 0.$$

Then, applying Parseval's equality to $f_n - f_m$,

$$\int_a^b |f_n(t) - f_m(t)|^2 dt = \int_{-\infty}^\infty |g_n(\lambda) - g_m(\lambda)|^2 d\rho(\lambda),$$

where

$$g_m(\lambda) = \int_a^b f_m(t) y_0(t,\lambda) dt.$$

Since the sequence $\{f_n\}_{n=1}^{\infty}$ converges in $L^2(a,b)$ to f, the sequence $\{g_n\}_{n=1}^{\infty}$ is a Cauchy sequence in $L^2(\rho)$. Thus there exists an element $g(\lambda)$ in $L^2(\rho)$ which is its limit. Since

$$\|g(\lambda) - \int_a^b f(t)y(t,\lambda)dt\| \leq \|g(\lambda) - g_n(\lambda)\|$$

$$+ \|\int_a^{b'} [f_n(t)-f(t)]y_0(t,\lambda)dt\|,$$

$$\leq \|g(\lambda) - g_n(\lambda)\|$$

$$+ [\int_a^{b'} |f_n(t)-f(t)|^2 dt]^{1/2}[\int_a^{b'} |y_0(t,\lambda)|^2 dt]^{1/2},$$

where b' is chosen sufficiently close to b so f vanishes when t is in $[b',b]$, we see that

$$g(\lambda) = \int_a^b f(t)y_0(t,\lambda)dt.$$

a.e. Then

$$\int_a^b |f(t)|^2 dt = \lim_{n\to\infty} \int_a^b |f_n(t)|^2 dt$$

$$= \lim_{n\to\infty} \int_{-\infty}^{\infty} |g_n(\lambda)|^2 d\rho(\lambda)$$

$$= \int_{-\infty}^{\infty} |g(\lambda)|^2 d\rho(\lambda).$$

Finally if f is completely arbitrary in $L^2(a,b)$, we let

$$
f_{b'}(t) = \begin{cases} f(t), & \text{if } t \in [a,b'], \\ \\ 0, & \text{if } t \in [b',b], \end{cases}
$$

and

$$
g_{b'}(\lambda) = \int_a^b f_{b'}(t)y_0(t,\lambda)dt,
$$

$$
= \int_a^{b'} f(t)y_0(t,\lambda)dt.
$$

Since

$$
\int_{-\infty}^{\infty} |g_c(\lambda) - g_d(\lambda)|^2 d\rho(\lambda) = \int_c^d |f(t)|^2 dt
$$

when $c < d$, we see that $\{g_{b'}\}$ is a Cauchy sequence as b' approaches b, converging to a limit $g(\lambda)$ in $L^2(\rho)$. Letting b' approach b in the previous result:

$$
\int_{-\infty}^{\infty} |g_{b'}(\lambda)|^2 d\rho(\lambda) = \int_a^{b'} |f(t)|^2 dt,
$$

we find

$$
\int_{-\infty}^{\infty} |g(\lambda)|^2 d\rho(\lambda) = \int_a^b |f(t)|^2 dt.
$$

The statement $g(\lambda)$ is the limit of $g_{b'}(\lambda)$ in $L^2(\lambda)$ as b' approaches b can be written in the form

$$\lim_{b'\to b} \int_{-\infty}^{\infty} |g(\lambda) - \int_a^{b'} f(t)y_0(t,\lambda)dt|^2 d\rho = 0.$$

XII.4.5. Theorem. If $g(\lambda)$ is the limit in $L^2(\rho)$ of $\int_a^{b'} f(t)y_0(t,\lambda)dt$ as b' approaches b, then $\int_{-\infty}^{\infty} g(\lambda)y_0(t,\lambda)d\rho(\lambda)$ converges in $L^2(a,b)$ to $f(t)$. That is,

$$\lim_{I\to(-\infty,\infty)} \int_a^b |f(t) - \int_I g(\lambda)y_0(t,\lambda)d\rho(\lambda)|^2 dt = 0.$$

Proof. Let $I = (\mu,\nu)$ and

$$f_I(t) = \int_I g(\lambda)y_0(t,\lambda)d\rho(\lambda).$$

If b' is in $[a,b)$, then

$$\int_a^{b'} f_I(t)\overline{[f(t) - f_I(t)]} \ dt = \int_a^{b'} [\int_I g(\lambda)y_0(t,\lambda)d\rho(\lambda)]\overline{[f(t)-f_I(t)]} \ dt,$$

$$= \int_I g(\lambda)[\int_a^{b'} \overline{[f(t)-f_I(t)]}y_0(t,\lambda)dt] \ d\rho(\lambda).$$

By a similar calculation

$$\int_a^{b'} f(t)\overline{[f(t)-f_I(t)]} \ dt = \int_{-\infty}^{\infty} g(\lambda)[\int_a^{b'} \overline{[f(t)-f_I(t)]}y_0(t,\lambda)dt] \ d\rho(\lambda).$$

Subtracting these yields

$$\int_a^{b'} |f(t)-f_I(t)|^2 dt = \int_{(-\infty,\infty)-I} g(\lambda)[\int_a^{b'} \overline{[f(t)-f_I(t)]}y_0(t,\lambda)dt] \, d\rho(\lambda).$$

Now

$$\int_a^{b'} [f(t)-f_I(t)]y_0(t,\lambda)dt$$

represents the transform of a function in $L^2(a,b)$ which vanishes in the interval $[b',b]$. Thus the integral is in $L^2(\rho)$. We apply Schwarz's inequality to the preceeding equation to find

$$[\int_a^{b'} |f(t)-f_I(t)|^2 dt]^2 \le \int_{(-\infty,\infty)-I} |g(\lambda)|^2 d\rho(\lambda) \int_{(-\infty,\infty)-I} |\int_a^{b'} [f(t)-f_I(t)]y_0(t,\lambda)dt|^2 d\rho(\lambda).$$

According to Theorem XII.4.4, the second integral on the right equals $\int_a^{b'} |f(t)-f_I(t)|^2 dt$. Thus

$$\int_a^{b'} |f(t)-f_I(t)|^2 dt \le \int_{(-\infty,\infty)-I} |g(\lambda)|^2 d\rho(\lambda).$$

We let b' approach b, then let (μ,ν) approach $(-\infty,\infty)$ to complete the proof.

The converse to Theorems XII.4.4 and XII.4.5 is also true. That is, for each g in $L^2(\rho)$ there exists a unique element f in $L^2(a,b)$ such that the preceeding theorems hold. We use four lemmas to show this is true.

XII.4.6. <u>Lemma</u>. <u>Let $g(\lambda)$ be in $L^2(\rho)$; let</u>

$$f_I(t) = \int_I g(\lambda)d\rho(\lambda).$$

<u>Then</u> $\lim\limits_{I \to (-\infty,\infty)} f_I(t) = f(t)$ <u>exists in $L^2(a,b)$.</u>

Proof. Let $I_1 \subset I_2$. Then

$$f_{I_2}(t) - f_{I_1}(t) = \int_{I_2 - I_1} g(\lambda)y_0(t,\lambda)d\rho(\lambda),$$

$$= \int_{-\infty}^{\infty} [K_{I_2-I_1}(\lambda)g(\lambda)]y_0(t,\lambda)d\rho(\lambda),$$

where K equals 1 when λ is in I_2-I_1 and 0 otherwise. Parseval's equality then shows

$$\int_a^b |f_{I_2}(t) - f_{I_1}(t)|^2 dt = \int_{-\infty}^{\infty} |K_{I_2-I_1}(\lambda)g(\lambda)|^2 d\rho(\lambda),$$

$$= \int_{I_2-I_1} |g(\lambda)|^2 d\rho(\lambda).$$

As I approaches $(-\infty,\infty)$, $\int_I |g(\lambda)|^2 d\rho(\lambda)$ forms a Cauchy sequence. Hence $f_I(t)$ does also in $L^2(a,b)$, and converges (to f(t)) in $L^2(a,b)$.

XII.4.7. <u>Lemma</u>. <u>Let $g(\lambda)$ be in $L^2(\rho)$; let</u>

$$f_I(t) = \int_I g(\lambda)y_0(t,\lambda)d\rho(\lambda),$$

$$f(t) = \lim_{I \to (-\infty, \infty)} \int g(t) y_0(t, \lambda) d\rho(\lambda)$$

in $L^2(a,b)$; let

$$g^*(\lambda) = \int_a^b f(t) y_0(t, \lambda) dt$$

in $L^2(\rho)$; and let

$$f_I^*(t) = \int_I g^*(\lambda) y_0(t, \lambda) d\rho(\lambda)$$

in $L^2(a,b)$. Then

$$\lim_{I \to (-\infty, \infty)} \int_a^b |f_I(t) - f_I^*(t)|^2 dt = 0.$$

Proof. From Theorem XII.4.5

$$f(t) = \lim f_I^*(t)$$

in $L^2(a,b)$. But by definition

$$f(t) = \lim_{I \to (-\infty, \infty)} f_I(t).$$

Since

$$[\int_a^b |f_I(t) - f_I^*(t)|^2 dt]^{1/2} \le [\int_a^b |f_I(t) - f(t)|^2 dt]^{1/2}$$

$$+ [\int_a^b |f(t) - f_I^*(t)|^2 dt]^{1/2},$$

the result follows immediately.

XII.4.8. Lemma. Let λ_0 have positive imaginary part, and let

$$F_I(t,\lambda_0) = \int_I \frac{g(\lambda)-g^*(\lambda)}{\lambda-\lambda_0} y_0(t,\lambda) d\rho(\lambda).$$

Then for all fixed λ_0 with positive imaginary part

$$\lim_{I \to (-\infty, \infty)} F_I(t,\lambda_0) = 0$$

in $L^2(a,b)$.

Proof. A simple calculation shows $F_I(t,\lambda_0)$ satisfies

$$LF_I = \lambda_0 F_I + (f_I - f_I^*),$$

as well as the same boundary condition at a as does y_0. Variation of parameters then shows that

$$F_I(t,\lambda_0) = \int_a^t [x_0(t)y_0(\tau) - y_0(t)x(\tau)][f_I(\tau) - f_I^*(\tau)]d\tau + c_I y_0(t).$$

Since $g(\lambda) - g^*(\lambda)$ is in $L^2(\rho)$, so is $[g(\lambda) - g^*(\lambda)]/[\lambda - \lambda_0]$, so, as I approaches $(-\infty, \infty)$, $F_I(t,\lambda_0)$ converges in $L^2(a,b)$ to an element $F(t,\lambda_0)$. Since by Lemma XII.4.7 $f_I(t) - f_I^*(t)$ approaches 0 in $L^2(a,b)$, an application of Schwarz's inequality shows the integral above approaches 0 as

I approaches $(-\infty,\infty)$. Hence $F(t,\lambda_0) = c\, y_0(t)$. Since y_0 is not in $L^2(a,b)$, $c = 0$, and $\lim_{I\to(-\infty,\infty)} F_I(t,\lambda_0) = 0$.

XII.4.9. **Lemma.** $g(\lambda) = g^*(\lambda)$ in $L^2(\rho)$.

Proof. Let

$$Y_s(\lambda) = \int_a^s y_0(t,\lambda)dt.$$

Since Y_s is the transform of the function which is 1 in $[a,s]$ and 0 in $(s,b]$, which is in $L^2(a,b)$, Y_s is in $L^2(\rho)$. Integrating $F_I(t,\lambda_0)$, we find

$$\int_a^s F_I(t,\lambda_0)dt = \int_I \frac{g(\lambda)-g^*(\lambda)}{\lambda-\lambda_0} Y_s(\lambda)d\rho(\lambda).$$

Letting I approach ∞, F_I approaches 0, so

$$\int_{-\infty}^{\infty} \frac{g(\lambda)-g^*(\lambda)}{\lambda-\lambda_0} Y_s(\lambda)d\rho(\lambda) = 0.$$

Taking the imaginary part with respect to $1/(\lambda-\lambda_0)$, we find

$$\int_{-\infty}^{\infty} \frac{\nu_0}{(\lambda-\mu_0)^2+\nu_0^2} (g(\lambda)-g^*(\lambda))Y_s(\lambda)d\rho(\lambda) = 0,$$

where $\lambda_0 = \mu_0 + i\nu_0$. If we integrate from α to β with respect to μ_0, after reversing the order of integration, we find

$$\int_{-\infty}^{\infty} [\tan^{-1}(\frac{\beta-\lambda}{\nu_0}) - \tan^{-1}(\frac{\alpha-\lambda}{\nu_0})](g(\lambda)-g^*(\lambda))Y_s(\lambda)d\rho(\lambda) = 0.$$

Taking the limit as ν_0 approaches 0 from above,

$$\int_{\alpha}^{\beta} (g(\lambda)-g^*(\lambda))Y_s(\lambda)d\rho(\lambda) = 0.$$

Since, with $I = [\alpha,\beta]$,

$$\int_I (g(\lambda)-g^*(\lambda))Y_s d\rho(\lambda) = \int_a^s [\int_I (g(\lambda)-g^*(\lambda))y_0(t,\lambda)d\rho(\lambda)]dt,$$

we find upon differentiation that $\int_I (g(\lambda)-g^*(\lambda))Y_s(\lambda)d\rho(\lambda)$ has a continuou

derivative with respect to t. Since the integral is 0, its derivative

must be also, and

$$\int_I (g(\lambda)-g^*(\lambda))y_0(t,\lambda)d\rho(\lambda) = 0.$$

Either by letting $t = 0$ directly or by differentiating and then letting

$t = 0$, we find

$$\int_I (g(\lambda)-g^*(\lambda))d\rho(\lambda) = 0.$$

Since I is arbitrary, by choosing suitable intervals and suitable

constant multipliers,

$$\int_{-\infty}^{\infty} G(\lambda)(g(\lambda)-g^*(\lambda))d\rho(\lambda) = 0$$

for all step functions (functions which are piecewise constant) which vanish outside some finite interval. Since according to Theorem XI.1.13 these are dense in $L^2(\rho)$. We conclude

$$\int_{-\infty}^{\infty} |g(\lambda) - g^*(\lambda)|^2 d\rho(\lambda) = 0$$

or $g(\lambda) = g^*(\lambda)$ in $L^2(\rho)$.

We summarize these results as follows:

XII.4.10. <u>Theorem.</u> <u>If $g(\lambda)$ is in $L^2(\rho)$, there exists a unique element</u> <u>$f(t)$ in $L^2(a,b)$ such that</u>

$$\lim_{I \to (-\infty, \infty)} \int_a^b \left| f(t) - \int_I g(\lambda) y_0(t,\lambda) d\rho(\lambda) \right|^2 dt = 0$$

<u>and</u>

$$g(\lambda) = \lim_{b' \to b} \int_a^{b'} f(t) y_0(t,\lambda) dt$$

<u>in $L^2(\rho)$.</u>

We have one final task for this section: to determine $\rho(\lambda)$ in terms of known functions.

XII.4.11. <u>Lemma.</u> <u>Let $\lambda = \mu + i\nu$, where $\nu \neq 0$. Then</u>

$$\int_a^b |\chi(t)|^2 dt = \mathrm{Im}(M_b)/\nu.$$

Proof. We fix λ and λ_0 with nonzero imaginary part, choose b' in $[a,$
and let $x_0(t,\lambda) + M(\lambda)y_0(t,\lambda)$ and $x_0(t,\lambda_0) + M(\lambda_0)y_0(t,\lambda_0)$ satisfy the
same boundary condition at b'. Then

$$W[x_0(b',\lambda) + M(\lambda)y_0(b',\lambda), \; x_0(b',\lambda_0) + M(\lambda_0)y_0(b',\lambda_0)] = 0.$$

Thus

$$W[X(b',\lambda), \; X(b',\lambda_0)]$$

$$+ \; (M(\lambda) - M_b(\lambda))W[y_0(b',\lambda), \; X(b',\lambda_0)]$$

$$+ \; (M(\lambda_0) - M_b(\lambda_0))W[X(b',\lambda), \; y_0(b',\lambda_0)]$$

$$+ \; (M(\lambda) - M_b(\lambda_0))(M(\lambda_0) - M_b(\lambda_0)W[y_0(b',\lambda), \; y_0(b',\lambda_0)] = 0.$$

Now

$$W[y_0(b',\lambda),X(b',\lambda)] = (\lambda_0-\lambda)\int_a^{b'} y_0(t,\lambda)X(t,\lambda_0)dt$$

$$+ \; W[y_0(a,\lambda),X(a,\lambda_0)].$$

So for some C_1 and C_2

$$\left|W[y_0(b',\lambda),X(b',\lambda_0)]\right| \le C_1\left[\int_a^{b'} |y_0(t,\lambda)|^2 dt\right]^{1/2} + C_2$$

as b′ approaches b.   Further

$$|M(\lambda) - M_b(\lambda)| \le 2r_{b'} = [|\nu| \int_a^{b'} |y_0(t,\lambda)|^2 dt]^{-1}.$$

Thus as b′ approaches b,   their product converges to 0, since the
denominator becomes arbitrarily large and dominant.   Similar estimates
show the other terms involving M and M_b approach 0.   So

$$\lim_{b' \to b} W[X(b',\lambda), X(b',\lambda_0)] = 0.$$

An application of Green's formula now shows

$$\int_a^b X(t,\lambda)X(t,\lambda_0)dt = \frac{M_b(\lambda) - M_b(\lambda_0)}{\lambda - \lambda_0}.$$

Letting $\lambda_0 = \bar{\lambda}$, we find

$$\int_a^b |X(t,\lambda)|^2 dt = \operatorname{Im}(M_b(\lambda))/\nu.$$

XII.4.12.   Lemma.   Let $\lambda_0 = \mu + i\nu$ where $\nu \ne 0$.   Then

$$\int_a^b |X(t,\lambda_0)|^2 dt = \int_{-\infty}^{\infty} \frac{d\rho(\lambda)}{|\lambda - \lambda_0|^2}.$$

Proof.   According to the proof of Theorem XII.4.3, the statement is true if
b′ replaces b.   Since according to Helly's second theorem

$$\lim_{b' \to b} \int_{-\infty}^{\infty} \frac{d\rho_{b'}(\lambda)}{|\lambda - \lambda_0|^2}$$

exists, and at each stage equals

$$\int_a^{b'} |\chi_{b'}(t, \lambda_0)|^2 dt,$$

this integral must also exist. Since $\chi_{b'}$ approaches χ, the result follows.

XII.4.13. <u>Theorem.</u> <u>If λ_1 and λ_2 are real valued, then</u>

$$\rho(\lambda_2) - \rho(\lambda_1) = \lim_{\nu \to 0} (1/\pi) \int_{\lambda_1}^{\lambda_2} \text{Im}(M_b(\mu + i\nu)) d\mu.$$

<u>Further when λ_1 and λ_2 have nonzero imaginary parts, then</u>

$$M_b(\lambda_2) - M_b(\lambda_1) = \int_{-\infty}^{\infty} [1/(\lambda - \lambda_2) - 1/(\lambda - \lambda_1)] d\rho(\lambda).$$

Proof. From Lemmas XII.4.11 and XII.4.12, we see that when $\lambda_0 = \mu + i\nu$ is not real,

$$\int_{-\infty}^{\infty} \frac{d\rho(\lambda)}{|\lambda - \lambda_0|^2} = \text{Im}(M_b(\lambda))/\nu.$$

Thus

$$\text{Im}(M_b(\mu + i\nu)) = \int_{-\infty}^{\infty} \frac{\nu \, d\rho(\lambda)}{(\lambda - \mu)^2 + \nu^2}.$$

Integrating both sides from λ_1 to λ_2, this becomes

$$\int_{\lambda_1}^{\lambda_2} \operatorname{Im}(M_b(\mu + i\nu))d\mu = \int_{\lambda_1}^{\lambda_2} \int_{-\infty}^{\infty} \frac{\nu \, d\rho(\lambda)}{(\lambda-\mu)^2+\nu^2} \, d\mu,$$

$$= \int_{-\infty}^{\infty} [\int_{\lambda_1}^{\lambda_2} \frac{\nu \, d\mu}{(\lambda-\mu)^2+\nu^2}]d\rho(\lambda),$$

$$= \int_{-\infty}^{\infty} [\tan^{-1}(\frac{\lambda_2-\lambda}{\nu}) - \tan^{-1}(\frac{\lambda_1-\lambda}{\nu})]d\rho(\lambda).$$

Taking the limit as ν approaches 0 on both sides,

$$\lim_{\nu \to 0} \int_{\lambda_1}^{\lambda_2} \operatorname{Im}(M_b(\mu + i\nu))d\mu = \pi \int_{\lambda_1}^{\lambda_2} d\rho(\lambda),$$

$$= \pi[\rho(\lambda_2)-\rho(\lambda_1)].$$

To prove the second part we again begin with the formula at the beginning of the proof. Letting $\lambda_1 = \mu_1 + i\nu_1$ and $\lambda_2 = \mu_2 + i\nu_2$, we find

$$\operatorname{Im}[M_b(\lambda_2)-M_b(\lambda_1)] = \int_{-\infty}^{\infty} [\frac{\nu_2}{|\lambda-\lambda_2|^2} - \frac{\nu_1}{|\lambda-\lambda_1|^2}]d\rho(\lambda),$$

$$= \operatorname{Im}[\int_{-\infty}^{\infty} [\frac{\overline{\lambda-\lambda_2}}{|\lambda-\lambda_2|^2} - \frac{\overline{\lambda-\lambda_1}}{|\lambda-\lambda_1|^2}]d\rho(\lambda)],$$

$$= \operatorname{Im}[\int_{-\infty}^{\infty} [1/(\lambda-\lambda_2)-1/(\lambda-\lambda_1)]d\rho(\lambda)].$$

Since both terms within the brackets are analytic in λ_1 and λ_2, their real parts differ at most by a constant. Further if $\lambda_1 = \lambda_2$, this constant is easily seen to be 0. Thus the result follows immediately.

XII.4.14.   <u>The Limit Point Case at Both Ends</u>. If the Sturm-Liouville problem is singular at both a and b, the same techniques can be applied to derive a spectral resolution. Rather than burden the reader with a lot of additional complicated details, we prefer at this point instead to merely outline what occurs.

As b′ approaches b, we find, as before, a solution $\chi(t,\lambda) = x_0(t,\lambda)$ $+ M_b(\lambda)y_0(t,\lambda)$ in $L^2(s,b)$ for all nonreal λ. As a′ approaches a, a solution $\eta(t,\lambda) = x_0(t,\lambda) + M_a(\lambda)y_0(t,\lambda)$ is similarly produced in $L^2(a,s)$ for all nonreal λ.

Now let $I = [a′,b′] \subsetneq [a,b]$. The regular Sturm-Liouville problem over I:

$$Lx = \lambda x,$$

$$\cos\theta\, x(a′) - \sin\theta\, x′(a′) = 0,$$

$$\cos\phi\, x(b′) - \sin\phi\, x′(b′) = 0,$$

possesses a countable sequence of eigenvalues $\{\lambda_{In}\}_{n=1}^{\infty}$ and normalized eigenfunctions

$$x_{In} = r_{In}\, x_0 + s_{In}\, y_0,$$

where x_0 and y_0 are the usual solutions satisfying the boundary conditions at s in I as described in the proof of Theorem XII.3.1, and r_{In} and s_{In} are appropriate coefficients.

If f is in $L^2(I)$ then Parseval's equality takes the form

$$\int_I |f(t)|^2 dt = \sum_{n=1}^{\infty} |\int_I x_{In}(t)\overline{f}(t)dt|^2,$$

$$= \sum_{n=1}^{\infty} |r_{In}\int_I x_0(t,\lambda_{In})\overline{f}(t)dt + s_{In}\int_I y_0(t,\lambda_{In})\overline{f}(t)dt|^2,$$

$$= \sum_{n=1}^{\infty} [|r_{In}|^2 |g_{In}|^2 + r_{In}\overline{s}_{In}g_{In}\overline{h}_{In}$$

$$+ \overline{r}_{In}s_{In}\overline{g}_{In}h_{In} + |s_{In}|^2 |h_{In}|^2],$$

where

$$g_{In} = g_I(\lambda)\Big|_{\lambda=\lambda_{In}} = \int_I x_0(t,\lambda)\overline{f}(t)dt\Big|_{\lambda=\lambda_{In}},$$

$$h_{In} = h_I(\lambda)\Big|_{\lambda=\lambda_{In}} = \int_I y_0(t,\lambda)\overline{f}(t)dt\Big|_{\lambda=\lambda_{In}}.$$

If ρ_{I11}, ρ_{I12} and ρ_{I22} are defined to be continuous from above, and to satisfy

$$\rho_{I11}(\lambda_{In} + 0) - \rho_{I11}(\lambda_{In} - 0) = |r_{In}|^2, \quad \rho_{I11}(0) = 0,$$

$$\rho_{I12}(\lambda_{In} + 0) - \rho_{I12}(\lambda_{In} - 0) = \overline{r}_{In}s_{In}, \quad \rho_{I12}(0) = 0,$$

$$\rho_{I22}(\lambda_{In} + 0) - \rho_{I22}(\lambda_{In} - 0) = |s_{In}|^2, \quad \rho_{I22}(0) = 0,$$

then the matrix

$$\rho_I(\lambda) = \begin{pmatrix} \rho_{I11}(\lambda) & \rho_{I12}(\lambda) \\ \overline{\rho}_{I12}(\lambda) & \rho_{I22}(\lambda) \end{pmatrix}$$

is self-adjoint (satisfies $\rho_I^* = \rho_I$), positive semidefinite and has finite total variation over finite intervals in λ. (We have implicity assumed the eigenvalues λ_{In} are simple. If not, then the jumps in ρ_I at λ_{In} must be increased to include the contributions of all eigenfunctions. That is, $|r_{In}|^2$ is to be replaced by $\sum_{\lambda=\lambda_{In}} |r_{In}|^2$, etc.) Parseval's equality then can be written as

$$\int_I |f(t)|^2 dt = \int_{-\infty}^{\infty} (\overline{g}_I(\lambda)\overline{h}_I(\lambda)) \begin{pmatrix} d\rho_{I11}(\lambda) & d\rho_{I12}(\lambda) \\ d\overline{\rho}_{I12}(\lambda) & d\rho_{I22}(\lambda) \end{pmatrix} \begin{pmatrix} g_I(\lambda) \\ h_I(\lambda) \end{pmatrix}.$$

If $G_I(\lambda) = \begin{pmatrix} g_I(\lambda) \\ h_I(\lambda) \end{pmatrix}$, then

$$\int_I |f(t)|^2 dt = \int_{-\infty}^{\infty} G_I(\lambda)^* d\rho_I(\lambda) G_I(\lambda)$$

The same procedure is now employed to show by Helly's theorems that as I approaches $[a,b]$, the functions g_I and h_I as well as the matrix ρ_I all converge in $L^2(\rho)$ to yield

$$\int_a^b |f(t)|^2 dt = \int_{-\infty}^{\infty} G(\lambda)^* d\rho(\lambda) G(\lambda),$$

where $g(\lambda) = \lim_{I\to[a,b]} g_I(\lambda)$, $h(\lambda) = \lim_{I\to[a,b]} h_I(\lambda)$, where the limits are taken in $L^2(\rho)$. Further if

$$X(t,\lambda) = \begin{pmatrix} x_0(t,\lambda) \\ y_0(t,\lambda) \end{pmatrix},$$

then

$$f(t) = \int_{-\infty}^{\infty} G^*(\lambda) d\rho(\lambda) X(t,\lambda).$$

As in the case with a singularity only at b, the matrix $\rho(\lambda)$ can be determined in terms of the limit points M_a and M_b. If λ_1 and λ_2 are points of continuity of $\rho(\lambda)$, then

$$\rho_{11}(\lambda_2) - \rho_{11}(\lambda_1) = \lim_{v\to 0} \frac{1}{\pi} \int_{\lambda_1}^{\lambda_2} \text{Im}[M_a(\mu + iv) - M_b(\mu + iv)] d\mu,$$

$$\rho_{12}(\lambda_2) - \rho_{12}(\lambda_1) = \lim_{v\to 0} \frac{1}{\pi} \int_{\lambda_1}^{\lambda_2} \text{Im}[\frac{M_a(\mu + iv) + M_b(\mu + iv)}{M_a(\mu + iv) - M_b(\mu + iv)}] d\mu,$$

$$\rho_{22}(\lambda_2) - \rho_{22}(\lambda_1) = \lim_{v\to 0} \frac{1}{\pi} \int_{\lambda_1}^{\lambda_2} \text{Im}[\frac{M_a(\mu + iv)M_b(\mu + iv)}{M_a(\mu + iv) - M_b(\mu + iv)}] d\mu.$$

To those interested in more details we recommend the books "Theory of Ordinary Differential Equations," by E.A. Coddington and N. Levinson, as well as their papers listed therein, and "Eigenfunction Expansions Associated With Second Order Differential Equations, part 1" by E.C. Titchmarsh.

XII.5.   The Limit Circle Case.   If the Sturm-Liouville problem

$$Lx = x'' + Qx,$$

$$\cos \theta \, x(a) - \sin \theta \, x'(a) = 0,$$

$$\cos \phi \, x(b') - \sin \phi \, x'(b') = 0,$$

where $0 \leq \theta < \pi$, $0 \leq \phi < \pi$, is in the limit circle case as b' approaches b, then there exists an infinite number of points M_b from which to choose. Having chosen one, the results of the preceeding section are all valid with only minor modifications in the proofs. The results, however can be made more precise. In particular the spectral resolution of Theorem XII.4.5 and Parseval's equality, Theorem XII.4.4 are representable as infinite series. Where Section XII.4 did not have a boundary condition at b, one now natural appears.

Throughout the remainder of this section we shall assume that the points M on the circles $C_{b'}$, converge to a specific fixed point M_b on the limit circle C_b for each fixed value of the complex variable λ, $\text{Im}(\lambda) \neq 0$.

XII.5.1.   Theorem.   If for $\lambda_0 = \mu_0 + i\nu_0$, $\nu_0 \neq 0$,

$$\lim_{b' \to b} M(\lambda_0) = M_b(\lambda_0),$$

then for all $\lambda = \mu + i\nu$, $\nu \neq 0$,

$$\lim_{b' \to b} M(\lambda) = M_b(\lambda).$$

Further, $M_b(\lambda)$ is a meromorphic function of λ, having poles which are all real and simple. $M_b(\lambda)$ is real valued when λ is real and is not one of its poles.

Proof. Let $X(t,\lambda) = x_0(t,\lambda) + M(\lambda)y_0(t,\lambda)$. If we apply Green's formula to $\overline{X}(t,\lambda)$ and $X(t,\lambda_0)$, we find

$$(\lambda_0-\lambda)\int_a^{b'} X(t,\lambda)X(t,\lambda_0)dt$$

$$= \int_a^{b'} [LX(t,\lambda_0)X(t,\lambda) - X(t,\lambda_0)LX(t,\lambda)]dt,$$

$$= X(a,\lambda_0)X'(a,\lambda) - X'(a,\lambda_0)X(a,\lambda),$$

$$= M(\lambda_0) - M(\lambda).$$

Replacing $X(t,\lambda)$ by $x_0(t,\lambda) + M(\lambda)y_0(t,\lambda)$ and solving for $M(\lambda)$,

$$M(\lambda) = \frac{M(\lambda_0) + (\lambda-\lambda_0)\int_a^{b'} x_0(t,\lambda)X(t,\lambda_0)dt}{1 - (\lambda-\lambda_0)\int_a^{b'} y_0(t,\lambda)X(t,\lambda_0)dt}.$$

If λ is restricted to some bounded portion of the complex plane, then as b' approaches b, the convergence in the quotient above is uniform. Since each component is meromorphic, the limits are all meromorphic, and so is $M_b(\lambda)$.

$$M_b(\lambda) = \frac{M_b(\lambda_0) + (\lambda-\lambda_0)\int_a^b x_0(t,\lambda)\chi(t,\lambda_0)dt}{1 - (\lambda-\lambda_0)\int_a^b y_0(t,\lambda)\chi(t,\lambda_0)dt} \quad ,$$

where now $\chi(t,\lambda_0) = x_0(t,\lambda_0) + M_b(\lambda_0)y_0(t,\lambda_0)$.

From the proof of Theorem XII.3.1, we see that

$$\int_a^b |\chi(t,\lambda)|^2 dt = \mathrm{Im}(M_b(\lambda))/\nu \le |M_b(\lambda)|/|\nu|.$$

Since

$$\int_a^b |\chi(t,\lambda)|^2 dt \ge \frac{1}{2}|M_b(\lambda)|^2 \int_a^b |y_0(t,\lambda)|^2 dt - \int_a^b |x_0(t,\lambda)|^2 dt,$$

we find, after solving for $|M_b(\lambda)|$, that

$$|M_b(\lambda)| \le [|\nu|\int_a^b |y_0(t,\lambda)|^2 dt]^{-1} + \left[\frac{2\int_a^b |x_0(t,\lambda)|^2 dt}{\int_a^b |y_0(t,\lambda)|^2 dt} + \frac{1}{|\nu|^2[\int_a^b |y_0(t,\lambda)|^2 d}\right.$$

So

$$M_b(\lambda) \le C/|\nu|$$

as ν approaches 0. This shows that $M_b(\lambda)$ can have poles only on the re axis, and these must be simple. Finally, since

$$\text{Im}(M_b(\lambda)) = \nu \int_a^b |\chi(t,\lambda)|^2 dt,$$

if λ approaches a real value μ which is not a pole of $M_b(\lambda)$, then

$$\text{Im}(M_b(\mu)) = 0.$$

XII.5.2.   Theorem.

$$\lim_{b'\to b} \rho_{b'}(\lambda) = \rho(\lambda)$$

exists. ρ is a step function, increasing only at the poles $\{\lambda_k\}_{k=1}^{\infty}$ of $M_b(\lambda)$. The jump at λ_k is equal to the negative of the residue of $M_b(\lambda)$ at λ_k.

Proof.   Let

$$M_b(\lambda) = a_k/(\lambda-\lambda_k) + f(\lambda)$$

where $f(\lambda)$ is analytic near λ_k. Then

$$\text{Im}(M_b(\lambda)) = \frac{-i\nu a_k}{(\mu-\lambda_k)^2 + \nu^2} + \text{Im}f(\lambda),$$

and

$$\rho(\lambda_k+\varepsilon) - \rho(\lambda_k-\varepsilon) = -\lim_{\nu\to 0} \frac{1}{\pi} \int_{\lambda_k-\varepsilon}^{\lambda_k+\varepsilon} \frac{\nu\, a_k\, d\mu}{(\mu-\lambda_k)^2 + \nu^2}$$

$$+ \lim_{\nu\to 0} \frac{1}{\pi} \int_{\lambda_k-\varepsilon}^{\lambda_k+\varepsilon} f(\mu + i\nu) d\mu.$$

The second integral becomes negligible if ε is chosen sufficiently small. The first can be integrated to yield

$$-\lim_{\nu \to 0} \frac{a_k}{\pi} \left[\tan^{-1}\left(\frac{\varepsilon}{\nu}\right) - \tan^{-1}\left(-\frac{\varepsilon}{\nu}\right) \right]$$

which equals $-a_k$.

At points where $M_b(\lambda)$ is analytic, $a_k = 0$, and $\rho(\lambda_k + \varepsilon) - \rho(\lambda_k - \varepsilon)$ approaches 0 as ε becomes small.

XII.5.3. <u>Theorem</u>. <u>The functions $\{y_0(t,\lambda_k)\}_{k=1}^{\infty}$, where $\{\lambda_k\}_{k=1}^{\infty}$ are the poles of $M_b(\lambda)$ form a complete orthogonal set in $L^2(a,b)$.</u>

Proof. This follows immediately from Theorem XII.4.5.

XII.5.4. <u>Theorem</u>. <u>Let $X(t,\lambda_0) = x_0(t,\lambda_0) + M_b(\lambda_0)y_0(t,\lambda_0)$, where λ_0 is not a pole of $M_b(\lambda)$. Then</u>

$$W[X(b,\lambda_0),y_0(b,\lambda_k)] = 0,$$

<u>$k=1,\ldots$.</u>

$$W[X(b,\lambda_0),y_0(b,\lambda)] \neq 0,$$

<u>when $\lambda \neq \lambda_k$, $k=1,\ldots$.</u>

Proof. We again employ Green's formula to derive

$$(\lambda-\lambda_0)\int_a^{b'} \chi(t,\lambda_0)y_0(t,\lambda)dt = 1 - W[\chi(b',\lambda_0),y_0(b',\lambda)]$$

or

$$W[\chi(b',\lambda_0),y_0(b',\lambda)] = 1 - (\lambda-\lambda_0)\int_a^{b'} \chi(t,\lambda_0)y_0(t,\lambda)dt.$$

If λ is a pole of $M_b(\lambda)$, then the expression for $M(\lambda)$, derived in the proof of Theorem XII.5.1, shows that the right side of the formula above approaches 0 as b' approaches b. If λ is not a pole, the right side does not approach 0.

The expression

$$W[\chi(b,\lambda_0),x(b)] = 0$$

is a boundary condition. In the limit point case it was not needed. Here, as with the regular Sturm-Liouville problem, it is.

In closing let us restate Parseval's equality and the spectral resolution in the limit circle case:

XII.5.5. Theorem. Let the singular Sturm-Liouville problem lie in the limit circle case at b. Let M_b be any point on the limit circle C_b. Then there exists a step function ρ, which increases only at the poles $\{\lambda_k\}_1^\infty$ of $M_b(\lambda)$, and which is given by

$$\rho(\lambda_2) - \rho(\lambda_1) = \lim_{\nu\to 0} \frac{1}{\pi} \int_{\lambda_1}^{\lambda_2} Im(M_b(\mu + i\nu))d\mu$$

at points of continuity of ρ, and which has the following additional pro-

perties.

1. Parseval's equality: If f is in $L^2(a,b)$, then

$$\int_a^b |f(t)|^2 dt = \int_{-\infty}^{\infty} |g(\lambda)|^2 d\rho(\lambda),$$

where

$$g(\lambda) = \int_a^b f(t)y_0(t,\lambda)dt.$$

2.

$$\lim_{I \to (-\infty,\infty)} \int_a^b \left| f(t) - \int_I g(\lambda)y_0(t,\lambda)d\rho(\lambda) \right|^2 dt = 0.$$

The first, of course states that

$$\|f\|_{L^2(a,b)} = \|g\|_{L^2(\rho)}.$$

The second states that

$$f = \lim_{I \to (-\infty,\infty)} \int_I g(\lambda)y_0(t,\lambda)d\rho(\lambda)$$

in $L^2(a,b)$. The right sides of both of these equations may be rewritten

as infinite series.

XII.6. Examples.

1. The Fourier Sine Transform. Perhaps the simplest situation to discuss

is the problem $Lx = x''$, $x(0) = 0$ over the interval $[0,\infty)$. According to

Exercise XII.3, this problem is in the limit point case at ∞, so no boundary

condition at ∞ is necessary, while 0 is regular.

The solutions to $x'' = \lambda x$ can be written as

$$x(t) = c_1 e^{i\sqrt{-\lambda}t} + c_2 e^{-i\sqrt{-\lambda}t}$$

or

$$x(t) = A \sin \sqrt{-\lambda}t + B \cos \sqrt{-\lambda}\ t.$$

in particular

$$x_0(t,\lambda) = \cos \sqrt{-\lambda}t,$$

$$y_0(t,\lambda) = - \sqrt{-\lambda} \sin \sqrt{-\lambda}t.$$

If $\text{Im}(\lambda) > 0$, then

$$\chi(t,\lambda) = e^{i\sqrt{-\lambda}t},$$

$$= \cos \sqrt{-\lambda}t + M_b(\lambda) \sin \sqrt{-\lambda}t.$$

Thus $M_b(\lambda) = i\sqrt{-\lambda} = \sqrt{\lambda}$, and

$$\rho(\lambda_2) - \rho(\lambda_1) = \lim_{\nu \to 0} \int_{\lambda_1}^{\lambda_2} \text{Im } \sqrt{\mu + i\nu} \, d\mu.$$

If $\mu > 0$, then as ν approaches 0, $\sqrt{\lambda}$ approaches $\sqrt{\mu}$, which is real, so

$$\rho(\lambda_2) - \rho(\lambda_1) = 0.$$

If $\mu < 0$, then $\sqrt{\lambda}$ approaches $i\sqrt{-\mu}$, and

$$\rho(\lambda_2) - \rho(\lambda_1) = \frac{1}{\pi} \int_{\lambda_1}^{\lambda_2} \sqrt{-\mu} \, d\mu.$$

$$d\rho(\lambda) = \frac{1}{\pi} (-\lambda)^{1/2} d\lambda.$$

Thus, when f is in $L^2(0,\infty)$,

$$g(\lambda) = \frac{1}{\pi} \int_0^\infty f(s) [(-\lambda)^{-1/2} \sin\sqrt{-\lambda}s] ds$$

where the integral converges in the sense of $L^2(0,\infty)$, and

$$f(t) = \frac{1}{\pi} \int_{-\infty}^0 g(\lambda) [(-\lambda)^{-1/2} \sin\sqrt{-\lambda}t] \sqrt{-\lambda} \, d\lambda.$$

If we let $\sqrt{-\lambda} = \mu$, then this can be rewritten as

$$f(t) = \frac{2}{\pi} \int_0^\infty [\int_0^\infty f(s) \sin \mu s \, ds] \sin \mu t \, d\mu.$$

2.   The Fourier Cosine Transform.   Here we consider the problem $Lx = x''$,

$x'(0) = 0$ over the interval $[0,\infty)$. The solutions to $x'' = \lambda x$ are the same

as in the previous case. However, now

$$x_0(t,\lambda) = (1/\sqrt{-\lambda})\sin \sqrt{-\lambda}t,$$

$$y_0(t,\lambda) = - \cos \sqrt{-\lambda}t.$$

Since $X(t,\lambda) = e^{i \sqrt{-\lambda}t}$ when $\text{Im}(\lambda) > 0$, we find $M_b(\lambda) = i/\sqrt{-\lambda}$.

In calculating $d\rho$, if λ approaches $\mu > 0$, then $i/\sqrt{-\lambda}$ approaches

$1/\sqrt{\mu}$, which is real. Thus $d\rho(\lambda) = 0$ when $\lambda > 0$.

If λ approaches $\mu < 0$, then $i/\sqrt{-\lambda}$ approaches $i/\sqrt{-\mu}$, and

$$\rho(\lambda_2) - \rho(\lambda_1) = \frac{1}{\pi} \int_{\lambda_1}^{\lambda_2} \frac{1}{\sqrt{-\mu}} d\mu,$$

$$d\rho(\lambda) = \frac{1}{\pi} (1/\sqrt{-\lambda})d\lambda.$$

Thus when f is in $L^2(0,\infty)$,

$$g(\lambda) = - \int_0^\infty f(s)\cos \sqrt{-\lambda}sds,$$

and

$$f(t) = - \frac{1}{\pi} \int_{-\infty}^0 g(\lambda)\cos \sqrt{-\lambda}t(1/\sqrt{-\lambda})d\lambda.$$

If $\mu = \sqrt{-\lambda}$, this can be rewritten as

$$f(t) = \frac{2}{\pi} \int_0^\infty [\int_0^\infty f(s)\cos \mu s\,ds]\cos \mu t\,d\mu.$$

Both the integrals involved converge in the sense of $L^2(0,\infty)$.

   3. <u>The Legendre Expansion.</u> According to Exercise XII.4 the problem
involving $Lx = ((1-t^2)x')'$ on $(-1,1)$ is in the limit circle case at both
1 and -1. Thus it would be necessary to find appropriate points on the
limit circles for 1 and -1 in order to generate the ordinary Legendre expansi
This is quite tedious. The details can be found in the book by Titchmarsh,
"Eigenfunction Expansions, part 1."

   Rather than derive the expansion directly, we prefer to draw upon what
we already know about the Legendre polynomials. We showed in Chapter X that
the Legendre polynomials $\{P_n\}_{n=1}^\infty$ form a complete orthogonal set in $L^2(-1,1)$
Since

$$\int_{-1}^1 P_n^2(t)\,dt = 2/(2n+1),$$

the set $\{[2/(2n+1)]^{-1/2}P_n\}_{n=0}^\infty$ is complete and orthonormal. Thus for each
f in $L^2(-1,1)$ Parseval's equality takes the form

$$\|f\|^2 = \sum_{n=0}^\infty |\int_{-1}^1 f(s)P_n(s)\,ds|^2 /[2/(2n+1)].$$

The expansion for f is given by

$$f(t) = \sum_{n=0}^\infty [2/(2n+1)]^{-1}[\int_{-1}^1 f(s)P_n(s)\,ds]P_n(t).$$

4. **The Laguerre Expansion.** Exercise XII.5 indicates that the problem
involving $Lx = (te^{-t}x')'$ on $(0,\infty)$ with weight e^{-t} is in the limit point
case at ∞, while it is in the limit circle case at 0. Thus to generate
the ordinary Laguerre expansion it is necessary to choose the appropriate
point on the limit circle at 0. The computations not only for this, but,
more generally, to find the solutions to $Lx = \lambda x$ are extremely tedious. We
therefore prefer again to write the expansions directly by drawing upon
previously derived material.

We showed in Chapter XI in Corollary XI.2.13 that the Laguerre poly-
nomials $\{L_n\}_{n=0}^{\infty}$ are dense in $L^2(e^{-t})$ over the interval $[0,\infty)$. This
enables us to write directly Parseval's equality:

$$\|f\|^2 = \sum_{n=0}^{\infty} (1/n!)^2 \left|\int_0^{\infty} f(s)L_n(s)e^{-s}ds\right|^2,$$

and the expansion in $L^2(e^{-t})$:

$$f(t) = \sum_{n=0}^{\infty} (1/n!)^2 \left[\int_0^{\infty} f(s)L_n(s)e^{-s}ds\right]L_n(t).$$

5. **The Hermite Expansion.** The Hermite operator $Lx = (e^{-t^2}x')'$ on
$(-\infty,\infty)$ with weight e^{-t^2} is in the limit point case at both ∞ and $-\infty$.
We can again write directly Parseval's equality

$$\|f\|^2 = \sum_{n=0}^{\infty} (1/2^n n!\sqrt{\pi}) \left|\int_{-\infty}^{\infty} f(s)H_n(s)e^{-s^2}ds\right|^2$$

and the expansion in $L^2(e^{-t^2})$

$$f(t) = \sum_{n=0}^{\infty} (1/2^n n! \sqrt{\pi}) [\int_{-\infty}^{\infty} f(s) H_n(s) e^{-s^2} ds] H_n(t),$$

by quoting Corollary XI.2.13, and, in so doing, save a considerable amount of computation.

XII. Exercises.

1. If

$$(px')' + qx = \lambda wx,$$

show that the transformations

$$x = yg,$$

$$s = \int^t hdt,$$

where g and h are unknown, imply

$$x' = (yg)^{\cdot}h,$$

$$(px')' = ((yg)^{\cdot}hp)^{\cdot}h,$$

where $( )^{\cdot} = \frac{d}{ds}$. Show then that y satisfies

$$y^{\cdot\cdot} + (1/hpg)[2(hp)g^{\cdot} + (hp)^{\cdot}g]y^{\cdot}$$

$$+ [(1/hpg)(hpg^{\cdot})^{\cdot} + q/hp]y = \lambda[w/h^2p]y.$$

If $w/h^2p = 1$ and

$$2(hp)g^{\cdot} + (hp)^{\cdot}g = 0,$$

then

$$h = (w/p)^{1/2},$$

$$g = 1/(wp)^{1/4},$$

and

$$y^{\cdot\cdot} + Qy = \lambda y,$$

where

$$Q = [-\frac{1}{2}\frac{1}{(wp)^{1/4}}\left(\frac{[(wp)^{1/2}]^{\cdot}}{(wp)^{1/4}}\right)^{\cdot} + \frac{q}{w}].$$

The equation in y is in Liouville normal form.

2. Show that the Liouville normal forms for the Legendre, Laguerre and
    Hermite equations

$$((1-t^2)x') = \lambda x,$$

$$(te^{-t}x')' = \lambda e^{-t}x,$$

$$(e^{-t^2}x')' = \lambda e^{-t^2}x$$

    are

Legendre: $y'' + (\frac{1}{2} + \frac{1}{4} \tan^2 t)y = \lambda y$

on $(-\frac{\pi}{2}, \frac{\pi}{2})$,

Laguerre: $y'' + (\frac{1}{4t^2} + \frac{1}{2} - \frac{t^2}{16})y = \lambda y$

on $[0,\infty)$,

Hermite: $y'' - t^2y = \lambda y$

on $(-\infty,\infty)$.

    In all three cases the problems are singular at both ends of the
    interval in question.

3. Prove the following:

Theorem. Consider the equation

$$x'' + Qx = \lambda x$$

on $[a,\infty)$. Suppose there exists a positive differentiable function $M(t)$, which satisfies

$$M(t) > M_0 > 0,$$

$$Q(t) \leq k_1 M(t),$$

for some constant k_1,

$$\left|M'(t)M^{-3/2}(t)\right| < k_2$$

for some constant k_2, and

$$\int_a^\infty M^{-1/2}(t)\,dt = \infty.$$

Then the differential equation does not possess two linearly independent solutions in $L^2(a,\infty)$.

Hint: Let $\lambda = 0$, and let x be a real solution of

$$x'' + Qx = 0$$

which satisfies

$$\int_a^\infty x^2 dt < \infty.$$

Then

$$\int_{t_0}^t (xx''/M)\,dt = -\int_{t_0}^t (Qx^2/M)\,dt.$$

Integrate by parts on the left to show

$$\int_{t_0}^t ((x')^2(M)\,dt = \frac{1}{2M}\frac{d}{dt}(x^2)\Big|_{t_0}^t + \int_{t_0}^t (xx'M'/M^2)\,dt$$

$$+ \int_{t_0}^t (Qx^2/M)\,dt.$$

Use Schwarz's inequality on the second term on the right.   If

$$F(t) = \int_{t_0}^t ((x')^2/M)\,dt,$$

show

$$F(t) - \frac{1}{2M}\frac{d}{dt}(x^2)\Big|_{t_0}^t - \sup|M'/M^{3/2}|(\int_{t_0}^t x^2 dt)^{1/2}F(t)^{1/2}$$

$$\leq \int_{t_0}^t (Qx^2/M)\,dt.$$

If t_0 is chosen large enough so

$$k_2 \int_{t_0}^{t} x^2 dt < 1/2,$$

then

$$F(t) - \frac{1}{2M_0} \frac{d}{dt} x(t)^2 - \frac{1}{2} F(t)^{1/2} < k_3.$$

Now if $F(t)$ approaches ∞, then ultimately

$$\frac{d}{dt} x(t)^2 > M_0 F(t)/2.$$

So x^2 is increasing, and x is not in $L^2(a,\infty)$. Thus $F(t)$ is bounded.

If x and y are independent solutions in $L^2(a,\infty)$ such that

$$xy' - x'y = \alpha,$$

then

$$\frac{xy'}{M^{1/2}} \frac{x'y}{M^{1/2}} = \frac{\alpha}{M^{1/2}}.$$

The left side is integrable, but the right side is not.

4. Show that the Legendre Sturm-Liouville problem on $(-1,1)$ is in the limit circle case at both 1 and -1.

Hint: Solve directly $((1-t^2)x')' = 0$.

5. Show that the Laguerre Sturm-Liouville problem on $(0,\infty)$ is in the limit circle case at 0 and the limit point case at ∞.

   Hint: Solve directly $(te^{-t}x')' = 0$ near 0. Use Exercise XII.3 near ∞

6. Show that the Hermite Sturm-Liouville problem on $(-\infty,\infty)$ is in the limit point case at both ∞ and $-\infty$.

7. Complete the details of Section XII.4 when the Sturm-Liouville problem is singular at both a and b.

8. Show that the Sturm-Liouville problem for $Lx = x''$ on $(-\infty,\infty)$ is in the limit point case at both ∞ and $-\infty$. Derive the spectral resolution for an arbitrary function in $L^2(-\infty,\infty)$. If the variables are chosen appropr the result should be the Fourier transform in $L^2(-\infty,\infty)$.

9. If the Sturm-Liouville problem is regular at a and in the limit circle case at b, let M_b be a point on C_b and let $X = x_0 + M_b y_0$.

   Define the domain D as those functions x which satisfy

   1. x is in $C^2(a,b)$ and $L^2(a,b)$.

   2. Lx is in $L^2(a,b)$.

   3. $\cos \theta\, x(a) - \sin \theta\, x'(a) = 0$.

   4. $W[X(b,\lambda),x(b)] = 0$, when λ is not a pole of $M_b(\lambda)$.

   Then define the operator A by

   $$Ax = x'' + Qx$$

   for all x in D.

Show for all λ, not a pole of $M_b(\lambda)$, and all f in $L^2(a,b)$ that the function defined by

$$Kf(t) = \int_a^t \chi(t,\lambda)y_0(\tau,\lambda)f(\tau)d\tau$$

$$+ \int_t^b \chi(\tau,\lambda)y_0(t,\lambda)f(\tau)d\tau$$

is in D. Show that

$$(A-\lambda)Kf = f$$

for all f in $L^2(a,b)$; that

$$K(A-\lambda)x = x$$

for all x in D. Show that A is self-adjoint, that is, that A satisfies

$$\int_a^b \overline{(Ay)}x\,dt = \int_a^b \overline{y}(Ax)dt$$

for all x and y in D.

9. Put Bessel's equation

$$(tx')' - (n^2/t)x = -\lambda^2 tx$$

into Liouville normal form. Discuss the resulting Sturm-Liouville

problem over the intervals [0,b), [a,∞) and [0,∞).

Hint: Consult Titchmarsh's book, "Eigenfunction Expansions, part 1,"

Chapter 4.

References

1. N.I. Akhiezer and I.M. Glazman, "Theory of Linear Operators in Hilbert
        Space, vols. I and II," Frederick Ungar, New York, 1961.

2. E.A. Coddington and N. Levinson, "Theory of Ordinary Differential Equati
        McGraw-Hill, New York, 1955.

3. N. Dunford and J.T. Schwartz, "Linear Operators, vol. II," Interscience,
        New York, 1964.

4. E.C. Titchmarsh, "Eigenfunction Expansions, part 1," Oxford University
        Press, Oxford, 1962.

XIII. AN INTRODUCTION TO PARTIAL DIFFERENTIAL EQUATIONS

The study of partial differential equations is an enormous field. If we recall that the theory of ordinary differential equations is subsumed within it, we can get a vague idea of its size. It goes almost without saying, therefore, that any short introduction, such as this chapter, can only be an introduction in a very limited sense. For that reason we can include here only one not too satisfactory existence theorem before turning our attention to some properties of first order and second order equations. We emphasize again that what follows is not in any sense complete.

XIII.1. <u>The Cauchy-Kowaleski Theorem</u>. The field of partial differential equations is considerably more complicated than that of ordinary differential equations. For example, a first order linear ordinary differential equation

$$a_0 x' + a_1 x = 0$$

has only one linearly independent solution, while the partial differential equation

$$a_0 \frac{\partial x}{\partial s} + a_1 \frac{\partial x}{\partial t} + a_2 x = 0$$

has an arbitrary number of linearly independent solutions. Furthermore, to prove that solutions exist or even to define what is a solution is more difficult. To calculate solutions is very much more complicated.

Of course, there are also a great many similarities, and we shall use them extensively.

There is a - more or less - fundamental existence theorem. As we shall show by example, it is not very satisfactory, largely because its use is so limited. We state it, nonetheless, in the interest of completeness.

XIII.1.1. **Definition.** Let $F(z_1,\ldots,z_m)$ be a function of m complex variables $z_1 = x_1 + iy_1, \ldots, z_m = x_m + iy_m$, and let $(z_1^0,\ldots,z_m^0)$ be a point in K^m. $F(z_1,\ldots,z_m)$ is analytic in a neighborhood of $(z_1^0,\ldots,z_m^0)$ if

$$F(z_1,\ldots,z_m) = \sum_{k_1,\ldots,k_m=0}^{\infty} A_{k_1\ldots k_m} (z_1-z_1^0)^{k_1} \ldots (z_m-z_m^0)^{k_m}$$

converges when $\|z-z_0\| = [\sum_{j=1}^{m} |z_j-z_j^0|^2]^{1/2}$ is sufficiently small.

Under these conditions it is easy to show that

$$A_{k_1\ldots k_m} = \frac{1}{k_1!\ldots k_m!} \frac{\partial^{k_1+\ldots+k_m}}{\partial z_1^{k_1}\ldots\partial z_m^{k_m}} F(z_1^0,\ldots,z_m^0)$$

for all $k_1,\ldots,k_m$.

XIII.1.2. **Theorem (Cauchy-Kowaleski Theorem).** Consider the system of equat

$$\frac{\partial^{n_i}}{\partial t^{n_i}} u_i = F_i(t,x_1,\ldots,x_n,u_1,\ldots,u_N,\ldots,\frac{\partial^k u_j}{\partial t^{k_0}\partial x_1^{k_1}\ldots\partial x_n^{k_n}},\ldots),$$

where $k_0 + \ldots + k_n = k \leq n_j$, and $k_0 \leq n_j$, $i, j=1,\ldots,N$, with initial data at t_0

$$\frac{\partial^k u_i}{\partial t^k} = \phi_i^{(k)}(x_1,\ldots,x_n),$$

$k = 0,\ldots,n_{i-1}$, $i=1,\ldots,N$. If $\partial^{n_i} u_i / \partial t^{n_i}$ is the derivative of highest order, if F_i is analytic in a neighborhood of $(t_0,x_1^0,\ldots,x_n^0,\ldots,\phi_j^0,k_0,\ldots,k_n,\ldots)$ and if $\phi_j^{(k)}$ are analytic in a neighborhood of $(x_1^0,\ldots,x_n^0)$, then the system with its initial data possesses a unique analytic solution in a neighborhood of $(t_0,x_1^0,\ldots,x_n^0)$.

The proof is a rather tedious computation which involves the use of power series to show that the coefficients of u_i, $i=1,\ldots,N$, are sufficiently restricted so the series converges. Since we will not depend upon the theorem for what is to follow. The proof is omitted.

XIII.1.3.   An Example.   In order to show the limited range of applicability of the Cauchy-Kowalesky Theorem, let us consider the following problem.

$$\frac{\partial u}{\partial t} = \frac{\partial^2 u}{\partial x^2},$$

$$u(0,x) = 1/(1-x)^{1/2},$$

where $|x| < 1$ and $t \geq 0$. Let us assume that

$$u(x,t) = \sum_{n=0}^{\infty} A_n(x) t^n.$$

Then

$$\frac{\partial u}{\partial t} = \sum_{n=0}^{\infty} n A_n(x) t^{n-1} = \sum_{m=0}^{\infty} (m+1) A_{m+1}(x) t^m,$$

and

$$\frac{\partial^2 u}{\partial x^2} = \sum_{n=0}^{\infty} A_n''(x) t^n.$$

Thus

$$\sum_{n=0}^{\infty} [A_n'' - (n+1)A_{n+1}] t^n = 0,$$

and $A_{n+1} = A_n''/(n+1)$, $n=0,1,\dots$. Further, if $t=0$, we see that $A_0 = 1/(1-$
An easy computation shows

$$A_1 = (\tfrac{1}{2})(\tfrac{3}{2})(1-x)^{-5/2},$$

$$A_2 = (\tfrac{1}{2})(\tfrac{1}{2})(\tfrac{3}{2})(\tfrac{5}{2})(\tfrac{7}{2})(1-x)^{-9/2},$$

$$\cdots$$

$$A_n = (4n)!/n! 2^{4n} (2n)! (1-x)^{\frac{4n+1}{2}}.$$

Thus

$$u(x,t) = \sum_{n=0}^{\infty} \frac{(4n)!\, t^n}{n!\, 2^{4n}(2n)!\,(1-x)^{\frac{4n+1}{2}}} \quad .$$

This series diverges for all $t \neq 0$.

In closing this section we remark that there are other existence proofs (although the Cauchy-Kowaleski Theorem is the best known). For our purposes, however, we shall use the simplest. Wherever possible we shall actually exhibit the solution! We shall not depend upon abstract existence theorems.

XIII.2. <u>First Order Equations.</u> In order to get some feeling for the solutions of partial differential equations, we first examine partial differential equations of the first order which are linear in the partial derivatives, but not necessarily linear in the dependent variable. Since higher dimensions add little in the way of insight, we consider only functions of two independent variables x and y. Thus, we consider the equation

$$P(x,y,u)\frac{\partial u}{\partial x} + Q(x,y,u)\frac{\partial u}{\partial y} = R(x,y,u),$$

where x and y are independent real variables, and $u = u(x,y)$. The functions P, Q, R are Lipschitzian (and, hence, continuous) in x, y and u.

XIII.2.1. <u>Theorem.</u> <u>The equation</u>

$$v(x,y,u) = c,$$

where   c   is constant, defines an implicit solution of

$$P(x,y,u)\frac{\partial u}{\partial x} + Q(x,y,u)\frac{\partial u}{\partial y} = R(x,y,u)$$

if and only if $v(x,y,u) = c$ satisfies

$$dx/P = dy/Q = du/R.$$

It is understood that if any of the functions   P,   Q   or   R   is zero, then the differential above is also zero.   The equations

$$dx/P = dy/Q = du/R$$

are called subsidiary equations.

Proof.   Suppose $v(x,y,u) = c$ defines a solution of $P \frac{\partial u}{\partial x} + Q \frac{\partial u}{\partial y} = R$.   Then $v_u = \frac{\partial v}{\partial u} \neq 0$.   Further $u_x = -v_x/v_u$,   and $u_y = -v_y/v_u$.   Thus

$$Pv_x + Qv_y + Rv_u = 0.$$

Now

$$dv = v_x dx + v_y dy + v_u du = 0,$$

is the equation of the tangent plane (in   dx,dy,du   coordinates) for all poi on the surface generated by $v(x,y,u) = c$. Thus (P,Q,R) is in the tangent plane. Hence, there exists a direction (dx,dy,du) such that

$$dx/P = dy/Q = du/R.$$

Conversely, suppose $v(x,y,u) = c$ satisfies

$$dz/P = dy/Q = du/R.$$

Then for some μ,

$$dx = \mu P, \; dy = \mu Q, \; du = \mu R.$$

Since $du = u_x dx + u_y dy$, substitution of the previous equations and cancellation of μ results in

$$P \frac{\partial u}{\partial x} + Q \frac{\partial u}{\partial y} = R.$$

The problem, of course, is to find solutions to the subsidiary equations.

XIII.2.2. Theorem. Let $v(x,y,u) = a$ and $w(x,y,u) = b$ be two independent solutions to the subsidiary equations (and thus to $P \frac{\partial u}{\partial x} + Q \frac{\partial u}{\partial y} = R$). Let $F(\cdot,\cdot)$ be an arbitrary differentiable function. Then $F(v,w) = 0$ is a solution to $P \frac{\partial u}{\partial x} + Q \frac{\partial u}{\partial y} = R$. It is called the general solution.

Proof. Differentiating F with respect to x and then y, we find

$$F_v v_x + F_w w_x + F_v u_x + F_w u_x = 0,$$

$$F_v v_y + F_w w_y + F_v u_y + F_w u_y = 0,$$

or

$$\dot{F}_v[v_x + v_u u_x] + F_w[w_x + w_u u_x] = 0,$$

$$F_v[v_y + v_u u_y] + F_w[w_y + w_u u_y] = 0.$$

Since both F_u and F_w are not identically 0, the determinant of their coefficients must vanish. Thus

$$\begin{vmatrix} v_x + v_u u_x & w_x + w_u u_x \\ v_y + v_u u_y & w_y + w_u u_y \end{vmatrix} = 0,$$

or

$$(v_u w_y - v_y w_u)u_x + (v_x w_u - v_u w_x)u_y = (v_y w_x - v_x w_y).$$

Now since

$$Pv_x + Qv_y + Rv_u = 0,$$

$$Pw_x + Qw_y + Rw_u = 0,$$

we find upon solving for P and Q,

$$P = \frac{v_y w_u - v_u w_y}{v_x w_y - v_y w_x}\, R,$$

$$Q = \frac{v_u w_x - v_x w_u}{v_x w_y - v_y w_x}\, R,$$

or

$$\frac{v_y w_u - v_u w_y}{P} = \frac{v_u w_x - v_x w_u}{Q} = \frac{v_x w_y - v_y w_x}{R}\,.$$

When these are substituted in the equation involving u,

$$P u_x + Q u_y = R$$

is the result.

We note in passing that the two subsidiary equations involve two arbitrary constants of integration. The general solution is a relation between them.

Solutions to the first order equation give additional insight when expressed in parametric form. If we set the subisidiary equations equal to a new differential ds, then

$$\frac{dx}{ds} = P(x,y,u),\ \frac{dy}{ds} = Q(x,y,u),\ \frac{du}{ds} = R(x,y,u),$$

a system of ordinary differential equations, which, when P, Q, and R are Lipschitzian, has a unique solution, called a characteristic curve, for each set of initial conditions.

If initial values $x(t)$, $y(t)$, $u(t)$ are introduced along a curve c depending upon the parameter t, then the result $x(s,t)$, $y(s,t)$, $u(s,t)$ is a family of characteristic solutions. If the equations

$$x(s,t) = x, \ y(s,t) = y$$

are solved to yield

$$s = s(x,y), \ t = t(x,y),$$

then the solution $u = u(s(x,y), t(x,y))$ can be expressed in terms of x and y.

The only problem in the elimination of s and t occurs when the Jacobian of the transformation

$$\begin{vmatrix} x_s & y_s \\ x_t & y_t \end{vmatrix} = 0.$$

In this case

$$\frac{x_t}{y_t} = \frac{x_s}{y_s} = \frac{P}{Q},$$

and

$$\frac{u_t}{R} = \frac{u_x x_t + u_y y_t}{Pu_x + Qu_y} = \frac{y_t}{Q} = \frac{x_t}{P} .$$

This implies that the initial value curve c is also characteristic, and an infinite number of solutions pass through c. Thus we conclude that in order to have a unique solution u, initial values must be specified along a non-characteristic curve c.

XIII.2.3. <u>Examples.</u> 1. Let $P(x,y,u) = x$, $Q(x,y,u) = -y$, and $R(x,y,u) = xy$. The differential equation is

$$x \frac{\partial u}{\partial x} - y \frac{\partial u}{\partial y} = xy.$$

The subsidiary equations are

$$\frac{dx}{x} = \frac{dy}{-y} = \frac{du}{xy} .$$

Using the first two parts, we find

$$ydx + xdy = 0,$$

or $xy = a$. To find a second solution to the subsidiary equations, we use the first and third parts. Letting $xy = a$, we have

$$\frac{dx}{x} = \frac{du}{a}.$$

or $u = a \ln x + b$.   So $u - xy \ln x = b$ is a second solution.   The general

solution is then

$$F(u - xy \ln x, xy) = 0.$$

    2.   In higher dimensions the equation

$$P_1 u_{x_1} + \ldots + P_n u_{x_n} = R$$

has subsidiary equations

$$\frac{dx_1}{P_1} = \ldots = \frac{dx_n}{P_n} = \frac{du}{R} .$$

If $v_1 = a_1, \ldots, v_n = a_n$ are independent solutions to the subsidiary equation

then the general solution is

$$F(v_1, \ldots, v_n) = 0.$$

XIII.3.   <u>Second Order Equations</u>.   In this section we turn our attention to

those equations of second order which are linear in the second derivatives.

Our goal at this point is to classify these problems into standard categories

and to put them into a standard form.   In the interest of simplicity we shall

only consider those equations in two independent variables.   We are more con-

cerned at the moment with the forms involved than in rigorous theory.

XIII.3.1.   **Definition.**   Let   A(x,y), B(x,y), C(x,y)   be continuous throughout
a region   D   in the plane E^2.   We define the formal differential operator
L   by letting

$$Lu = A(x,y)\frac{\partial^2 u}{\partial x^2} + B(x,y)\frac{\partial^2 u}{\partial x \partial y} + C(x,y)\frac{\partial^2 u}{\partial y^2} + \text{lower order terms.}$$

L   is hyperbolic, parabolic, or elliptic whenever B^2-4AC is positive, zero,
or negative.

   We will assume that B^2-4AC remains in the same category throughout our
discussion.

XIII.3.2.   **Theorem.**   Let   L   be hyperbolic throughout a region   D   in E^2.
Then for each point (x_0, y_0) in   D   there exists a transformation $\xi = \xi(x,y)$,
$\eta = \eta(x,y)$ under which the resulting coefficients of $\dfrac{\partial^2 u}{\partial \xi^2}$ and $\dfrac{\partial^2 u}{\partial \eta^2}$ in   L
are both zero near (x_0, y_0).

Proof.   If $\xi = \xi(x,y)$ and $\eta = \eta(x,y)$,   then

$$u_x = u_\xi \xi_x + u_\eta \eta_x,$$

$$u_y = u_\xi \xi_y + u_\eta \eta_y.$$

$$u_{xx} = u_{\xi\xi}(\xi_x)^2 + 2u_{\xi\eta}\xi_x\eta_x + u_{\eta\eta}(\eta_x)^2 + u_\xi\xi_{xx} + u_\eta\eta_{xx},$$

$$u_{xy} = u_{\xi\xi}\xi_x\xi_y + u_{\xi\eta}(\xi_x\eta_y + \xi_y\eta_x) + u_{\eta\eta}\eta_x\eta_y + u_\xi\xi_{xy} + u_\eta\eta_{xy},$$

$$u_{yy} = u_{\xi\xi}(\xi_y)^2 + 2u_{\xi\eta}\xi_y\eta_y + u_{\eta\eta}(\eta_y)^2 + u_\xi\xi_{yy} + u_\eta\eta_{yy}.$$

Thus, computing Lu,   we find

$$Lu = u_{\xi\xi} [A(\xi_x)^2 + B\xi_x\xi_y + C(\xi_y)^2]$$

$$+ u_{\xi\eta} [2A\xi_x\eta_x + B\xi_x\eta_y + B\xi_y\eta_x + 2C\xi_y\eta_y]$$

$$+ u_{\eta\eta} [A(\eta_x)^2 + + B\eta_x\eta_y + C(\eta_y)^2]$$

$$+ \text{ lower order terms.}$$

If we temporarily assume that $A \neq 0$ near (x_0,y_0) and set the first and th

coefficients equal to zero, then

$$\xi_x/\xi_y = (-B\pm \sqrt{B^2-4AC})/2A,$$

$$\eta_x/\eta_y = (-B\pm \sqrt{B^2-4AC})/2A.$$

Now, the curves where ξ is constant satisfy

$$\xi(x,y) = c.$$

Differentiating, we find

$$\xi_x dx + \xi_y dy = 0,$$

or

$$dy/dx = - \xi_x/\xi_y.$$

Thus, the curves $\xi(x,y) = c$ may be found by solving the ordinary differential equation

$$dy/dx = (B - \sqrt{B^2-4AC})/2A,$$

where we have made a specific choice (+) in the arbitrary sign. We find as its solution $f_1(x,y) = c_1$. Similarly, solving

$$dy/dx = (B + \sqrt{B^2-4AC})/2A,$$

we find $f_2(x,y) = c_2$. This solution corresponds to $\eta(x,y)$ being held constant. We then choose

$$\xi = f_1(x,y) , \quad \eta = f_2(x,y)$$

as our transformation. The derivation is completed by dividing by the coeffients of $u_{\xi\eta}$. The case where $A = 0$, but $B \neq 0$ is similar. If $A = 0$, $B = 0$, there is no transformation necessary. The final form of the equation $Lu = 0$ is

$$u_{\xi\eta} + \text{lower order terms} = 0.$$

The curves $f_1(x,y) = c_1$, $f_2(x,y) = c_2$ are called characteristic curve
If initial values are arbitrarily assigned when y is fixed, then these valu
are propogated along the characteristic curves.

XIII.3.3. <u>An Example</u>. Let

$$Lu = u_{xx} + (5 + 2y^2)u_{xy} + (1+y^2)(4+y^2)u_{yy}.$$

The differential equations determining the characteristics are

$$dy/dx = [(5 + 2y^2) \pm 3]/2.$$

First fixing the arbitrary sign at (-) we find

$$dy/(1+y^2) = dx,$$

or

$$\tan^{-1}y - x = c_1.$$

If the sign is fixed at (+), then

$$dy/(4 + y^2) = dx,$$

and

$$(1/2) \tan^{-1}(y/2) - x = c_2.$$

Thus the transformation needed to reduce the first and third coefficients to zero is

$$\xi = \tan^{-1} y - x,$$

$$\eta = (1/2) \tan^{-1}(y/2) - x.$$

The coefficient of $u_{\xi\eta}$ is $-9/(y^2+1)(y^2+4)$.

XIII.3.4. Theorem. Let L be parabolic throughout a region D in E^2. Then for each point (x_0, y_0) in D there exists a transformation $\xi = \xi(x,y)$, $\eta = \eta(x,y)$ under which the resulting coefficients of $\dfrac{\partial^2 u}{\partial \xi^2}$ and $\dfrac{\partial^2 u}{\partial \xi \partial \eta}$ in L are both zero near (x_0, y_0).

Proof. Setting the coefficient of $\dfrac{\partial^2 u}{\partial \xi^2}$ equal to zero, we find

$$\xi_x / \xi_y = -B/2A,$$

since L is parabolic. Since $B^2 - 4AC = 0$, this implies that the coefficients of $\dfrac{\partial^2 u}{\partial \xi \partial \eta}$ is also zero. We therefore choose ξ in the same way as was done in Theorem XIII.3.2. η may be chosen any way, so long as the Jacobian $J(\xi, \eta) \neq 0$.

This, of course, is only possible when $A \neq 0$. If $A = 0$, then $B = 0$ also, and no transformation is necessary. The final form of the equation $Lu = 0$ may be reduced to

$$u_{\eta\eta} + \text{lower order terms} = 0$$

by dividing by the coefficient of $u_{\eta\eta}$.

The curve $\xi = f_1(x,y) = c_1$, a constant, is again a characteristic curve.

XIII.3.5. <u>An Example.</u> Let

$$Lu = u_{xx} + 2e^y u_{xy} + e^{2y} u_{yy}.$$

The differential equation determining ξ is then

$$dy/dx = e^y,$$

which has as its solution

$$e^{-y} + x = c_1.$$

We therefore let $\xi = e^{-y} + x$ and $\eta = y$. Then $Lu = e^{2\eta} u_{\eta\eta}$.

XIII.3.6. <u>Theorem. Let L be elliptic throughout a region D in E^2. Then</u>

<u>for each point (x_0,y_0) in D there exists a transformation $\xi = \xi(x,y)$,</u>

<u>$\eta = \eta(x,y)$ under which the resulting coefficients of $\dfrac{\partial^2 u}{\partial \xi^2}$ and $\dfrac{\partial^2 u}{\partial \eta^2}$ in L</u>

<u>are equal and the coefficient of $\dfrac{\partial^2 u}{\partial \xi \partial \eta}$ is zero near (x_0,y_0).</u>

Proof. As was accomplished in the hyperbolic case, we find the differential

equations which determine the characteristics. In this case, however, since

$B^2-4AC < 0$, the equations and their solutions will be complex. Let $F(x,y) = c$

be the solution to

$$dy/dx = (B - i\sqrt{4AC-B^2})/2A,$$

and let $\xi = \mathrm{Re}(F)$, $\eta = \mathrm{Im}(F)$:

$$F = \xi + i\eta.$$

Then, since

$$A(F_x)^2 + BF_xF_y + C(F_y)^2 = 0,$$

we find, upon separating real imaginary parts, that

$$A(\xi_x)^2 + B\xi_x\xi_y + C(\xi_y)^2 =$$

$$A(\eta_x)^2 + B\eta_x\eta_y + C(\eta_y)^2,$$

and

$$2A\xi_x\eta_x + B\xi_x\eta_y + B\xi_y\eta_x + 2C\xi_y\eta_y = 0.$$

Thus in L the first and third coefficients are equal; the middle coefficient
is 0.

The characteristics $F(x,y) = c_1$ and $\overline{F}(x,y) = c_2$ are complex in nature
in the elliptic case, and do not cause any difficulty.

XIII.3.7. <u>An Example.</u> Let

$$Lu = (x^2+1)u_{xx} + 4xu_{xy} + (x^2+1)u_{yy}.$$

Then the differential equation satisfied by one of the characteristics is

$$dy/dx = (2x + i(x^2-1))/(x^2+1),$$

so

$$y - \ln(x^2+1) - ix + 2i \tan^{-1}x = C.$$

Thus we let

$$\xi = y - \ln(x^2+1)$$

$$\eta = 2 \tan^{-1}x - x.$$

Under this transformation we find

$$Lu = \frac{(x^2-1)^2}{x^2+1} [u_{\xi\xi} + u_{\eta\eta}].$$

We remark that in general the reduction to a standard form is possible when the number of independent variables is less than or equal to three. When it is greater than three, the differential equations which must be satisfied overdetermine the coefficients.

However, when the coefficients are all constant, there exists a linear transformation of the coordinates which transforms any second order partial differential equation, linear in the second derivatives, into the form

$$\sum_{j=1}^{n} A_j \frac{\partial^2 u}{\partial \xi_j^2} + \text{lower order terms} = 0,$$

where $A_j = 1, 0,$ or -1, $j=1,\ldots,n$.

If all the coefficients in the transformed equation are nonzero and have the same sign, the equation is elliptic. If just one is of opposite sign, the equation is hyperbolic. If several are of opposite sign, such as in $u_{x_1 x_1} + u_{x_2 x_2} - u_{x_3 x_3} - u_{x_4 x_4} = 0$, the equation is ultrahyperbolic. If any of the coefficients A_j, $j=1,\ldots,n$, is zero the equation is parabolic. Ultrahyperbolic equations are almost never encountered in physical systems.

XIII.4. <u>Green's Formula.</u> In the last three chapters we shall need various theorems from elementary vector calculus. It is convenient at this point to give the appropriate definitions and state these theorems. Our setting is E^n.

XIII.4.1. Definition. Let $u = u(x_1,...,x_n)$ be differentiable in each variable. Then the gradient of u is

$$\text{grad } u = \nabla u = (\frac{\partial u}{\partial x_1},...,\frac{\partial u}{\partial x_n}).$$

This is an n-dimensional vector.

XIII.4.2. Definition. Let $v = (v_1,...,v_n)$ be an n-dimensional vector in which each component v_j, $j=1,...,n$, is a function of $x_1,...,x_n$. Then if $\frac{\partial v_j}{\partial x_j}$, $j=1,...,n$, exist, the divergence of v is

$$\text{div } v = \nabla \cdot v = \frac{\partial v_1}{\partial x_1} + ... + \frac{\partial v_n}{\partial x_n}.$$

As inferred in these definitions, the operator ∇ denotes the vector differential operator.

$$\nabla = (\frac{\partial}{\partial x_1},...,\frac{\partial}{\partial x_n}).$$

XIII.4.3. Theorem (Divergence Theorem). Let D be an open region in E^n containing an open region R and its piecewise smooth boundary S. Let n denote the outward unit normal vector of S with respect to R. If v is a continuously differentiable vector, then

$$\int_R \text{div } v \, dx = \int_S v \cdot n \, ds,$$

where   dx   indicates increments of volume and   ds   indicates increments

of surface area.

The proof usually consists of showing that

$$\int_S \cdots \int v_i dx_1 \cdots dx_{i-1} dx_{i+1} \cdots dx_n = \int_R \cdots \int \frac{\partial v_i}{\partial x_i} dx_1 \cdots dx_n$$

for suitable regions   R,   then extending this formula to more general regions

through addition and limit processes.   It can be found in any standard vector

analysis text.

XIII.4.4.   Definition.   1.   Let $k = (k_1, \ldots, k_p)$ be a multivariable summing

index.   We define $|k|$ by

$$|k| = \sum_{j=1}^p k_j .$$

2.   We define the partial differential operator D^k by

$$D^k = \partial^{|k|} / \partial x_1^{k_1} \cdots \partial x_n^{k_n} .$$

XIII.4.5.   Definition.   Let the function $u = u(x_1, \ldots, x_n)$ be sufficiently

differentiable in an open region   D   of E^n.   We then define the partial

differential operator   L   by

$$Lu = \sum_{|k| \leq p} a_k(x) D^k u ,$$

where $a_k(x) = a_{k_1 \cdots k_n}(x_1, \ldots x_n)$ are appropriate coefficients.

XIII.4.6. Definition. Let the function $v = v(x_1,\ldots,x_n)$ be sufficiently
differentiable in an open region D of E^n, and let the functions $a_k(x)$
be sufficiently differentiable such that $D^k a_k$ exists for $|k| \leqq p$. The
formal adjoint of L, denoted by L^+, is defined by

$$L^+ v = \sum_{|k|\leqq p} (-1)^{|k|} D^k (\overline{a_k(x)} v).$$

The following theorem is the multidimensional analog of Green's formula
Corollary V.2.3.

XIII.4.7. Theorem (Green's Formula). Let D be an open region in E^n con
taining an open region R and its piecewise smooth boundary S. Let n
denote the outward unit normal vector of S with respect to R. Then for
sufficiently differentiable functions u and v and appropriately defined
partial differential operators L and L^+,

$$\int_R [\overline{v}(Lu) - \overline{(L^+ v)}u]dx = \int_S J \cdot n\, ds,$$

where $J = J(u,v)$ is an appropriate vector function of u and v.

We shall prove this theorem only in the case $p = 2$, since higher order
operators yield little in the way of enlightment.

Proof. When $p = 2$,

$$Lu = [\sum_{j=1}^{n} \sum_{k=1}^{n} A_{jk} \frac{\partial^2}{\partial x_j \partial x_k} + \sum_{j=1}^{n} B_j \frac{\partial}{\partial x_j} + C]u.$$

Now note that

$$
\overline{vA}_{jk} \frac{\partial^2 u}{\partial x_j \partial x_k} = \frac{\partial}{\partial x_j} [(\overline{vA}_{jk}) \frac{\partial u}{\partial x_k}] - [\frac{\partial}{\partial x_j}(\overline{v}(A_{jk}))] \frac{\partial u}{\partial x_k} ,
$$

$$
= \frac{\partial}{\partial x_j} [(\overline{vA}_{jk}) \frac{\partial u}{\partial x_k}] - \frac{\partial}{\partial x_k}[[\frac{\partial}{\partial x_j}(\overline{vA}_{jk})]u] + [\frac{\partial}{\partial x_j \partial x_k}(\overline{vA}_{jk})]u.
$$

Further

$$
\overline{vB}_j \frac{\partial u}{\partial x_j} = \frac{\partial}{\partial x_j} [\overline{vB}_j u] - u \frac{\partial}{\partial x_j} [\overline{vB}_j].
$$

There is no loss of generality in assuming that $A_{jk} = A_{kj}$, $j, k = 1, \ldots, n$. Then

$$
\overline{v}Lu = \{ \sum_{j=1}^{n} \sum_{k=1}^{n} \frac{\partial^2}{\partial x_j \partial x_k} [\overline{vA}_{jk}] - \sum_{j=1}^{n} \frac{\partial}{\partial x_j} [\overline{vB}_j] + \overline{vC}\}u
$$

$$
+ \sum_{j=1}^{n} \frac{\partial}{\partial x_j} \{\overline{v} \sum_{k=1}^{n} A_{jk} \frac{\partial u}{\partial x_k} - \sum_{k=1}^{n} \frac{\partial}{\partial x_k} [\overline{vA}_{jk}]u + \overline{vB}_j u\}.
$$

The first expression is $\overline{(L^+v)}u$. If we define the vector $J = (J_1, \ldots, J_n)$

by

$$
J_j = \{\overline{v} \sum_{k=1}^{n} A_{jk} \frac{\partial u}{\partial x_k} - \sum_{k=1}^{n} \frac{\partial}{\partial x_k} [\overline{vA}_{jk}]u + \overline{vB}_j u\},
$$

then the Divergence Theorem shows

$$
\int_R [\overline{v}(Lu) - \overline{(L^+v)}u]dx = \int_S J \cdot n \ ds.
$$

The following examples are the subjects of the last three chapters.
The application of Green's formula will be essential, especially when discussing Green's functions, which produce solutions of not only nonhomogeneou
problems, but also homogeneous problems with nonhomogeneous boundary conditi

XIII.4.8. Examples. 1. Let L be the Laplace operator

$$L = \nabla^2 = \frac{\partial^2}{\partial x_1^2} + \ldots + \frac{\partial^2}{\partial x_n^2} \, .$$

Then an easy computation shows that $L^+ = L$.   L is formally self-adjoint.
Green's formula is

$$\int_R [\overline{v}(\nabla^2 u) - \overline{(\nabla^2 v)}u]\,dx = \int_S J \cdot n \, ds,$$

where $J = (J_1, \ldots, J_n)$ and

$$J_j = (\overline{v} \frac{\partial u}{\partial x_j} - \frac{\partial \overline{v}}{\partial x_j} u),$$

$j=1,\ldots,n$. The right side may also be written as $\int_S [\overline{v} \frac{\partial u}{\partial n} - \frac{\partial \overline{v}}{\partial n} u]\,ds$,   where
$(n_1, \ldots, n_n)$ is the unit outward normal, and

$$\frac{\partial u}{\partial n} = \sum_{j=1}^{n} \frac{\partial u}{\partial x_j} n_j \, .$$

2. Let L be the heat conduction operator

$$L = \frac{\partial}{\partial t} - \nabla^2 = \frac{\partial}{\partial t} - (\frac{\partial^2}{\partial x_1^2} + \ldots + \frac{\partial^2}{\partial x_n^2}).$$

Then

$$L^{+} = -\frac{\partial}{\partial t} - (\frac{\partial^2}{\partial x_1^2} + \ldots + \frac{\partial^2}{\partial x_n^2}),$$

and

$$\bar{v}(Lu) - (L^{+}\bar{v})u = \frac{\partial}{\partial t}(\bar{v}u) + \sum_{j=1}^{n} \frac{\partial}{\partial x_j} [\frac{\partial \bar{v}}{\partial x_j} u - \bar{v} \frac{\partial u}{\partial x_j}].$$

If we let

$$J = (\bar{v}u, \frac{\partial \bar{v}}{\partial x_1} u - \bar{v} \frac{\partial u}{\partial x_1}, \ldots, \frac{\partial \bar{v}}{\partial x_n} u - \bar{v} \frac{\partial u}{\partial x_n}),$$

we can write Green's formula in the standard way, or, if we let e_t denote a unit vector in the t direction, then

$$J = e_t(\bar{v}u) + u \text{ grad}_x \bar{v} - \bar{v} \text{ grad}_x u,$$

where grad_x denotes the gradient taken over the x coordinates only. Then

$$\int_t \int_R [\bar{v}(Lu) - \overline{(L^{+}v)}u]dxdt =$$

$$\int_S [e_t(\bar{v}u) + u \text{ grad}_x \bar{v} - \bar{v} \text{ grad}_x u] \cdot n \text{ ds}.$$

3.   Let L be the wave operator

$$L = \frac{\partial^2}{\partial t^2} - \nabla^2 = \frac{\partial^2}{\partial t^2} - (\frac{\partial^2}{\partial x_1^2} + \ldots + \frac{\partial^2}{\partial x_n^2}).$$

Then L is self-adjoint, and

$$\int_t \int_R [\bar{v}(Lu) - \overline{(Lv)}u]dxdt =$$

$$\int_S [e_t(\bar{v} \frac{\partial u}{\partial t} - \frac{\partial \bar{v}}{\partial t} u) + u \; grad_x \; \bar{v} - \bar{v} \; grad_x \; u] \cdot n \; ds.$$

XIII. Exercises

1. Solve the partial differential equation

$$a_0 u_x + a_1 u_y + a_2 u = 0,$$

where a_0, a_1, a_2 are constants. Find the solution passing through $x =$ █
$y = t^2$, $u = t^3$.

2. Show that if the operator L is hyperbolic in a region D in E^2, the█
there exists a transformation $\xi = \xi(x,y)$, $\eta = \eta(x,y)$ such that

$$Lu = u_{\xi\xi} - u_{\eta\eta} + \text{lower order terms.}$$

3. Reduce the following operators to standard form:

$$Lu = u_{xx} + 5u_{xy} + 4u_{yy},$$

$$Lu = u_{xx} + 4u_{xy} + 4u_{yy},$$

$$Lu = u_{xx} + 2u_{xy} + 2u_{yy}.$$

4. Show that the type of second order partial differential operator does not change under a real transformation of coordinates.

5. Compute Green's formula for Exercise XIII.3 and for Examples XIII.3.3, XIII.3.5, and XIII.3.7.

References

1. R. Courant and D. Hilbert, "Methods of Mathematical Physics, vol. II, Interscience, New York, 1962.

2. P.R. Garabedian, "Partial Differential Equations," John Wiley and Sons, New York, 1967.

3. F. Miller, "Partial Differential Equations," John Wiley and Sons, New York, 1947.

4. I.G. Petrovskii, "Partial Differential Equations," Scripta Technica, London, 1967.

5. I. Stakgold, "Boundary Value Problems of Mathematical Physics, vol. II," Macmillan, New York, 1968.

6. H.F. Weinberger, "A First Course in Partial Differential Equations," Blaisdell, New York, 1965.

XIV. DISTRIBUTIONS

Unlike ordinary differential equations, where practice and theory seem t
be quite compatable, the physical applications of partial differential
equations are frequently too idealized to possess solutions. For instance,
the theoretical description of the plucked string will not satisfy the partia
differential equation from which it came. Although the solution seems to be
quite reasonable intuitively, it can not be differentiated often enough to
satisfy the idealized wave equation (See Example XVII.4., no.2).

In order to circumvent this difficulty mathematicians have invented two
different but related settings in which these idealized solutions make good
sense. One is called Operational Calculus and is due primarily to the Polish
mathematician Jan Mikusinski. It is quite similar but substantially more
general than the Laplace transform. The other, perfected by the Frenchman
Laurent Schwartz, is called Distributions. Because it is much more widely
used in the theory of partial differential equations, we have chosen to use a
distributional setting for a detailed look at Laplace's equation, the heat
equation, and the wave equation in the following chapters.

Our setting is E^n, although any open region could be used just as easi
with only minor modifications.

XIV.1.   Test Functions and Distributions.

XIV.1.1.   Definition.   A measurable function $f = f(x_1, \ldots, x_n)$ is locally integrable if

$$\int_R |f(x)|\, dx = \int_R \cdots \int |f(x_1,\ldots,x_n)|\, dx_1 \cdots dx_n$$

exists for every bounded measurable set R in E^n.

$f(x)$ is locally integrable on a hypersurface S in E^n if $\int_\Sigma |f(x)|\, ds$

exists for every bounded measurable set Σ in S.

XIV.1.2. Definition. The support of a function $f(x)$ is the set of all x

such that $f(x) \neq 0$.

We have already encountered this concept before when dealing with

infinitely differentiable functions with compact support in the spaces

$L^p(-\infty,\infty)$, $1 \leq p < \infty$.

XIV.1.3. Definition. A function $\phi(x) = \phi(x_1,\ldots,x_n)$ is called a test

function if

1. ϕ is infinitely differentiable. That is, $D^k\phi$ exists for all

indices $k = (k_1,\ldots,k_n)$,

2. ϕ has compact support. That is, there exists a number $N > 0$

such that $\phi(x) = 0$ when $\|x\| = [\sum_{j=1}^{n} x_j^2]^{1/2} > N$.

We denote the set of all test functions by D.

As an example of a test function we present

$$\phi(x) = \begin{cases} 0, & \text{when} \quad \|x\| \geq a, \\[2em] \exp[(\|x\|^2 - a^2)^{-1}], & \text{when} \quad \|x\| \leq a. \end{cases}$$

It is evident that D is a linear space, that $D^k\phi$ is in D when ϕ is, and that if ϕ is in D and f is infinitely differentiable, then $f\phi$ is in D.

XIV.1.4. <u>Definition</u>. <u>A sequence</u> $\{\phi_j\}_{j=1}^{\infty}$ <u>converges to</u> ϕ_0 <u>in D if</u>

  1. <u>All ϕ_j, j=1,..., and ϕ_0 vanish outside a common region.</u>

  2. $\underline{D^k\phi_j}$ <u>approaches</u> $\underline{D^k\phi_0}$ <u>as j approaches ∞ for all indices k.</u>

Let us recall that a linear functional is a linear function which takes on real or complex values.

XIV.1.5. <u>Definition</u>. <u>A distribution is a continuous linear functional on</u> <u>the linear space D. We denote the set of all distributions by D'.</u>

There is a special set of distributions which are most useful: those generated by locally integrable functions.

XIV.1.6. <u>Theorem</u>. <u>Every locally integrable function f defines a dis-</u> <u>tribution through the formula</u>

$$<f,\phi> = \int_{E^n} f(x)\phi(x)\,dx.$$

Proof. Let the support of ϕ be denoted by R_ϕ. Then

$$|<f,\phi>| \leq \sup_{x \in R_\phi} |\phi(x)| \int_{R_\phi} |f(x)| dx < \infty.$$

If ϕ approaches 0 in D, then $<f,\phi>$ also approaches 0, so

$<f,\phi>$ is continuous.

As we shall see, there are distributions other than those generated by

locally integrable functions. These are, nonetheless, important enough to

give them a special name.

XIV.1.7. <u>Definition.</u> <u>A distribution is regular if it can be written in the</u>

<u>form</u> $\int_{E^n} f(x)\phi(x) dx.$ <u>All other distributions are singular.</u>

Just as with the test functions it is evident that the space of dis-

tributions D′ is a linear space. Furthermore a topology, a collection of

open sets, etc., can be induced on the space of distributions D′. Under

this topology it can be shown that the subset of regular distributions is

dense in D′. Hence every distribution is the limit of regular distributions.

XIV.1.8. <u>Examples.</u> 1. Let R be a fixed region in E^n, and define the

distribution H_R by

$$<H_R,\phi> = \int_R \phi(x) dx.$$

H_R is generated by the characteristic function of R,

$$K_R(x) = \begin{cases} 1, & \text{when } x \text{ is in } R, \\ \\ 0, & \text{when } x \text{ is not in } R. \end{cases}$$

Thus H_R is a regular distribution. It is called the Heaviside distribution after its inventor.

  2. Let ξ be a fixed point in E^n, and define the distribution δ_ξ by

$$<\delta_\xi, \phi> = \phi(\xi).$$

If ϕ approaches 0 in D, then clearly $\phi(\xi)$ also approaches 0, so δ_ξ is in D'. Frequently δ_ξ is symbollically written as

$$<\delta_\xi, \phi> = \int_{E^n} \delta(x-\xi)\phi(x)\,dx,$$

where

$$\delta(x-\xi) = \begin{cases} 0, & \text{when } x \neq \xi, \\ \\ \infty, & \text{when } x = \xi, \end{cases}$$

and $\delta(x-\xi)$ also has the property

$$\int_R \delta(x-\xi)\,dx = \begin{cases} 1, & \text{when } \xi \text{ is in } R, \\ 0, & \text{when } \xi \text{ is not in } R. \end{cases}$$

Of course, there is no such function as $\delta(x-\xi)$, but the symbolism works quite well. δ_ξ is a perfectly valid distribution. It is also named after its inventor. It is called the Dirac distribution or the Dirac delta function.

3. If f is locally integrable, the translate of the distribution f is defined by

$$<f(x-a),\phi(x)> = \int_{E^n} f(x-a)\phi(x)\,dx.$$

Since

$$\int_{E^n} f(x-a)\phi(x)\,dx = \int_{E^n} f(x)\phi(x+a)\,dx,$$

$$= <f(x),\phi(x+a)>,$$

we use this as the basis of a definition for all distributions:

XIV.1.9. <u>Definition</u>. <u>The translate of a distribution t by a is given by</u>

$$<t(x-a),\phi(x)> = <t(x),\phi(x+a)>.$$

The product of two distributions can also be defined, provided one is infinitely differentiable:

XIV.1.10. Definition. Let t be a distribution and let g be infinitely

differentiable. Then the distribution tg is defined by

$$<tg,\phi> = <t,g\phi>.$$

The extra conditions imposed upon g are necessary. For instance,

if t is generated by the locally integrable function $1/\|x\|^{1/2}$ and g als

equals $1/\|x\|^{1/2}$, then tg would have to be generated by $1/\|x\|$, which

cannot generate a distribution because of its nonintegrable behavior near x

Now suppose that f and its first partial derivative $\frac{\partial f}{\partial x_1}$ are locally

integrable. Then

$$<\frac{\partial f}{\partial x_1},\phi> = \int_{E^n} \frac{\partial f}{\partial x_1}(x)\,\phi(x)\,dx$$

is well defined. the integral can be integrated by parts to yield

$$<\frac{\partial f}{\partial x_1},\phi> = - \int_{E^n} f(x)\,\frac{\partial \phi}{\partial x_1}(x)\,dx = - <f,\frac{\partial \phi}{\partial x_1}>.$$

Again we use this as a basis for a more general definition:

XIV.1.11. Definition. The derivative of a distribution t, $\frac{\partial t}{\partial x_1}$ is

given by

$$<\frac{\partial t}{\partial x_1},\phi> = - <t,\frac{\partial \phi}{\partial x_1}>$$

More generally, $D^k t$ is defined by

$$<D^k t, \phi> = (-1)^{|k|} <t, D^k \phi>.$$

This is one of the most important properties of distributions. They all inherit infinite differentiability from the test functions, even though they may not be differentiable in the ordinary sense.

XIV.1.12. Examples. 1. The derivative $\dfrac{\partial \delta_\xi}{\partial x_1}$ is calculated as follows:

$$<\frac{\partial \delta_\xi}{\partial x_1}, \phi> = - <\delta_\xi, \frac{\partial \phi}{\partial x_1}>,$$

$$= - <\delta(x-\xi), \frac{\partial \phi}{\partial x_1}(x)>,$$

$$= - \frac{\partial \phi}{\partial x_1}(\xi).$$

2. If f is infinitely differentiable,

$$<f\delta_\xi, \phi> = <f(x)\delta(x-\xi), \phi(x)>,$$

$$= <\delta(x-\xi), f(x)\phi(x)>,$$

$$= f(\xi)\phi(\xi),$$

$$= f(\xi)<\delta_\xi, \phi>.$$

This may be used as a basis for defining the distribution $f(x)\delta_\xi(x)$ by $f(\xi)\delta_\xi(x)$, when f is only continuous at $x = \xi$.

3. If f is infinitely differentiable

$$<f\ \frac{\partial\delta_\xi}{\partial x_1},\phi> \ = \ <\frac{\partial\delta_\phi}{\partial x_1},f\phi>,$$

$$= \ -<\delta_\xi\ ,\frac{\partial(f\phi)}{\partial x_1}>,$$

$$= \ - \ <\delta(x-\xi),f(x)\ \frac{\partial\phi}{\partial x_1}(x)\ + \ \frac{\partial f}{\partial x_1}(x)\phi(x)>,$$

$$= \ -f(\xi)\ \frac{\partial\phi}{\partial x_1}(\xi)\ - \ \frac{\partial f}{\partial x_1}(\xi)\phi(\xi),$$

$$= \ f(\xi)<\frac{\partial\delta_\xi}{\partial x_1},\phi> \ - \ \frac{\partial f}{\partial x_1}(\xi)<\delta_\xi,\phi>.$$

This may also be extended to those functions f which are merely continuous differentiable to yield

$$f(x)\ \frac{\partial\delta_\xi}{\partial x_1}\ = \ f(\xi)\ \frac{\partial\delta_\xi}{\partial x_1}\ - \ \frac{\partial f}{\partial x_1}(\xi)\delta_\xi.$$

4. If $L = \sum_{|k|\leq p} a_k(x)D^k$ has infinitely differentiable real coefficien[t]s

then we may define Lt, where t is an arbitrary distribution, by

$$<Lt,\phi> \ = \ <t,L^+\phi>.$$

When f is sufficiently differentiable,

$$<Lf,\phi> = \int_{E^n} Lf(x)\phi(x)dx,$$

$$= \int_{E^n} f(x)L^+\phi(x)dx,$$

$$= <f,L^+\phi>,$$

so the definition is consistent when applied to these functions. There is a similar definition available when the coefficients are complex.

Finally, we can extend the concept of support to distributions, again through test functions.

XIV.1.13. <u>Definition</u>. <u>A distribution t vanishes in a region R if $<t,\phi> = 0$ for all test functions ϕ with support in R. t has support in R if $<t,\phi> = 0$ for all test functions ϕ with support in the compliment of R.</u>

For example we note that the delta function δ_ξ has support only $x = \xi$. The delta function $\delta(=\delta_0)$ has support only at $x = 0$.

XIV.2. <u>Limits of Distributions</u>. If a collection of distributions depends upon a parameter, we may define limit processes just as with ordinary functions.

XIV.2.1. <u>Definition</u>. <u>Let the distribution t_α depend upon the real valued parameter α. The</u>

$$\lim_{\alpha \to \alpha_0} t_\alpha = t_{\alpha_0}.$$

<u>if</u>

$$\lim_{\alpha \to \alpha_0} <t_\alpha,\phi> = <t_{\alpha_0},\phi>$$

<u>for all ϕ in D.</u>

   <u>If α takes on integer values only,</u>

$$\lim_{n \to \infty} t_n = t_0$$

<u>if</u>

$$\lim_{n \to \infty} <t_n,\phi> = <t_0,\phi>$$

for all ϕ in D.

   Now comes a most remarkable theorem. Its importance to partial differential equations is hard to overemphasize. It is one of the main reasons distributions make an advantageous setting for them!!! <u>Attention,</u> <u>please</u>!!!

XIV.2.2. <u>Theorem</u>. <u>Every convergent sequence or convergent series of distributions can be differentiated termwise with respect to any combinations of the variables $x_1,\ldots,x_n$ ($x = (x_1,\ldots,x_n)$) without affecting the convergence.</u>

   This theorem is not true in the ordinary sense. For instance, the series $\sum\limits_{n=1}^{\infty} (1/n^2) \cos nx$ converges uniformly for all x. Its second

derivative termwise, however, would be $- \sum\limits_{n=1}^{\infty} \cos nx$, which fails to converge

in the usual sense. (See Example XVII.4, no.2).

Proof. Clearly we need only show that convergence of a sequence is unaffected

by one differentiation. Other steps will be of a similar nature.

Suppose $\lim\limits_{\alpha \to \alpha_\phi} t_\alpha = t_{\alpha_0}$. Then

$$\lim_{\alpha \to \alpha_0} \langle \frac{\partial t_\alpha}{\partial x_1}, \phi \rangle = - \lim_{\alpha \to \alpha_0} \langle t_\alpha, \frac{\partial \phi}{\partial x_1} \rangle ,$$

$$= - \langle t_{\alpha_0}, \frac{\partial \phi}{\partial x_1} \rangle ,$$

$$= \langle \frac{\partial t_{\alpha_0}}{\partial x_1}, \phi \rangle .$$

Thus $\lim\limits_{\alpha \to \alpha_0} \dfrac{\partial t_\alpha}{\partial x_1} = \dfrac{\partial t_{\alpha_0}}{\partial x_1}$.

The convergence of differentiated series follows from the convergence of

the differentiated sequences of partial sums.

XIV.2.3. Examples. 1. The series

$$f(x) = \sum_{n=1}^{\infty} \frac{1}{n^2} \cos nx$$

converges uniformly for all x, is continuous, and is, therefore, also

a regular distribution. As a distribution it is infinitely differentiable.

Thus

$$f'(x) = - \sum_{n=1}^{\infty} \frac{1}{n} \sin nx,$$

$$f''(x) = - \sum_{n=1}^{\infty} \cos nx,$$

. . .

$$f^{(2j-1)}(x) = (-1)^j \sum_{n=1}^{\infty} n^{2j-3} \sin nx,$$

$$f^{(2j)}(x) = (-1)^j \sum_{n=1}^{\infty} n^{2j-2} \cos nx,$$

. . .

also converge as distributions for all $j=1,\dots$.

2. The function

$$f(x) = \begin{cases} 1, & \text{when } 2j \le x < 2j+1, \\ \\ -1, & \text{when } 2j+1 \le x < 2j+2, \end{cases} \quad j \text{ an integer,}$$

is piecewise continuous and is thus a distribution. It can be represented

by the series

$$f(x) = \sum_{j=-\infty}^{\infty} 2[H(x-2j) - H(x-[2j+1])] - 1,$$

where

$$H(x) = H_{[0,\infty)}(x) = \begin{cases} 1, & \text{when } x \in [0,\infty), \\ \\ 0, & \text{when } x \notin [0,). \end{cases}$$

Now $H'(x) = \delta(x)$, so

$$f'(x) = \sum_{j=-\infty}^{\infty} 2[\delta(x-2j) - \delta(x-[2j+1])].$$

The derivative of a distribution t_α with respect to the parameter

α is defined in the usual way:

$$dt_\alpha/d\alpha = \lim_{h \to 0} [t_{\alpha+h} - t_\alpha]/h,$$

when the limit exists.

XIV. An Example. Let us compute $\dfrac{\partial \delta(x-\xi)}{\partial \xi_1}$, where $\xi = (\xi_1, \ldots, \xi_n)$. If e_1 denotes a unit vector in the x_1 direction, then

$$<\frac{\partial \delta(x-\xi)}{\partial \xi_1}, \phi> = \lim_{h \to 0} [<\delta(x-\xi-he_1), \phi> - <\delta(x-\xi), \phi>]/h,$$

$$= \lim_{h \to 0} [\phi(\xi_1+he_1, \xi_2, \ldots, \xi_n) - \phi(\xi_1, \ldots, \xi_n)]/h$$

$$= \frac{\partial \phi}{\partial x_1} (\xi).$$

But

$$- <\frac{\partial \delta(x-\xi)}{\partial x_1}, \phi> = \frac{\partial \phi}{\partial x_1}(\xi).$$

Thus $\dfrac{\partial}{\partial \xi_1} \delta(x-\xi) = - \dfrac{\partial}{\partial x_1} \delta(x-\xi).$

Another example of distribution convergence concerns those which converge to the δ function.

XIV.2.6. Theorem. Let f be a nonnegative integrable function satisfying $\int_{E^n} f(x)dx = 1,$ and let

$$f_\alpha(x) = (1/\alpha^n) f(x/\alpha),$$

$$= (1/\alpha^n) f(x_1/\alpha, \ldots, x_n/\alpha).$$

__Then__ $\lim_{\alpha \to 0} f_\alpha(x) = \delta(x).$

Proof. By letting $u = x/\alpha,$ we find

1. $\int_{E^n} f_\alpha(x) dx = 1,$

2. $\lim_{\alpha \to 0} \int_{|x|>A} f_\alpha(x) dx = 0,$

3. $\lim_{\alpha \to 0} \int_{|x|<A} f_\alpha(x) dx = 1.$

Now let ϕ be an arbitrary test function, and let $\eta(x) = \phi(x) - \phi(0).$ Then

$$\int_{E^n} f_\alpha(x)\eta(x) dx = \int_{|x|\leq A} f_\alpha(x)\eta(x) dx + \int_{|x|>A} f_\alpha(x)\eta(x) dx.$$

Since ϕ is bounded, so is η by some number M. Further let

$N(A) = \sup_{|x|\leq A} |\eta(x)|.$ We now have

$$\left| \int_{E^n} f_\alpha(x)\eta(x) dx \right| \leq N(A) + M \int_{|x|>A} f_\alpha(x) dx.$$

Now η is continuous and $\eta(0) = 0,$ so $\lim_{A \to 0} N(A) = 0.$ We already know

that $\lim_{\alpha \to 0} \int_{|x|>A} f_\alpha(x) dx = 0.$ Thus if $\varepsilon > 0$ is arbitrary, we first choose

A sufficiently small to have $N(A) < \varepsilon/2$, then α sufficiently small to

have $\int_{|x|>A} f_\alpha(x)dx < \varepsilon/2M$. Then

$$\left| \int_{E^n} f_\alpha(x)\eta(x)dx \right| < \varepsilon,$$

or

$$\lim_{\alpha\to 0} \int_{E^n} f_\alpha(x)\phi(x)dx = \lim_{\alpha\to 0} \int_{E^n} f_\alpha(x)\phi(0)dx,$$

$$= \phi(0),$$

$$= \int_{E^n} \delta(x)\phi(x)dx.$$

XIV.2.7. **Examples.** 1. In E, the functions

$$f_\alpha(x) = \begin{cases} 1/(2\alpha), & \text{when } -\alpha \le x \le \alpha, \\ \\ 0, & \text{when } |x| > \alpha, \end{cases}$$

converge to $\delta(x)$.

2. In E^2, the functions $\alpha/[2\pi(x_1^2 + x_2^2 + a^2)^{3/2}]$ converge to $\delta(x_1,x_2$

3. In E^2, the functions $\alpha/[\pi^2(x_1^2 + x_2^2 + x_3^2 + a^2)^2]$ converge to

$\delta(x_1,x_2,x_3)$.

4. In E^n, the functions $(1/\sqrt{\pi}\ a)^n e^{-(x_1^2 + \ldots + x_n^2)/a^2}$ converge

to $\delta(x_1,\ldots,x_n)$.

For higher dimensions distributions may be composed of products of several lower dimensional distributions:

XIV.2.8. __Definition.__ __Let__ $t_1(x_1)$ __and__ $t_2(x_2)$ __be distributions on__ E^{n_1} __and__ E^{n_2}. __The product distribution__ $t(x_1,x_2) = t_1(x_1)t_2(x_2)$ __on__ $E^{n_1+n_2}$ __is given by__ $<t,\phi(x_1,x_2)> = <t(x_1),<t_2(x_2),\phi(x_1,x_2)>>$.

For example the two dimensional δ-function $\delta(x_1,x_2) = \delta(x_1)\delta(x_2)$, the product of two one-dimensional δ-functions. In general

$$\delta(x_1,\ldots,x_n) = \prod_{j=1}^{n} \delta(x_j).$$

Finally, we note that in E, if f_1 and f_2 are integrable functions, then the convolution

$$f_1 * f_2(x) = (1/\sqrt{2\pi})\int_{-\infty}^{\infty} f_1(t)f_2(t-x)dt$$

is also. In general, however the convolution for distributions cannot be defined by an iteration, since

$$<f_1 * f_2,\phi> = \int_{-\infty}^{\infty} (1/\sqrt{2\pi})\int_{-\infty}^{\infty} f_1(t)f_2(x-t)dt\phi(x)dx,$$

$$= \int_{-\infty}^{\infty} f_1(t)[(1/\sqrt{2\pi})\int_{-\infty}^{\infty} f_2(x-t)\phi(x)dx]dt,$$

and the expression within the brackets does not have compact support. If, however, f_2 had compact support, then so does $f_2 * \phi$, and we may use this iteration as a definition:

XIV.2.9. <u>Definition</u>. <u>Let f_1 and f_2 be one-dimensional distributions and let f_2 have compact support. Then we define the convolution $f_1 * f_2$ by</u>

$$<f_1 * f_2, \phi> = <f_1, f_2 * \phi> .$$

XIV.2.10. <u>Examples</u>. 1. Let t be an arbitrary distribution. Then $t * \delta$ is defined by

$$<t * \delta, \phi> = <t(x), (1/\sqrt{2\pi})<\delta(y), \phi(x-y)>>,$$

$$= <t(x), (1/\sqrt{2\pi})\phi(x)>,$$

$$= <(1/\sqrt{2\pi})t, \phi>.$$

Or $t * \delta = (1/\sqrt{2\pi})t$. This implies that $\sqrt{2\pi}\, \delta$ is a sort of identity element among distributions so long as the convolution is defined. This is the nonexistent identity in the Banach algebra $L^1(-\infty, \infty)$. Since $L^1(-\infty, \infty)$ may also be considered as a set of regular distributions, we note that its identi belongs within a larger class of elements. $L^1(-\infty, \infty)$ is simply to restricti

2. If L is an arbitrary differential operator with infinitely differer able coefficients, then

$<L\delta*t,\phi> = <t(x),(1/\sqrt{2\pi})<L\delta(y),\phi(x-y)>>,$

$= <t(x),(1/\sqrt{2\pi})<\delta(y),L^+\phi(x-y)>>,$

$= <t(x),(1/\sqrt{2\pi})L^+\phi(x)>,$

$= <(1/\sqrt{2\pi})Lt,\phi>.$

Or $L\delta*t = (1/\sqrt{2\pi})Lt.$

The factor $1/\sqrt{2\pi}$ was originally introduced into the convolution so that the Fourier transform of a convolution would be an ordinary product. It has reappeared above a bit awkwardly. It can, of course, be removed from the definition of a convolution, which makes the formulas above neater. But then the Fourier transform formula is more complicated. The reader can make his choice and adjust the formulas accordingly.

XIV.3.   Fourier Transforms of Distributions.   If we attempt to generalize the formula

$$Ff(x) = (1/\sqrt{2\pi})\int_{-\infty}^{\infty} e^{-ixy}f(x)dx$$

to define the Fourier transform of a distribution, we are in trouble because e^{-ixy} is not a test function. The formula may not have meaning. If we attempt to define it by

$$<Ff,\phi> = <f,F\phi>,$$

we again are in trouble since $F\phi$ is not necessarily a test function.

To circumvent these difficulties a change of setting has been found to be most satisfactory: we enlarge the space of test functions. In so doing we also restrict the space of distributions, but this will cause us no problem. We shall work in 1 dimension for convenience. The generalizatio to higher dimensions is obvious.

XIV.3.1. Definition. A function $\phi(x)$ belongs to S, the space of test functions of rapid decay, if

1. $\phi(x)$ is infinitely differentiable.

2. $\phi(x)$, as well as all its derivatives, vanish at $|x| = \infty$ faster than the recriprocal of any polynomial. Thus for each j, k,

$$\lim_{|x|\to\infty} x^j \frac{d^k\phi}{dx^k} = 0, \quad j, \ k=0,1,\dots .$$

It is evident that $D \subset S$, but not the reverse. e^{-x^2} is in S, but not in D.

XIV.3.2. Definition. A sequence $\{\phi_j\}_{j=1}^{\infty}$ converges to ϕ_0 in S if, for all k, $\ell=0,\dots,$

$$\lim_{j\to\infty} \sup_{x\in E} \left| x^k \left[\frac{d^\ell \phi_j(x)}{dx^\ell} - \frac{d^\ell \phi_0(x)}{dx^\ell} \right] \right| = 0.$$

This is slightly different than convergence in D.

XIV.3.3. Definition. A distribution of slow growth is a continuous linear

functional on the space S. The space of all continuous linear functionals

on S is denoted by S'.

 Just as in D' the regular distributions formed a special subset, in

S' the functions of slow growth are especially important:

XIV.3.4. Definition. A function f(x) is of slow growth if

 1. f(x) is locally integrable. That is, $\int_I |f(x)| dx < \infty$ for all finite

 intervals I.

 2. There exist constants c,n,r such that $|f(x)| < c|x|^n$, when

 $|x| > r$.

XIV.3.5. Theorem. Every function of slow growth generates a distribution

of slow growth through the formula

$$<f,\phi> = \int_{-\infty}^{\infty} f(x)\phi(x)dx.$$

Proof. It is sufficient to show that if ϕ_j approaches 0 in S, then

$<f,\phi_j>$ also approaches 0. For each j,

$$\int_{-\infty}^{\infty} f(x)\phi_j(x)dx = \int_{-\infty}^{\infty} \frac{f(x)}{(1+x^2)^p} [(1+x^2)^p \phi_j(x)]dx.$$

If p is sufficiently large, then $|f(x)|/(1+x^2)^p$ is integrable. Thus

$$\left|\int_{-\infty}^{\infty} f(x)\phi_j(x)dx\right| \leq \sup_{x \in E} |[1+x^2]^p \phi_j(x)| \cdot \int_{-\infty}^{\infty} \frac{|f(x)|}{(1+x^2)^p} dx.$$

This approaches 0 as ϕ_j approaches 0 in S.

We reiterate that not every distribution in D′ is also in S′, but nearly all of the important ones are. In particular, the Dirac δ-function and the Heaviside distributions are in S′.

We now turn our attention to the main reason for introducing S and S′ The Fourier Transform (See Chapter XI.).

XIV.3.6. <u>Theorem</u>. <u>If ϕ is in S, then $F(\phi)$ exists and is also in S.</u>

Proof. the rapid decay at ±∞ guarantees the convergence of the integral

$$\int_{-\infty}^{\infty} (-ix)^k e^{-ixy} \phi(x)\, dx$$

for all k=0,... . This represents $\dfrac{d^k}{dy^k} F(\phi)(y)$. Now for each p ≥ 0,

$$(-iy)^p \frac{d^k}{dy^k} F(\phi)(y) = \frac{1}{\sqrt{2\pi}} \int_{-\infty}^{\infty} (-ix)^k p(x) \frac{d^p}{dx^p} (e^{-ixy})\, dx.$$

Integrating by parts, we find

$$(-iy)^p \frac{d^k}{dy^k} F(\phi)(y) = \frac{(-1)^p}{\sqrt{2\pi}} \int_{-\infty}^{\infty} e^{-ixy} \frac{d^p}{dx^p} [(-ix)^k \phi(x)]\, dx.$$

Since $\dfrac{d^p}{dx^p}[(-ix)^k \phi(x)]$ is in S, the integral is well defined. Thus $\dfrac{d^k}{dy^k} F(\phi)(y)$ decreases faster than any polynomial in y. Since k and p are arbitrary, $F(\phi)$ is in S.

We note since $FF\phi(x) = \phi(-x)$, that each function ϕ in S is the Fourier transform of some function $(F(\phi)(-x))$ also in S.

XIV.3.7. Definition. Let t be a distribution of slow growth and ϕ be in S. Then Ft is defined by

$$<Ft,\phi> = <t,F\phi>.$$

XIV.3.8. Theorem. Definition XIV.3.8 is consistent with Definition XII.2.1 when the distribution t is generated by an integrable function.

Proof.

$$<Ft,\phi> = <t,F\phi>,$$

$$= \int_{-\infty}^{\infty} t(x)[(1/\sqrt{2\pi})\int_{-\infty}^{\infty} e^{-ixy}\phi(y)dy]dx,$$

$$= \int_{-\infty}^{\infty} [(1/\sqrt{2\pi})\int_{-\infty}^{\infty} e^{-ixy}t(x)dx]\phi(y)dy,$$

$$= \int_{-\infty}^{\infty} F(t)(y)\phi(y)dy.$$

Thus

$$F(t)(y) = (1/\sqrt{2\pi})\int_{-\infty}^{\infty} e^{-ixy}t(x)dx.$$

Since distributions are differentiable, we can calulate the derivatives of the Fourier transform.

XIV.3.9. Theorem. **If t is an arbitrary distribution in S′, then**

1. $F(\frac{d^k t}{dx^k})(y) = (iy)^k F(t)(y)$, $k=0,\ldots$.

2. $F((-ix)^k t)(y) = \frac{d^k}{dy^k} F(t)(y)$, $k=0,\ldots$.

3. $F(F(t))(x) = t(-x)$.

Proof. 1. Let k be arbitrary. Then

$$<F(\frac{d^k t}{dx^k})(y),\phi(y)> = <\frac{d^k t}{dx^k},F(\phi)(x)>,$$

$$= (-1)^k <t(x), \frac{d^k}{dx^k}(F(\phi))(x)>,$$

$$= (-1)^k <t(x), F((-iy)^k \phi)(x)>,$$

$$= <t(x),F((iy)^k \phi)(x)>,$$

$$= <(iy)^k F(t)(y),\phi(y)>,$$

2 and 3 are similar.

XIV.3.10. Examples. 1. We compute $F(\delta)(y)$.

$$<F(\delta)(y),\phi(y)> = <\delta(x),F(\phi)(x)>,$$
$$= (1/\sqrt{2\pi})\int_{-\infty}^{\infty} \phi(x)dx,$$
$$= <1/\sqrt{2\pi},\phi(x)>.$$

Thus $F(\delta)(y) = 1/\sqrt{2\pi}$. The reader should compare this result with Theorem

XII.2.7 and Example XIV.2.10.

  2. To calculate $F(1)(y)$, we have

$$<F(1)(y),\phi(y)> = <1,F(\phi)(x)>,$$

$$= \int_{-\infty}^{\infty} F(\phi)(x)\,dx,$$

$$= \int_{-\infty}^{\infty} F(\phi)(x)e^{ixy}\,dx\Big|_{y=0},$$

$$= \sqrt{2\pi}\ \phi(0),$$

$$= <\sqrt{2\pi}\ \delta(y),\phi(y)>\ .$$

So $F(1)(y) = \sqrt{2\pi}\ \delta(y)$. This gives considerable insight into why constants

do not have Fourier transforms in the classical sense. Their transforms are

singular distributions.

     The extension of the Fourier transform to higher dimensional distributions

is straight forward.

XIV.4. <u>Applications of Distributions to Ordinary Differential Equations</u>.

Having built up a rather elaborate theory for distributions, including derivatives,

let us now see how they affect differential equations. We first consider or-

dinary differential equations. We recall that differential operators are defined

for distribution only when the coefficients are infinitely differentiable, and

that every n-th order equation can be written as a first order system. Although we have not expressedly discussed distributional systems, this extensic
is also straight forward. We first consider homogeneous equations, then
follow it with their nonhomogeneous counterparts. We again remind the reader
we are assuming that all coefficients are real valued. Solutions in a distributional setting are called generalized solutions, while those discussed
in earlier chapters are called classical solutions.

XIV.4.1. Theorem. The only generalized solution to the equation $dt/dx = 0$
is $t = c,$ a constant.

Proof. First note that, when ψ is a test function,

$$\phi(x) = \int_{-\infty}^{x} \psi(u)\,du$$

is also a test function if and only if

$$\int_{-\infty}^{\infty} \psi(u)\,du = 0.$$

In that case $\phi' = \psi$.

The distributional equation $dt/dx = 0$ is equivalent to $<t',\phi> = 0$,
or $<t,\psi> = 0$, where $\psi = \phi'$. Let ϕ_0 be an arbitrary test function
satisfying

$$\int_{-\infty}^{\infty} \phi_0(u)\,du = 1.$$

Then write

$$\phi(x) = \phi_0(x)\int_{-\infty}^{\infty} \phi(u)du + [\phi(x) - \phi_0(x)\int_{-\infty}^{\infty} \phi(u)du].$$

The function

$$\psi(x) = \phi(x) - \phi_0(x)\int_{-\infty}^{\infty} \phi(u)du$$

has the property

$$\int_{-\infty}^{\infty} \psi(y)du = 0.$$

Thus if $dt/dx = 0$, then $<t,\psi> = 0$. We therefore conclude

$$<t,\phi> = <t,\phi_0\int_{-\infty}^{\infty} \phi(u)du>,$$

$$= \int_{-\infty}^{\infty} \phi(u)du<t,\phi_0>.$$

Since $<t,\phi_0>$ is independent of ϕ, we let $<t,\phi_0> = c$. Then

$$<t,\phi> = \int_{-\infty}^{\infty} c \phi(u)du,$$

or $t = c$.

Let us now turn our attention to first order vector systems.

XIV.4.2.   <u>Theorem.</u>   <u>In each of the problems</u>

   a. $a_0 t^n + \ldots + a_n t = 0,$

       <u>where t is a scalar distribution, a_j, j=0,...,n, are infinitely</u>
       <u>differentiable, $a_0 \neq 0$, or</u>

   b. $T' + AT = 0$

       <u>where T is an n-dimensional vector distribution, and A is an</u>
       <u>infinitely differentiable n × n matrix,</u>

<u>the generalized solution is identical with the classical solution, calculated</u>
<u>in Chapters III and IV.</u>

Proof.   Clearly it is sufficient to consider the matrix equation.   Since
$T' + AT = 0$ has a fundamental matrix solution X, and X is nonsingular,
we make the invertable transformation

$$T = XY$$

to find $XY' = 0$, or

$$Y' = 0.$$

From the matrix extension of the previous theorem we see that

$$Y = C,$$

and

$$T = XC,$$

the classical solution.

Let us next examine the nonhomogeneous counterparts.

XIV.4.3. Theorem. The only generalized solution of the equation $dt/dx = f$,
where f is continuous, is identical with the classical solution calculated
in Chapters III and IV.

Proof. The differential equation is equivalent to

$$<t',\phi> = <f,\phi>,$$

or

$$<t,\phi'> = <f,-\phi>.$$

Using the decomposition of ϕ in Theorem XIV.4.1,

$$\phi(x) = \phi_0(x)\int_{-\infty}^{\infty} \phi(u)\,du + [\phi(x) - \phi_0(x)\int_{-\infty}^{\infty} \phi(u)\,du],$$

we have

$$<t,\phi> = <t,\phi_0\int_{-\infty}^{\infty} \phi(u)\,du> + <t,\psi>,$$

where

$$\psi(x) = [\phi(x) - \phi_0(x)\int_{-\infty}^{\infty} \phi(u)du]$$

is the derivative of

$$\phi_1(x) = \int_{-\infty}^{x} \phi(u)du - \int_{-\infty}^{x}\phi_0(x)du \int_{-\infty}^{\infty} \phi(u)du.$$

From the differential equation we see that $<t,\psi>$ is known:

$$<t,\psi> = <f,-\phi_1>.$$

We define the distribution t_0 by

$$<t_0,\phi> = <f,-\phi_1>.$$

Then

$$<t,\phi> = <t,\phi_0>\int_{-\infty}^{\infty} \phi(u)du + <t_0,\phi>,$$

$$= <c_t,\phi> + <t_0,\phi>,$$

where $c_t = <t,\phi_0>$.

Now let y represent a classical solution of $dy/dx = f$. By performing the same operations again, we find

$$<y,\phi> \; = \; <c_y,\phi> + <t_0,\phi>,$$

where $c_y = <y,\phi_0>$. Then

$$<t,\phi> \; = \; <y,\phi> + <c_t - c_y,\phi>,$$

or

$$t = c + y,$$

which is the classical general solution, defined in Theorem V.1.3.

XIV.4.4. Theorem. In each of the problems

a. $a_0 t^{(n)} + \ldots + a_n t = f,$

where t is a scalar distribution, a_j, j=0,...,n, are infinitely
differentiable, $a_0 \neq 0$, and f is continuous, or

b. $T' + AT = F,$

where T is an n-dimensional vector distribution, A is an infinitely
differentiable n × n matrix, and F is a continuous n × 1
dimensional vector,

the generalized solution is identical with the classical solution, calculated
in Chapters III and IV.

Proof. As in Theorem XIV.4.2, only the vector problem needs to be considered.

The transformation

$$T = XY,$$

where X is a fundamental matrix for the homogeneous problem, reduces

$$T' + AT = F$$

to

$$Y' = X^{-1}F = F_1 .$$

From the extension of Theorem XIV.4.3 to vector equations, we see that

$$Y = C + Y_0 ,$$

where Y_0 is a classical solution and C is constant. Then

$$T(x) = X(x)C + X(x)\int_{\xi}^{x} X^{-1}(u)F(u)\,du,$$

the classical solution.

The implication of these theorems is that in dealing with classical problems in ordinary differential equations it is pointless to use distributions as a setting. Nothing new is gained. The solutions are the same as previously calculated.

It is only when considering new types of problems, such as with functions which have discontinuities, that distributions prove to be an asset.

XIV.4.5. Definition. Generalized solutions fall into three categories:

1. The solution corresponds to a function which is sufficiently differentiable in the classical sense. These are classical (and also distributional) solutions.

2. The solution corresponds to a function which is not sufficiently differentiable, but satisfies the equation as a distribution. These are called weak solutions.

3. The solution is a singular distribution and does not correspond to a locally integrable function. These are generalized solutions.

XIV.4.6. Examples. 1. As we have seen, the equation $dx/dt = 0$ has $x = c$ as its solution. This is a classical solution.

2. The equation $x(dt/dx) = 0$ has as a solution

$$t(x) = c_1 + c_2 H(x),$$

where $H(x)$ is the Heaviside distribution,

$$H(x) = \begin{cases} 0, & \text{when } x < 0, \\ 1, & \text{when } x \geq 0. \end{cases}$$

Then $dt/dx = \delta(x)$,   and $x(dt/dx) = 0$.   This is a weak solution.

   3.   The equation $x^2(dt/dx) = 0$ has a solution

$$t(x) = c_1 + c_2 H(x) + c_3 \delta(x),$$

since

$$t'(x) = c_2 \delta(x) + c_3 \delta'(x),$$

and

$$x^2 t'(x) = c_2 x^2 \delta(x) + c_3 x^2 \delta'(x),$$

$$= 0.$$

This is a generalized solution.

   4.   Consider the differential equation

$$Lt = a_0 t^{(n)} + \ldots + a_n t = \delta(x-\xi),$$

where a_j, $j=0,\ldots,n$,   are infinitely differentiable, $a_0 > 0$,   and ξ is a fixed parameter.   Since $\delta(x-\xi) = 0$ when $x \neq \xi$,

$$t(x,\xi) = \sum_{j=1}^{n} a_j(\xi) y_j(x), \quad \text{when}\quad x > \xi,$$

$$t(x,\xi) = \sum_{j=1}^{n} \beta_j(\xi) y_j(x), \quad \text{when} \quad x < \xi,$$

where $\{y_j(x)\}_{j=1}^{n}$ are solutions to the homogeneous equation.

If we integrate the differential equation from $\xi - \varepsilon$ to $\xi + \varepsilon$,

where $\varepsilon > 0$ is small, we find

$$\sum_{j=0}^{n} \int_{\xi - \varepsilon}^{\xi + \varepsilon} a_{n-j}(x) y^{(j)}(x) \, dx = \int_{\xi - \varepsilon}^{\xi + \varepsilon} \delta(x - \xi) \, dx = 1.$$

Since any jump in $\int_{\xi - \varepsilon}^{\xi + \varepsilon} a_{n-j}(x) y^{(j)}(x) \, dx$ produces a higher discontinuity in

$t^{(j+1)}$ at $x = \xi$, we conclude that all $t^{(j)}$ must be continuous at $x = \xi$

for $j = 0, \ldots, n-2$, and that

$$\lim_{\varepsilon \to 0} \int_{\xi - \varepsilon}^{\xi + \varepsilon} a_0(x) t^{(n)} \, dx = a_0(\xi) [t^{(n-1)}|_{x=\xi+} - t^{(n-1)}|_{x=\xi-}],$$

$$= 1.$$

This implies

$$\sum_{j=1}^{n} (\alpha_j - \beta_j) y_j(\xi) = 0,$$

$$\sum_{j=1}^{n} (\alpha_j - \beta_j) y_j'(\xi) = 0,$$

$$\cdots$$

$$\sum_{j=1}^{n} (\alpha_j - \beta_j) y_j^{(n-2)}(\xi) = 0,$$

$$\sum_{j=1}^{n} (\alpha_j - \beta_j) y_j^{(n-1)}(\xi) = 1/a_0(\xi).$$

The determinant of the coefficients of $\{(\alpha_j - \beta_j)\}_{j=1}^{n}$ is the Wronskian of

$y_1, \ldots, y_n$, and is nonzero when $y_1, \ldots, y_n$ are linearly independent. Assum:

this is so, we can solve these equations simultaneously to find $\{(\alpha_j - \beta_j)\}_{j=}^{n}$

If t is prescribed at a particular point (for example at $x = 0$), then

$\{\alpha_j\}_{j=1}^{n}$ and $\{\beta_j\}_{j=1}^{n}$ are determined. The solution is called a one sided

Green's function.

If t is prescribed at two points α and β as in a boundary value

problem, then n additional equations in α_j and β_j, $j=1,\ldots,n$, result,

determining α_j and β_j, $j=1,\ldots,n$. In this case the solution is the stan-

dard Green's function as was found in Chapter V: If

$$Lt = \delta(x-\xi),$$

then

$$y(x) = \int_{-\infty}^{\infty} t(x,\xi)f(\xi)d\xi$$

satisfies

$$Ly(x) = \int_{-\infty}^{\infty} Lt(x,\xi)f(\xi)d\xi$$

$$= \int_{-\infty}^{\infty} \delta(x-\xi)f(\xi)d\xi$$

$$= f(x).$$

So $Ly = f$, and t generates solutions to the nonhomogeneous equation.

XIV.5. Applications of Distributions to Partial Differential Equations. We
now turn our attention to the standard second order partial differential
equations as classified in Section XIII.3. While the classification made
there is still valid, we cannot say as much about the solutions to such
equations as was said in the previous section about ordinary differential
equations. The primary reason is the lack of a fundamental matrix for
linear partial differential equations. We can, however, show that classical
solutions are also generalized solutions.

XIV.5.1. Theorem. Let R be an open region in E^n with piecewise smooth
boundary S, and let the partial differential operator L, defined by

$$Lu = \sum_{|k| \leq p} a_k D^k u,$$

have infinitely differentiable coefficients in R.   Then

    1. If u is a classical solution of Lu = 0 then it is also a
generalized solution.

    2. If u is a generalized solution of Lu = 0 and has continuous
derivatives of order p, then u is also a classical solution.

Proof. 1. Let ϕ be an arbitrary test function with support in the interior
of R.   Then

$$<u, L^+\phi> = \int_R u \, L^+\phi \, dx.$$

Since ϕ vanishes on S, by Green's formula,

$$\int_R u \, L^+\phi dx = \int_R Lu\phi dx,$$

$$= 0.$$

So $<u,L^+\phi> = 0$, and u is a generalized solution.

2. We now assume that $<u,L^+\phi> = 0$. Since

$$\int_R Lu\phi dx = <u,L^+\phi>,$$

by Green's formula,

$$\int_R Lu\phi dx = 0$$

for any test function ϕ. If Lu $\neq$ 0 in the neighborhood of any point x_0 in R (Suppose Lu > 0), then we choose ϕ such that $\phi > 0$ near x_0 and is zero elsewhere. Then

$$\int_R Lu \, \phi dx > 0,$$

a contradiction.

We now turn our attention to each of the standard linear equations of elliptic, parabolic and hyperbolic type: Laplaces equation, the heat equation and the wave equation. In each case we wish to solve the equation

$$Lu = \delta(x-\xi),$$

or

$$<u,L^+\phi> = \phi(\xi).$$

Such solutions are called fundamental, since, if $u(x,\xi)$ is a solution, then

$$U(x) = \int_{E^n} u(x,\xi)f(\xi)d\xi$$

satisfies $LU = f$ when f is sufficiently well behaved. In all three cases $u(x,\xi)$ is a type of Green's function. By finding it we can, in fact, solve much more than just the nonhomogeneous equation.

XIV.5.2. **Theorem.** In E^n a fundamental solution to

$$- \left(\frac{\partial^2}{\partial x_1^2} + \ldots + \frac{\partial^2}{\partial x_n^2}\right)u = \delta(x-\xi)$$

is

$$u(x,\xi) = - (1/2)|x-\xi|,$$

when $n = 1$;

$$u(x,\xi) = (-1/2\pi)\ell n[(x_1-\xi_1)^2 + (x_2-\xi_2)^2]^{1/2},$$

$$= (-1/2\pi)\ell n\|x-\xi\|,$$

when n = 2;

$$u(x,\xi) = (1/4\pi)/[\sum_{j=1}^{3} (x_j-\xi_j)^2]^{1/2},$$

$$= (1/4\pi)/\|x-\xi\|,$$

when n = 3.

Higher dimensions are similar, but more complicated. Since we are primarily interested in those situations which have applications, we have little to gain from considering arbitrary n-dimensions.

Proof. One dimension is trivial. The technique for 2-dimensions is similar to that for 3-dimensions. Therefore we shall only consider the 3-dimensional case.

Let us temporarily set $\xi = 0$. We, therefore, consider

$$-(\frac{\partial^2}{\partial x_1^2} + \frac{\partial^2}{\partial x_2^2} + \frac{\partial^2}{\partial x_3^2})u = \delta(x),$$

or

$$- \operatorname{div}(\operatorname{grad} u) = \delta(x).$$

We assume that u depends only on $r = \sqrt{x_1^2 + x_2^2 + x_3^2}$ because of the symmetry of the equation. Our equation is thus reduced to

$$- (1/r^2) \frac{\partial}{\partial r} (r^2 \frac{\partial u}{\partial r}) = \delta(r).$$

If $r > 0$, the solution to this equation is

$$u(r) = A/r + B.$$

Since $u = B$ is a solution to the homogeneous equation, even when $r = 0$, we let $B = 0$. To find A, we let $\varepsilon > 0$ be arbitrarily chosen and integrate over a sphere of radius ε, centered at the origin. We find

$$- \int_{r \leq \varepsilon} \text{div}(\text{grad } u)\,dx = \int_{r \leq \varepsilon} \delta(r)\,dx = 1.$$

By the Divergence Theorem (Theorem XII.4.3)

$$- \int_{r = \varepsilon} [\frac{\partial u}{\partial x_1}, \frac{\partial u}{\partial x_2}, \frac{\partial u}{\partial x_3}] \cdot [x_1, x_2, x_3]/r \; ds = 1,$$

or

$$- \int_{r = \varepsilon} \frac{\partial u}{\partial r} \; ds = 1.$$

Letting $u = A/r$, we have $\frac{\partial u}{\partial r} = - A/r^2$, and

$$- \int_{r = \varepsilon} (- A/r^2)\,ds = (A/\varepsilon^2)(4 \pi \varepsilon^2) = 1.$$

Thus   A = 1/(4π),   and

$$u(x) = 1/(4\pi r).$$

Finally, replacing x by x-ξ, we have

$$u(x,\xi) = 1/4\pi \|x-\xi\|.$$

Having now found u(x,ξ), we must now verify that u(x,ξ) is, in fact, a solution. In view of Green's formula, we must show

$$\int_{E^3} (1/4\pi r)\nabla^2\phi\, dx = -\phi(0),$$

and then replace x by x-ξ.

Now, $\int_{E^3} = \lim_{\delta\to 0} \int_{E^3-R_\delta}$, where R_δ is the sphere of radius δ, centered at the origin. If φ vanishes when r is sufficiently large, then

$$\int_{E^3-R_\delta} (1/4\pi r)\nabla^2\phi\, dx = \int_{R^3-R_\delta} \nabla^2(1/4\pi r)\phi\, dx$$

$$+ \int_{r=\delta} [(-1/4\pi r)\frac{\partial\phi}{\partial r} + \frac{\partial}{\partial r}(1/4\pi r) \cdot \phi]\, ds.$$

Since $\nabla^2(1/4\pi r) = 0$ when r > 0,

$$\int_{E^3-R_\delta} (1/4\pi r)\nabla^2\phi\, dx = - \int_{r=\delta} [1/4\pi r) \frac{\partial\phi}{\partial r} + (1/4\pi r^2)\phi]\, ds$$

Now since φ is a test function, $|\frac{\partial\phi}{\partial r}| < M$ for all x. Thus

$$\left| \int\limits_{r=\delta} (1/4\pi r)\, \frac{\partial\phi}{\partial r}\, ds \right| < (1/4\pi\delta)M4\pi\delta^2,$$

$$= M\delta.$$

Further

$$\int\limits_{r=\delta} (1/4\pi r^2)\phi ds = \int\limits_{r=\delta} [\phi(0)/4\pi r^2]ds$$

$$+ \int\limits_{r=\delta} [(\phi(x) - \phi(0))/4\pi r^2]ds.$$

Now let $\varepsilon > 0$ be arbitrary. We choose δ sufficiently small such that $\delta < \varepsilon/2M$ and $|\phi(x) - \phi(0)| < \varepsilon/2$ when $|x| \leqq \delta$. Then

$$\left| \int\limits_{r=\delta} (1/4\pi r)\, \frac{\partial\phi}{\partial r}\, ds \right| < \varepsilon/2,$$

and

$$\int\limits_{r=\delta} [(\phi(x) - \phi(0))/4\pi r^2]ds < \varepsilon/2.$$

Thus

$$\left| \int\limits_{E^3-R_\delta} (1/4\pi r)\nabla^2\phi dx + \int\limits_{r=\delta} [\phi(0)/4\pi r^2]ds \right| =$$

$$\left| \int\limits_{E^3-R_\delta} (1/4\pi r)\nabla^2\phi dx + \phi(0) \right| < \varepsilon.$$

As ε approaches 0, δ approaches 0 also, and

$$\lim_{\delta \to 0} - \int_{E^3 - R_\delta} (1/4\pi r)\nabla^2 \phi\, dx = \phi(0).$$

Finally, replacing x by x-ξ we find

$$- \int_{E^3} (1/4\pi \|x-\xi\|)\nabla^2\phi\, dx = \phi(\xi),$$

or

$$<- \nabla^2(1/4\pi\|x-\xi\|), \phi> = <\delta(x-\xi), \phi>.$$

The general result in n-dimensions is

$$u(x,\xi) = c/\|x-\xi\|^{n-2},$$

where c is an appropriate constant.

We next consider the standard linear equation of parabolic type, the heat equation. Specifically we shall study the differential equation

$$\frac{\partial u}{\partial t} - \nabla^2 u = \delta(x-\xi)\delta(t-\tau),$$

where x, ξ are in E^n, and t, τ are one-dimensional. We temporarily replace x-ξ by x and t-τ by τ.

XIV.5.3.   Definition.   The causal fundamental solution for the heat equation

is the one which satisfies

$$\frac{\partial u}{\partial t} - \nabla^2 u = \delta(x)\delta(t),$$

where   x   is in E^n,   and   t   is real valued, and which vanishes when $t < 0$.

XIV.5.4.   Theorem.   For $t \geq 0$,   the causal fundamental solution coincides

with the solution of the initial value problem

$$\frac{\partial u}{\partial t} - \nabla^2 u = 0,$$

$$u(x,0+) = \delta(x).$$

Proof.   Let $u(x,t)$ be the solution of the initial value problem and let

$$v(x,t) = H(t)u(x,t),$$

where $H(t)$ is the Heaviside function,

$$H(t) = \begin{cases} 1, & \text{when } t \geq 0, \\ 0, & \text{when } t < 0. \end{cases}$$

Then

$$\nabla^2 v = H \nabla^2 u$$

and

$$\frac{\partial v}{\partial t} = \frac{\partial H}{\partial t} u + H \frac{\partial u}{\partial t},$$

$$= \delta(t)u + H\nabla^2 u,$$

$$= \nabla^2 v + \delta(t)\delta(x).$$

It is the initial value problem we actually solve to find the causal fundamental solution.

XIV.5.5. Theorem. Let $u(x,t)$ denote the causal fundamental solution to the heat equation in E^n. Then for $t \geq 0$,

$$u(x,t) = (4\pi t)^{n/2} e^{-r^2/4t},$$

where $r = (x_1^2 + \ldots + x_n^2)^{1/2}$.

Proof. Let $U(y,t)$ denote the n-dimensional Fourier transform of $u(x,t)$,

$$U(y,t) = (1/2\pi)^{n/2} \int_{E^n} e^{-ix \cdot y} u(x,t)\, dx,$$

where $x \cdot y = \sum_{j=1}^{n} x_j \cdot y_j$. Then $U(y,t)$ satisfies

$$\frac{dU}{dt} + \|y\|^2 U = 0,$$

$$U(y, 0+) = (1/2\pi)^{n/2},$$

where $(1/2\pi)^{n/2}$ is the n-dimensional Fourier transform of $\delta(x)$. Thus

$$U(y, t) = e^{-\|y\|^2 t}/(2\pi)^{n/2}$$

when $t \geq 0$. Inverting, we find

$$u(x, t) = (1/2\pi)^n \int_{E^n} e^{ix \cdot y} e^{-\|y\|^2 t} dy,$$

$$= \prod_{j=1}^{n} [(1/2\pi) \int_{-\infty}^{\infty} e^{ix_j y_j} e^{-y_j^2 t} dy_j].$$

From Lemma XII.2.10, we find

$$(1/2\pi) \int_{-\infty}^{\infty} e^{ix_j y_j} e^{-y_j^2 t} dy_j = e^{-x_j^2/4t}/(4\pi t)^{1/2}.$$

Thus

$$u(x, t) = e^{-r^2/4t}/(4\pi t)^{n/2},$$

when $t \geq 0$.

Finally we examine the causal fundamental solution for the wave equation.

XIV.5.6. Definition. The causal fundamental solution for the wave equation
is the one which satisfies

$$\frac{\partial^2 u}{\partial t^2} - \nabla^2 u = \delta(x)\delta(t),$$

where x is in E^n, and t is real valued, and which vanishes when $t < 0$.

XIV.5.7. Theorem. For $t \geq 0$, the causal fundamental solution coincides
with the solution of the initial value problem

$$\frac{\partial^2 u}{\partial t^2} - \nabla^2 u = 0,$$

$$u(x,0+) = 0,$$

$$\frac{\partial u}{\partial t}(x,0+) = \delta(x).$$

The proof is similar to that of Theorem XIV.5.4 and is left to the reader.

XIV.5.8. Theorem. Let $u(x,t)$ denote the causal fundamental solution to the
wave equation in E^n. Then for $t \geq 0$,

$$u(x,t) = (1/2\pi)^n \int_{E^n} e^{ix \cdot y} \frac{\sin \|y\|t}{\|y\|} \, dy.$$

Again, the proof is similar to that of Theorem XIV.5.5 and is left to the
reader.

In some instances $u(x,t)$ can be found more easily by other techniques.

In particular

$$u(x,t) = (1/2)H(t-|x|)$$

in one-dimension;

$$u(x,t) = (1/2\pi)H(t-r)/(t^2-r^2)^{1/2}$$

in two-dimensions;

$$u(x,t) = (1/4\pi r)\delta(t-r)$$

in three dimensions. We leave it to the reader to verify these results

(See Chapter XVII.)

XIV. Exercises

1. Define the ordinary differential operator L for distributions when L
    has complex coefficients.

2. Verify in one dimension that $H'(x) = \delta(x)$, where H is the Heaviside
    distribution and δ is the Dirac delta function.

3. Verify Example XIV.2.7.

4. In view of Example XIV2.10, no.1, what do the approximate identities in
    $L^1(-\infty,\infty)$ converge to as distributions? Prove your result.

5. Show that for t in S,

1. $F((-ix)^k t)(y) = \dfrac{d^k}{dy^k} F(t)(y),$

2. $F(F(t))(x) = t(-x).$

6. Show $F(\delta(x-a))(y) = (1/\sqrt{2\pi})e^{-iay}$ in one-dimension.

7. Show in detail how the notation for distributions should be extended from one-dimension to n-dimensions as a function of one independent variable.

8. Show

$$x\delta(x) = 0,$$

$$x^2\delta'(x) = 0,$$

$$\cdots$$

$$x^n\delta^{(n-1)}(x) = 0,$$

as distributions.

9. Solve the distributional equation

$$x^3 t''(x) = 0.$$

10. Solve the boundary value problem

$$y'' + \lambda x = \delta(x-\xi),$$

$$a_1 y(\alpha) + a_2 y'(\alpha) = 0,$$

$$\beta_1 y(\beta) + \beta_2 y'(\beta) = 0.$$

For what values of λ does a solution exist? Show the result is the Green's function for the regular Sturm-Liouville problem of Chapter V.

11. Show that if $u = u(r)$, then the equation

$$- \nabla^2 u = \delta(x)$$

is equivalent to

$$- (1/r^2) \frac{\partial}{\partial r} (r^2 \frac{\partial u}{\partial r}) = \delta(r).$$

12. Show that in one-dimension

$$- \frac{\partial^2 u}{\partial x_1^2} = \delta(x-\xi)$$

has

$$u(x,\xi) = - (1/2)|x-\xi|$$

as a solution. In two-dimensions

$$- (\frac{\partial^2}{\partial x_1^2} + \frac{\partial^2}{\partial y_2^2})u = \delta(x-\xi)$$

has

$$u(x,\xi) = - (1/2\pi)\ell n\|x-\xi\|$$

as a   solution.

13. Show that in n-dimensions, $n \geq 3$, the equation

$$- \nabla^2 u = \delta(x-\xi)$$

has

$$u(x,\xi) = c/\|x-\xi\|^{n-2}$$

as a   solution. Find   c.

14. Prove Theorems XIV.5.7 and XIV.5.8.

15. Show that the distributions given after Theorem XIV.5.8 are causal fundamental solutions to the wave equation in one, two, and three dimensions.

References

1. A. Friedman, "Generalized Functions and Partial Differential Equations," Prentice-Hall, Englewood Cliffs, N.J., 1963.

2. I.M. Gelfand and G.E. Shilov, "Generalized Functions, vol. 1," Academic Press, New York, 1964.

3. L. Schwartz, "Theorie des Distributions," Hermann, Paris, 1966.

4. I. Stakgold, "Boundary Value Problems of Mathematical Physics, vol. II,"
        Macmillan, New York, 1968.

5. A.H. Zemanian, "Distribution Theory and Transform Analysis," McGraw-Hill,
        New York, 1965.

XV.   LAPLACE'S EQUATION

In this chapter we shall study the standard examples of elliptic equation
which bear the name of the French mathematician Laplace.   Our setting will
be, in general, a region in E^n. We shall, however, restrict ourselves to
E^1, E^2, or E^3 in various instances for computational purposes, since the
techniques to be used are easily extended to higher dimensions.

We shall primarily consider all solutions as distributions.   However,
we shall take note when the solution is also a solution in the classical
sense.

Specifically we shall consider the equation

$$\nabla^2 u = (\frac{\partial^2}{\partial x_1^2} + \ldots + \frac{\partial^2}{\partial x_n^2})u = 0,$$

or its nonhomogeneous form (Poisson's equation)

$$-\nabla^2 u = -(\frac{\partial^2}{\partial x_1^2} + \ldots + \frac{\partial^2}{\partial x_n^2})u = F,$$

where F is locally integrable.

XV.1.   <u>Introduction, Well Posed Problems</u>.   The theorem immediately following
is needed for computational purposes.

XV.1.1.   <u>Theorem</u>.   Let R be a bounded region in E^n with piecewise smooth
boundary S.   Let u satisfy $\nabla^2 u = 0$ and have continuous first order

derivatives on S. Then

$$\int_S \frac{\partial u}{\partial n}\, ds = 0,$$

where $\frac{\partial}{\partial n}$ denotes the derivative in the outward normal direction, and

$$0 \le \int_R [(\frac{\partial u}{\partial x_1})^2 + \dots + (\frac{\partial u}{\partial x_n})^2]\, dx = \int_S u\, \frac{\partial u}{\partial n}\, ds.$$

Proof. The first follows from Green's formula for ∇^2, letting $u = u$, $v = 1$ (See Example XIII.4.8).

To show the second, we note that

$$\mathrm{div}[u\, \frac{\partial u}{\partial x_1}, \dots, u\, \frac{\partial u}{\partial x_n}] = (\frac{\partial u}{\partial x_1})^2 + \dots + (\frac{\partial u}{\partial x_n})^2 + u\nabla^2 u.$$

Since $\nabla^2 u = 0$, the divergence theorem shows

$$0 \le \int_R [(\frac{\partial u}{\partial x_1})^2 + \dots + (\frac{\partial u}{\partial x_n})^2]\, dx = \int_S [u\, \frac{\partial u}{\partial x_1}, \dots, u\, \frac{\partial u}{\partial x_n}]\cdot n\, ds = \int_S u\, \frac{\partial u}{\partial n}\, ds \; .$$

There is a remarkable property shared by all harmonic functions, the solutions of $\nabla^2 u = 0$:

XV.1.2. Theorem (Mean Value Theorem). Let R be any bounded spherical region in E^n with boundary S. Let $\nabla^2 u = 0$ in a region containing R. Then

$$u\Big|_{\text{center of } R} = (1/A) \int_S u(x)\, ds,$$

<u>where</u> $A = \int_S 1\ ds.$

Proof.   We may assume without loss of generality that the center of R is at the origin. Let $G_0(x,\xi)$ be the solution of

$$- \nabla^2 u = \delta(x-\xi),$$

found in Theorem XIV.5.2 and Exercise XII.13.   Then

$$- u(x)\nabla_x^2 G_0(x,\xi) = \delta(x-\xi)u(x),$$

$$G_0(x,\xi)\nabla_x^2 u(x) = 0,$$

so

$$[G_0(x,\xi)\nabla_x^2 u(x) - u(x)\nabla_x^2 G_0(x,\xi)] = \delta(x-\xi)u(x).$$

Thus integrating over R and applying Green's formula,

$$\int_S [G_0(x,\xi)\frac{\partial u}{\partial n_x} - u(x)\frac{\partial G_0(x,\xi)}{\partial n_x}]ds_x = u(\xi).$$

Interchanging x and ξ, this becomes

$$u(x) = \int_S [G_0(\xi,x)\frac{\partial u(\xi)}{\partial n_\xi} - u(\xi)\frac{\partial G_0(\xi,x)}{\partial n_\xi}]ds_\xi.$$

Now R is a sphere. Thus the normal direction is in the direction of the increasing radius r. Further $G_0(\xi,x)$ and $\dfrac{\partial G_0(\xi,x)}{\partial n_\xi} = \dfrac{\partial G_0(\xi,r)}{\partial r}$ are functions of r alone when x = 0. Thus if x = 0,

$$u(0) = G_0(r) \int_S \frac{\partial u(\xi)}{\partial n_\xi}\, ds_\xi - \frac{\partial G_0(r)}{\partial r} \int_S u(\xi)\, ds_\xi .$$

By Theorem XV.1.1 the first integral is zero. By using the technique employed in Theorem XIV.5.2 we find

$$- \int_S \frac{\partial G_0(r)}{\partial r}\, ds = 1,$$

or

$$- \frac{\partial G_0(r)}{\partial r} = 1/[\int_S ds] = 1/A.$$

Thus

$$u(0) = (1/A)\int_S u(\xi)\, ds.$$

In layman's terms this says that every solution to Laplace's equation is the average of the values near by. From this result we can also derive another rather interesting fact:

XV.1.3. <u>Theorem (The Maximum Principle).</u> <u>Let R be a region in E^n with piecewise smooth boundary S. If $\nabla^2 u = 0$ inside R, then the maximum and minimum values of u are attained on S. If u is not constant, they are</u>

not attained in R.

Proof. Suppose u is maximal at x_0 in R. Suppose further that there
exists a point x_1 in R such that $u(x_0) \neq u(x_1)$. That is, u is not
constant. Then $u(x_1) < u(x_0)$. We let x_1 be connected to x_0 a simple
arc in R, and let x_2 be the first point along the arc from x_1 to x_0
where $u(x_2) = u(x_0)$. Since u is continuous, on any sphere Σ, centered
at x_2, there exists a region σ such that for x in σ, $u(x) < M < u(x_0)$.
Then, by the Mean Value Theorem

$$u(x_2) = (1/A)\int_{\Sigma} u(x)ds,$$

$$= (1/A)\int_{\Sigma-\sigma} u(x)ds + (1/A)\int_{\sigma} u(x)ds,$$

$$\leq (1/A)[u(x_0)(\text{Area of } \Sigma-\sigma) + M(\text{Area of } \sigma)],$$

$$< u(x_0),$$

and we have a contradiction.

Before we actually begin to consider problems associated with Laplace's
equation, we need to say specifically what sort of problems we wish to consider.

XV.1.4. Definition. A problem is well posed if

1. There exists a solution.

2. The solution is unique.

3. The solution depends continuously upon the boundary data and the
    nonhomogeneous term (if any).

Most problems which accurately reflect physical situations are well
posed. However, where simplifications (such as linearization) are made for
mathematical reasons, the resulting idealized problem may not be well posed.
These situations are usually easily recognized.

As an example of a problem that is not well posed,   consider

$$\frac{\partial^2 u}{\partial x^2} + \frac{\partial^2 u}{\partial y^2} = 0,$$

$$u(0,y) = 0,$$

$$\frac{\partial u}{\partial x}(0,y) = (1/n)\sin ny.$$

Its solution is

$$u(x,y) = (1/n^2)\sinh nx \sin ny.$$

As x approaches ∞, so does u(x,y), no matter how small the initial data
is made by choosing n large.

XV.2. Dirichlet, Neumann, and Mixed Boundary Value Problems. There are

three standard problems which arise naturally in the discussion of Laplace's

equation. They are in the form of boundary value problems.

XV.2.1. Definition. Let R be a bounded region in E^n with piecewise

smooth boundary S. Let $f = f(x_1,...,x_n)$ be continuous on S. The (interio

Dirichlet problem for R is to determine the function $u = u(x_1,...,x_n)$

which satisfies

  1. u is defined and continuous on R + S.

  2. $\nabla^2 u = 0$ in R.

  3. u = f on S.

There are, of course, similarly defined exterior problems as well as

problems over infinite regions.

XV.2.2. Theorem. If the solution to the Dirichlet problem exists, then it

is unique.

Proof. If u_1 and u_2 are two solutions, then $u_1 - u_2$ is zero on S, a

$\nabla^2(u_1 - u_2) = 0$ in R. The maximum principle then guarantees that $u_1 = u_2$

on R + S.

XV.2.3. Theorem. The solution to the Dirichlet problem is continuous with

respect to the boundary function. If u_f, u_g are two solutions correspondi

to boundary functions f, g, and $|f-g| < \varepsilon$, then $|u_f - u_g| < \varepsilon$.

Proof. If

$$|f-g| < \varepsilon$$

on S, then by the maximum principle

$$|u_f-u_g| \leq \sup_{x \in S} |f-g| < \varepsilon .$$

The second boundary value problem is called the Neumann problem.

XV.2.4. Definition. Let R be a bounded region in E^n with piecewise smooth boundary S. Let $f = f(x_1,\ldots,x_n)$ be continuous on S. The (interior) Neumann problem for R is to determine the function $u = u(x_1,\ldots,x_n)$ which satisfies

1. u is defined and continuous on R + S.

2. $\nabla^2 u = 0$ in R.

3. The outward normal derivative $\frac{\partial u}{\partial n}$ exists on S wherever S is smooth, and $\frac{\partial u}{\partial n} = f$ at these points.

Again, there exist exterior problems, as well as problems over infinite regions.

XV.2.5. Theorem. If u_1 and u_2 are solutions to the Neumann problem, then $u_1 - u_2$ is constant.

Proof. Let $u = u_1 - u_2$. Then on S, $\frac{\partial u}{\partial n} = 0$. By Theorem XV.1.1,

$\frac{\partial u}{\partial x_1} = 0, \ldots, \frac{\partial u}{\partial x_n} = 0$ in R. Thus $du = 0$, and u is constant.

The Neumann problem as stated is not well posed. ·

XV.2.5. <u>Theorem.</u> <u>If the solution to the Neumann problem exists, then</u>

$$\int_S f(x) ds = 0.$$

Proof. This follows immediately from Theorem XV.1.1.

The third boundary value problem which arises naturally is a mixed bound value problem.

XV.2.7. <u>Definition.</u> <u>Let R be a bounded region in E^n with piecewise smoo</u>
<u>boundary S. Let $f = f(x_1, \ldots, x_n)$ be continuous on S, and let α be a</u>
<u>real number. The (interior) mixed boundary value problem for R is to deter</u>
<u>mine the function $u = u(x_1, \ldots, x_n)$ which satisfies</u>

1. <u>u is defined and continuous on $R + S$.</u>

2. <u>$\nabla^2 u = 0$ in R.</u>

3. <u>The outward normal derivative $\frac{\partial u}{\partial n}$ exists on S wherever S is smo</u>
   <u>and $\frac{\partial u}{\partial n} + \alpha u = f$ on S.</u>

In this instance also there are similarly defined exterior and infinite problems.

XV.2.8.   Theorem.   Let $\alpha \geq 0$.   If u_1 and u_2 are two solutions of the mixed boundary value problem, then $u_1 - u_2$ is constant.   If $\alpha \neq 0$,   then the solution u is unique.

We leave the proof to the reader.

Sometimes the operator

$$Lu = [\nabla^2 - P(x)]u,$$

where $P \geq 0$,   is considered instead of $\nabla^2 u$.   An additional term Pu^2 is then introduced into the integral of $[(\frac{\partial u}{\partial x_1})^2 + \ldots + (\frac{\partial u}{\partial x_n})^2]$ of Theorem XV.1.1.   This results in uniqueness for both the Neumann problem and the mixed boundary value problem.   Continuity with respect to the boundary functions f   then follows, and the problems are well posed.

When an exterior problem or a problem over an infinite region is considered, an auxilliary condition that the solution u vanish at ∞ is usually imposed.

There are existence theorems for the Dirichlet, Neumann and mixed boundary value problems.   The classical existence theorem for the Dirichlet problem involves the use of harmonic, subharmonic and superharmonic functions (solutions of $\nabla^2 u = 0$, $\nabla^2 u \leq 0$, $\nabla^2 u \geq 0$).   Although it would be possible to present it, the proof is not constructive (as was the case with ordinary differential equations), and so there seems little reason to do so.   Rather we shall prove existence by actually exhibiting the solutions by using a Green's function.

The Neumann problem must be modified slightly, as just indicated, in order to prove the existence of solutions. In this case as well as the mixed

boundary value problem, the proofs are substantically more difficult. For

those interested we recommend the book "Partial Differential Equations", by

P.R. Garabedian. We shall again be content to prove existence by actually

exhibiting the solutions for the examples we consider.

XV.3. <u>The Dirichlet Problem</u>. Associated with the interior Dirichlet problem

there is also a nonhomogeneous counterpart. These problems can be solved

simultaneously by a Green's function, similar to those encountered in

ordinary differential boundary value problems. We adopt the Green's function

technique as the method of attack because of its high versitility.

XV.3.1. <u>Definition</u>. Let R be a bounded region in E^n with piecewise smooth

boundary S. By the Green's function for the Dirichlet problem on R we

mean the generalized function $G(x,\xi)$ which satisfies

   1. $- \nabla_x^2 G = \delta(x-\xi)$, when x is in R.

   2. $G = 0$, when x is on S.

XV.3.2. <u>Theorem</u>. Let R be a bounded region in E^n with piecewise smooth

boundary S, and let $G_0(x,\xi)$ denote the solution to

$$- \nabla^2 u = \delta(x-\xi)$$

found in Theorem XIV.5.2. If the problem

$$\nabla_x^2 v(x,\xi) = 0, \quad \text{when } x \text{ is in } R,$$

$$v(x,\xi) = - G_0(x,\xi), \quad \text{when } x \text{ is on } S$$

has a solution for all ξ in R, then the Green's function for the Dirichlet problem $G(x,\xi)$ exists, and

$$G(x,\xi) = G_0(x,\xi) + v(x,\xi).$$

The proof is an exercise.

G has the following property, which it shares with the Green's function for the regular Sturm-Liouville problem.

XV.3.3. Theorem. G is symmetric. That is, for all x, ξ in R,

$$G(x,\xi) = G(\xi,x).$$

Proof. G satisfies

$$- \nabla_x^2 G(x,\xi) = \delta(x-\xi), \quad \text{when } x \text{ is in } R,$$

$$G(x,\xi) = 0, \quad \text{when } x \text{ is on } S,$$

$$- \nabla_x^2 G(x,\eta) = \delta(x-\eta), \quad \text{when } x \text{ is in } R,$$

$$G(x,\eta) = 0, \quad \text{when } x \text{ is on } S,$$

Thus

$$\int_R [G(x,\xi)\nabla_x^2 G(x,\eta) - G(x,\eta)\nabla_x^2 G(x,\xi)]dx$$

$$= \int_R [G(x,\eta)\delta(x-\xi) - G(x,\xi)\delta(x-\eta)]dx,$$

$$= G(\xi,\eta) - G(\eta,\xi).$$

But applying Green's formula to the original integral

$$\int_R [G(x,\xi)\nabla_x^2 G(x,\eta) - G(x,\eta)\nabla_x^2 G(x,\xi)]dx$$

$$= \int_S [G(x,\xi)\frac{\partial G(x,\eta)}{\partial \eta_x} - G(x,\eta)\frac{\partial G(x,\xi)}{\partial \eta_x}]ds = 0.$$

Thus

$$G(\xi,\eta) = G(\eta,\xi).$$

In order to properly define an integral by using G as its kernel we need the following result.

XV.3.4. <u>Lemma.</u> <u>Let R be a bounded region in E^n with piecewise smooth</u> <u>boundary S. Then for all x, ξ in R,</u>

$$0 < G(x,\xi) < (1/2\pi) \ell n [\{ \sup_{x,\xi \in S} |x-\xi|\}/|x-\xi|],$$

when   n = 2,

$$0 < G(x,\xi) < G_0(x,\xi),$$

when   n ≧ 3.

Proof.   First let us delete the point ξ by surrounding it by a sphere R_ε,

centered at ξ with radius ε and surface S_ε On S_ε,   in all cases, G_0

is very large, and positive, and   v   is bounded.   Thus if ε is sufficiently

small,   G   is positive.

Since   G   satisfies $\nabla_x^2 G(x,\xi) = 0$ in $R - R_\varepsilon$,   and   G > 0   on S_ε,

G = 0   on   S,   the maximum principle shows that   G > 0   in $R - R_\varepsilon$.   Letting

ε approach   0,   we find   G > 0   in   R.

If   n = 2,   then

$$G(x,\xi) = (1/2\pi) \ln [1/|x-\xi|] + v(x,\xi),$$

where

$$\nabla_x^2 v(x,\xi) = 0$$

in   R,   and

$$v(x,\xi) = - (1/2\pi) \ln [1/|x-\xi|]$$

when   x   is on   S.   Now

$$G(x,\xi) = (1/2\pi) \, \ell n \, [\{ \sup_{x,\xi \in S} |x-\xi| \} / |x-\xi|] + v(x,\xi)$$

$$- (1/2\pi) \, \ell n [\sup_{x,\xi \in S} |x-\xi|].$$

If

$$u(x,\xi) = v(x,\xi) - (1/2\pi) \, \ell n [\sup_{x,\xi \in S} |x-\xi|],$$

then $\nabla_x^2 u = 0.$ On   S

$$u(x,\xi) = (1/2\pi) \, \ell n [|x-\xi| / \{ \sup_{x,\xi \in S} |x-\xi| \}] < 0,$$

so $u(x,\xi) \leqq 0$ for all   x, ξ in   R + S.   Since   u   is not constant, the

maximum principle shows that $u < 0$ in   R.   Thus if   x, ξ are in   R,

$$G(x,\xi) < (1/2\pi) \, \ell n [\{ \sup_{x,\xi \in S} |x-\xi| \} / |x-\xi|].$$

If $n \geqq 3$, $v(x,\xi) = -c/\|x-\xi\|^{n-2} < 0$.   Thus $v(x,\xi) < 0$ when   x, ξ are

in   R,   and the result is obvious.

We now have sufficient machinery to develop a theory that very closely

parallels the development of the regular Sturm-Liouville problem.

XV.3.5.   Definition.   Let   R   be a bounded region in E^n with piecewise smoo boundary S. We denote by $L^2(R)$ the Hilbert space of all measurable

functions defined on R generated by the inner product

$$(f,g) = \int_R \overline{g}(x)f(x)dx$$

and norm

$$\|f\| = [\int_R |f(x)|^2 dx]^{1/2}.$$

XV.3.6. Theorem. Let the operator K be defined on the Hilbert space $L^2(R)$

by

$$Kf(x) = \int_R G(x,\xi)f(x)dx,$$

where G is the Green's function for the Dirichlet problem for R. Then K

is compact and self-adjoint.

For the sake of computational convenience we shall only prove the result

for n = 2 and n = 3. In the case n = 3, we shall present two proofs.

The first is short. The second is somewhat longer, but can be extended to

higher dimensions.

Proof. Let n = 2. From Lemma XV.3.4, we have

$$0 < G(x,\xi) < (1/2\pi)\, \ell n\, [\{ \sup_{x,\xi \in S} |x-\xi| \}/|x-\xi|].$$

Let $M = \sup_{x,\xi \in S} |x-\xi|$, and let S_ρ be a circle of radius ρ, centered at ξ.

We assume that ρ is sufficiently large such that $R \subset S_\rho$. Then

$$\int_R \int_R |G(x,\xi)|^2 dx d\xi \leq (1/4\pi^2) \int_R \int_{S_\rho} |\ell n \, [M/|x-\xi|]|^2 dx d\xi.$$

Now

$$\int_{S_\rho} |\ell n \, [M/|x-\xi|]|^2 dx = \int_0^{2\pi} \int_0^\rho |\ell n \, (M/r)|^2 r \, dr \, d\theta.$$

If we let $u = \ell n \, (M/r)$, then $r = Me^{-u}$, $dr = -Me^{-u} du$, and

$$\int_{S_\rho} |\ell n \, [M/|x-\xi|]|^2 dx d\xi = -\int_0^{2\pi} \int_{\ell n \, (M/\rho)}^\infty u^2 M^2 e^{-2u} du \, d\theta,$$

$$= N,$$

a finite number.   Thus

$$\int_R \int_R |G(x,\xi)|^2 dx d\xi \leq NV_R/4\pi^2,$$

where V_R is the area of R. By Theorem IX.2.4, K is compact.

Let $n = 3$. Here

$$\int_R \int_R |G(x,\xi)|^2 dx d\xi < c^2 \int_R \int_R \|x-\xi\|^{-2} dx d\xi.$$

Now

$$\int_R \|x-\xi\|^{-2} dx \quad < \quad \int_0^{2\pi} \int_0^\rho \int_0^\pi \sin\phi \, d\phi \, dr \, d\theta,$$

$$= 4\pi\rho,$$

where the integration is over a sphere S_ρ with center at ξ and radius sufficiently large such that $R \subset S_\rho$. Thus

$$\int_R \int_R |G(x,\xi)|^2 dx \, d\xi < 4\pi \rho c^2 V_R,$$

where V_R is the volume of R. Again we employ Theorem IX.2.4.

Unfortunately for $n > 3$, this technique of proof fails. $|G(x,\xi)|^2$ is not integrable over $R \times R$. To illustrate a technique which can be extended when $n > 3$, we present a second proof for the case $n = 3$.

We note that

$$G(x,\xi) = h(x,\xi)/\|x-\xi\|,$$

where $|h(x,\xi)|$ is bounded by some number M. We then write

$$Kf = K_1 f + K_2 f,$$

where

$$K_1 f(x) = \int_{R \cap \|x-\xi\| > a} G(x,\xi) f(\xi) d\xi,$$

$$K_2 f(x) = \int_{R \cap \|x-a\| \leq a} G(x,\xi) f(\xi) d\xi.$$

Clearly K_1 is compact, since its Kernel $G(x,\xi)$ is square integrable. On the other hand

$$|K_2 f(x)|^2 = |\int_{R \cap \|x-\xi\| \leq a} [h(x,\xi)/\|x-\xi\|] f(\xi) d\xi|^2,$$

$$\leq M^2 |\int_{R \cap \|x-\xi\| \leq a} [|f(\xi)|/\|x-\xi\|] d\xi|^2,$$

$$\leq M^2 |\int_{R \cap \|x-\xi\| \leq a} [|f(\xi)|/\|x-\xi\|^{1/2}][\|x-\xi\|]^{-1/2} d\xi|^2,$$

$$\leq M^2 \int_{\|x-\xi\| \leq a} \|x-\xi\|^{-1} d\xi \int_{R \cap \|x-\xi\| \leq a} [|f(\xi)|^2/\|x-\xi\|] d\xi,$$

where the last follows from Schwarz's inequality. In spherical coordinates with center at ξ, the first integral is

$$\int_0^a \int_0^{2\pi} \int_0^{\pi} r \sin \phi \, d\phi \, d\theta \, dr = 2\pi a^2.$$

Thus

$$|K_2 f(x)|^2 \leq 2\pi a^2 M^2 \int_{R \cap \|x-\xi\| \leq a} [|f(\xi)|^2/\|x-\xi\|] d\xi,$$

and

$$\|K_2 f\|^2 \leq 2\pi a^2 M^2 \int_R \int_{R \cap \|x-\xi\| \leq a} [|f(\xi)|^2/\|x-\xi\|] d\xi \, dx,$$

$$= 2\pi a^2 M^2 \int\limits_{R \cap \|x-\xi\| \le a} |f(\xi)|^2 \int\limits_R \|x-\xi\|^{-1} dx d\xi.$$

In the inner integral we enlarge the limits of integration to cover a sphere R_A, centered at the point ξ with radius $A = \sup\limits_{x,\xi \in R} \|x-\xi\|$. Then

$$\|K_2 f\|^2 \le 2\pi a^2 M^2 \int\limits_R |f(\xi)|^2 [\int\limits_{R_A} \|x-\xi\|^{-1} dx] d\xi$$

$$\le (4\pi a M A)^2 \|f\|^2,$$

or

$$\|K_2\| \le 4\pi a A M.$$

Thus we find that

$$\|K-K_1\| = \|K_2\| \le 4\pi a A M.$$

As a approaches 0, K_1 approaches K in the operator norm. By Theorem IX.1.5, K, being the limit of compact operators K_1, is also compact.

Self-adjointness follows from Theorem XV.3.3.

XV.3.7. Definition. We denote by D those functions f which satisfy

1. f is in $L^2(R)$.

2. f is differentiable; $\nabla^2 f$ exists almost everywhere.

3. $\nabla^2 f$ is in $L^2(R)$.

4. $f = 0$ on S.

XV.3.8. Definition. We define the partial differential operator A by
letting

$$Af = - \nabla^2 f$$

for all f in D.

XV.3.9. Theorem. Let R be a bounded region with piecewise smooth boundary
S. Let the Green's function $G(x,\xi)$ for the Dirichlet problem for R exist,
and let the compact, self-adjoint operator K be defined by

$$Kf(x) = \int_R G(x,\xi)f(\xi)d\xi.$$

Then

1. 0 is not in $\sigma_p(K)$.

2. $\sigma(K) = \{1/\lambda, \quad \lambda \text{ is an eigenvalue of } A\} \cup \{0\}$.

3. Each eigenfunction of K corresponding to the eigenvalue $1/\lambda$
is also an eigenfunction of A corresponding to the eigenvalue λ.

4. The eigenvalues of A (and K) are all positive.

Proof.   We have already shown that

$$AKf = f$$

for all f in $L^2(R)$.   Furthermore, Green's formula shows that when f is
in D,

$$\int_R [G(x,\xi)(-\nabla^2 f(x)) - (-\nabla_x^2 G(x,\xi))f(x)]dx = 0.$$

Since $-\nabla_x^2 G(x-\xi) = \delta(x-\xi)$,

$$\int_R G(x,\xi)(-\nabla^2 f(x))dx = \int_R \delta(x-\xi)f(x)dx,$$

$$= f(\xi).$$

Using $G(x,\xi) = G(\xi,x)$ and interchanging x and ξ,   we find

$$f(x) = \int_R G(x,\xi)(-\nabla^2 f(\xi))d\xi,$$

or

$$KAf = f.$$

Part 1 then is proved in a manner similar to part 2 of Theorem IX.5.1,
just as parts 2 and 3 follow in a manner similar to parts 3 and 5 of Theorem
IX.5.1.   Part 4 follows from an application of the divergence theorem,
similar to that in Theorem XV.1.1.

XV.3.10.   Theorem.   If the eigenfunctions of A corresponding to the same
eigenvalue have been made orthogonal, and all eigenfunctions are normal, then
the eigenfunctions of A form a complete, orthonormal set in $L^2(R)$.   If
$\{u_j\}_{j=1}^{\infty}$ denotes these eigenfunctions, then for all f in $L^2(R)$

$$f = \sum_{j=1}^{\infty} a_j u_j,$$

where

$$a_j = \int_R u_j(\xi) f(\xi) d\xi.$$

Proof.   Since 0 is not in $\sigma_p(K)$, it follows from Theorem IX.4.3 that
$P_0 = 0$ and that the eigenfunctions of K, and thus of A are complete.

    Since an infinite series of distributions can be differentiated term
by term, an easy computation shows that for all f in $L^2(R)$,

$$Af = \sum_{j=1}^{\infty} \lambda_j a_j u_j,$$

where λ_j is the eigenvalue of u_j with respect to A. This series will
not converge in $L^2(R)$, however, unless

$$\sum_{j=1}^{\infty} \lambda_j^2 |a_j|^2 < \infty,$$

which holds if and only if f is in D.

    If Af = g, then comparing coefficients, we find

$$\lambda_j \int_R u_j(\xi) f(\xi) d\xi = \int_R u_j(\xi) g(\xi) d\xi,$$

and

$$f = \sum_{j=1}^{\infty} (1/\lambda_j) [\int_R u_j(\xi) g(\xi) d\xi] u_j.$$

This shows that

$$G(x,\xi) = \sum_{j=1}^{\infty} (1/\lambda_j) u_j(x) u_j(\xi).$$

This formula is interesting and sometimes useful. We will derive G in one example to follow in precisely this form.

As the final subject of this section we derive the solution to the nonhomogeneous Laplace equation (the Poisson equation) with Dirichlet boundary conditions.

XV.3.11. <u>Theorem. Let R be a bounded region in E^n with piecewise smooth boundary S. Let F be locally integrable in R and f be continuous on S. Then the solution to</u>

$$- \nabla^2 u = F,$$

<u>when x is in R,</u>

$$u = f,$$

when   x   is on   S,   is given by

$$u(x) = \int_R G(x,\xi)F(\xi)d\xi - \int_S \frac{G(x,\xi)}{\partial n_\xi} f(\xi)ds_\xi.$$

Proof.   We have

$$- G(x,\xi)\nabla_x^2 u(x) = G(x,\xi)F(x),$$

$$- u(x)\nabla_x^2 G(x,\xi) = u(x)\delta(x-\xi).$$

Then

$$\int_R [G(x,\xi)(\nabla_x^2 u(x)) - u(x)(\nabla_x^2 G(x,\xi))]dx =$$

$$\int_R [u(x)\delta(x-\xi) - G(x,\xi)F(x)]dx ,$$

or, according to Green's formula,

$$\int_S [G(x,\xi)\frac{\partial u(x)}{\partial n_x} - u(x)\frac{\partial G(x,\xi)}{\partial n_x}]ds_x = u(\xi) - \int_R G(x,\xi)F(x)dx.$$

Since on   S,   G = 0,   and   u = f,   we find

$$u(x) = \int_R G(x,\xi)F(\xi)d\xi - \int_S \frac{\partial G(x,\xi)}{\partial n} f(\xi)d\xi.$$

The first integral solves the nonhomogeneous problem

$$- \nabla^2 u = F,$$

when x is in R,

$$u = 0,$$

when x is on S. The second solves the problem

$$- \nabla^2 u = 0,$$

when x is in R,

$$u = f,$$

when x is on S, the Dirichlet problem.

We cannot help remarking as we continue that this last formula is almost too slick. The result comes so easily that the power it possesses may escape notice. It says: Given the Green's function for the Dirichlet problem for a region R, all Dirichlet problems and their nonhomogeneous counterparts may be solved merely by a substitution in the formula above!

XV.4. <u>The Dirichlet Problem on the Unit Circle</u>. Before actually attempting to solve the Dirichlet problem on the unit circle, a word is in order about one of the principal techniques, known as <u>separation of variables</u>, which is not only used for the solution of Laplace's equation, but also for linear

partial differential equations in general. The procedure is as follows:
We first assume that the solution can be written as a product of functions,
one for each of the independent variables. Upon substitution, these functions
are found to satisfy certain linear ordinary differential equations, which
are inherited from the original partial differential equation. In addition,
they inherit the boundary conditions to a degree. These products, and there
are many, are then added together to build the final form of the required
solution. Separation of variables is the principal technique known for
solving elementary linear partial differential equations. There are others,
of course, but none so extensively used.

Let us now turn our attention to the Dirichlet problem on the unit
circle. We wish to solve

$$(1/r)\frac{\partial}{\partial r}(r\,\frac{\partial u}{\partial r}) + (1/r^2)\frac{\partial^2 u}{\partial \phi^2} = 0,$$

$$u(1,\phi) = f(\phi),$$

where $f(\phi)$ is continuous on the unit circle. We employ separation of
variables by assuming that

$$u(r,\phi) = R(r)\Phi(\phi).$$

Then

$$\frac{r^2\nabla^2 u}{u} = \frac{r\,\frac{d}{dr}(r\,\frac{dR}{dr})}{R} + \frac{\frac{d^2}{d\phi^2}(\Phi)}{\Phi} = 0,$$

or

$$r \frac{d}{dr}(r \frac{dR}{dr})/R = - \frac{d^2\Phi}{d\phi^2}/\Phi.$$

Since the left side is a function of r alone, while the right side is a function of ϕ alone, both sides must equal the same constant. If λ is that constant, then

$$r \frac{d}{dr} (r \frac{dR}{dr}) - \lambda R = 0,$$

$$\frac{d^2\Phi}{d\phi^2} + \lambda\Phi = 0.$$

Now since u must be smooth for all (r,ϕ) within the unit circle, we require R to be bounded, and Φ to be continuous and have continuous derivatives within the unit circle. Since

$$\Phi(\phi) = \begin{cases} c_1 + c_2\phi & , \text{ when } \lambda = 0, \\ \\ c_1 e^{i\sqrt{\lambda}\phi} + c_2 e^{-i\sqrt{\lambda}\phi}, & \text{ when } \lambda \neq 0, \end{cases}$$

the only choices possible for $\sqrt{\lambda}$ are the integers $0, 1, \ldots, \pm n, \ldots$. For arbitrary n, the equation in R becomes

$$r^2 \frac{d^2R}{dr^2} + r \frac{dR}{dr} - n^2 R = 0,$$

which has solutions

$$R(r) = \begin{cases} A + B \log r, & , \quad \text{when } n = 0, \\[2em] Ar^{|n|} + Br^{-|n|}, & , \quad \text{when } n \neq 0. \end{cases}$$

Since R is bounded, $B = 0$ in all cases.

$$R(r) = Ar^{|n|},$$

$n = 0, \pm 1, \ldots$. The solutions u therefore have the form

$$u = r^{|n|} e^{in\phi},$$

where $n = 0, \pm 1, \ldots$. Since any linear combination is also a solution, we attempt to find the solution which also equals f on the boundary by writing

$$u(r,\phi) = \sum_{n=-\infty}^{\infty} c_n r^{|n|} e^{in\phi}.$$

Imposing the boundary condition when $r = 1$, we have

$$f(\phi) = \sum_{n=-\infty}^{\infty} c_n e^{in\phi}.$$

Thus the coefficients c_n are the Fourier coefficients. We state the following theorem in summary.

XV.4.1. Theorem. The formal solution to the Dirichlet problem on the unit

circle is given by

$$u(r,\phi) = \sum_{n=-\infty}^{\infty} [(1/2\pi)\int_{-\pi}^{\pi} e^{-in\theta}f(\theta)d\theta]r^{|n|}e^{in\phi}.$$

This formula can be somewhat simplified.

XV.4.2. Theorem (Poisson's Formula). The formal solution to the Dirichlet

problem on the unit circle is given by

$$u(r,\phi) = (1/2\pi)\int_{-\pi}^{\pi} \frac{(1-r^2)f(\theta)d\theta}{1-2r\cos(\phi-\theta)+r^2}.$$

Proof. Formally interchanging the integration and summation of the previous

formula,

$$u(r,\phi) = (1/2\pi)\int_{-\pi}^{\pi} f(\theta)[\sum_{n=-\infty}^{\infty} e^{in(\phi-\theta)}r^{|n|}]d\theta.$$

When $r < 1$,

$$\sum_{n=-\infty}^{\infty} e^{in(\phi-\theta)}r^{|n|} = \sum_{n=1}^{\infty} e^{-in(\phi-\theta)}r^{n} + 1 + \sum_{n=1}^{\infty} e^{in(\phi-\theta)}r^{n},$$

$$= \frac{re^{-i(\phi-\theta)}}{1-re^{-i(\phi-\theta)}} + 1 + \frac{re^{i(\phi-\theta)}}{1-re^{i(\phi-\theta)}},$$

$$= \frac{1-r^2}{1-2r\cos(\phi-\theta)+r^2}.$$

The last expression is known as Poisson's kernel. Thus

$$u(r,\phi) = (1/2\pi)\int_{-\pi}^{\pi} \frac{(1-r^2)f(\theta)d\theta}{1-2r\,\cos(\phi-\theta)+r^2} \ .$$

XV.4.3. <u>Corollary</u>. <u>The formal solution to the Dirichlet problem on a circle</u>

<u>centered at the origin with radius</u>   a   <u>is given by</u>

$$u(r,\phi) = (1/2\pi)\int_{-\pi}^{\pi} \frac{(a^2-r^2)f(\theta)d\theta}{a^2-2ar\,\cos(\phi-\theta)+r^2} \ .$$

We now show that Poisson's formula yields more than just a formal solu-

tion. When f is continuous, it yields a classical solution. This is some-

what surprising, since mere continuity is not enough to guarantee the converge

of the Fourier series for f to converge pointwise to f.

XV.4.4. <u>Lemma</u>. <u>When</u> r < 1,

$$(1/2\pi)\int_{-\pi}^{\pi} \frac{(1-r^2)d\psi}{1-2r\,\cos\psi+r^2} = 1.$$

Proof. Let $z = e^{i\psi}$. Then $\psi = (1/i)\ell n\ \ z$, and $d\psi = (1/iz)dz$. Thus

$$(1/2\pi)\int_{-\pi}^{\pi} \frac{(1-r^2)d\psi}{1-2r\,\cos\psi+r^2} = \int_{|z|=1} \frac{-(1-r^2)dz}{2\pi ir(z-r)(z-1/r)} \ ,$$

where the integration in the   z   plane is counterclockwise around the unit

circle. The only residue is at   z = r,   and is $(1/2\pi i)$. Thus, according

to the residue theorem, the integral is   1.

XV.4.5. <u>Lemma</u>. <u>Let f be continuous on the unit circle. Then</u>

$$\lim_{r \to 1} u(r, \phi_0) = f(\phi_0)$$

<u>for each ϕ_0 in $[-\pi, \pi]$</u>.

Proof. Let $\phi_0 + \psi = \theta$ in Poisson's formula. Then

$$u(r, \phi_0) = (1/2\pi) \int_{-\pi}^{\pi} \frac{(1-r^2) f(\phi_0 + \psi) d\psi}{1 - 2r \cos \psi + r^2} .$$

By using Lemma XV.4.4, we find

$$u(r, \phi_0) - f(\phi_0) = (1/2\pi) \int_{-\pi}^{\pi} \frac{(1-r^2)(f(\phi_0 + \psi) - f(\phi_0)) d\psi}{1 - 2r \cos \psi + r^2} .$$

Since $1 - r^2 > 0$, $1 - 2r \cos \psi + r^2 > 0$,

$$|u(r, \phi_0) - f(\phi_0)| \leq (1/2\pi) \int_{-\pi}^{\pi} \frac{(1-r^2) |f(\phi_0 + \psi) - f(\phi_0)| d\psi}{1 - 2r \cos \psi + r^2} .$$

We now choose $\varepsilon > 0$. Since f is continuous, there exists a $\delta > 0$ such that

$$|f(\phi_0 + \psi) - f(\phi_0)| < \varepsilon/3,$$

when $|\psi| < \delta$. Keeping this in mind, we split the integral above in to three pieces, from $-\pi$ to $-\delta$, from $-\delta$ to δ, and from δ to π. The middle integral satisfies

$$(1/2\pi)\int_{-\delta}^{\delta} \frac{(1-r^2)|f(\phi_0+\psi)-f(\phi_0)|d\psi}{1-2r\cos\psi+r^2} < (\epsilon/3)(1/2\pi)\int_{-\pi}^{\pi} \frac{(1-r^2)d\psi}{1-2r\cos\psi+r^2}$$

$$= \epsilon/3.$$

If   f < M,   then the first integral satisfies

$$(1/2\pi)\int_{\delta}^{\pi} \frac{(1-r^2)|f(\phi_0+\psi)-f(\phi_0)|d\psi}{1-2r\cos\psi+r^2} \qquad \frac{(1-r^2)2M(\pi-\delta)}{2\pi(1-2r\cos\delta+r^2)}$$

If r is sufficiently close to 1, this also will be less than $\epsilon/3$. In a

similar manner, if r is sufficiently close to 1, the third integral will

be less than $\epsilon/3$. Thus if r is sufficiently close to 1,

$$|u(r,\phi_0)-f(\phi_0)| < \epsilon.$$

XV.4.6. Theorem. If f is continuous on the unit circle, then the formal

solution to the Dirichlet problem given by Poisson's formula satisfies

$$\lim_{(r,\phi)\to(1,\phi_0)} u(r,\phi) = f(\phi_0).$$

Proof.

$$|u(r,\phi)-f(\phi_0)| \le |u(r,\phi)-f(\phi)| + |f(\phi)-f(\phi_0)|.$$

For $\varepsilon > 0$,

$$|f(\phi) - f(\phi_0)| < \varepsilon/2,$$

when $|\phi - \phi_0|$ is sufficiently small. Further, from Lemma XV.4.5, when $|r-1|$ is sufficiently small,

$$|u(r,\phi) - f(\phi)| < \varepsilon/2.$$

Thus if $|\phi - \phi_0|$ and $|1-r|$ are both sufficiently small,

$$|u(r,\phi) - f(\phi_0)| < \varepsilon.$$

XV.4.7.   Theorem.   If   f   is continuous on the unit circle, then

$$u(r,\phi) = (1/2\pi) \int_{-\pi}^{\pi} \frac{(1-r^2) f(\theta) d\theta}{1 - 2r \cos(\phi-\theta) + r^2}$$

provides a classical solution to the Dirichlet problem on the unit circle.

Proof. It is only necessary verify that $\nabla^2 u = 0$. This is left as an exercise.

In closing, we would like to make two remarks. First, the solution above, given by Poisson's formula, converges to   f,   even through the Fourier series for   f,   from which it was derived, may not converge to   f.   Second, if   f   is in $L^2(-\pi, \pi)$,   but not necessarily continuous, it is possible to show

$$\lim_{r \to 1} \left[\int_{-\pi}^{\pi} |u(r,\phi) - f(\phi)|^2 d\phi \right]^{1/2} = 0.$$

That is, $u(r,\phi)$ converges to $f(\phi)$ in the sense of $L^2(-\pi,\pi)$ as r approaches 1. For further details we recommend the book by I. Stakgold, "Boundary Value Problems of Mathematical Physics, vol. 2," page 95.

XV.5.   Other Examples.   The Neumann Problem on the Unit Circle.   We wish to solve

$$(1/r)\frac{\partial}{\partial r}(r \frac{\partial u}{\partial r}) + (1/r^2)\frac{\partial^2 u}{\partial \phi^2} = 0,$$

$$\frac{\partial u}{\partial r}(1,\phi) = f(\phi),$$

where $f(\phi)$ is continuous on the unit circle. As in the Dirichlet problem, we find

$$u(r,\phi) = \sum_{n=-\infty}^{\infty} c_n r^{|n|} e^{in\phi}.$$

Since the normal directions is that of increasing r,

$$\frac{\partial u}{\partial n} = \frac{\partial u}{\partial r} = \sum_{n=-\infty}^{\infty} |n| c_n r^{|n|-1} e^{in\phi}.$$

Letting $r = 1$, and imposing the boundary condition,

$$f(\phi) = \sum_{n=-\infty}^{\infty} |n| c_n e^{in\phi}.$$

This is only possible if the Fourier coefficient of the constant term in the

Fourier series for f is 0. This is guaranteed by Theorem XV.2.6 if the

problem is to have a solution. Thus we assume that

$$\int_{-\pi}^{\pi} f(\theta)d\theta = 0.$$

Therefore, equating $|n|c_n$ to the nth Fourier coefficient for f, we have

$$c_n = (1/2\pi|n|)\int_{-\pi}^{\pi} e^{-in\theta}f(\theta)d\theta,$$

and

$$u(r,\phi) = \sum_{\substack{n=-\infty \\ n\neq 0}}^{\infty} [(1/2\pi|n|)\int_{-\pi}^{\pi} e^{-in\theta}f(\theta)d\theta]r^{|n|}e^{in\phi}+C,$$

where C is an arbitrary constant. (Recall the interior Neumann is not well

posed.).

Like the Dirichlet problem, this can be simplified.

XV.5.1. Theorem. The formal solution to the Neumann problem on the unit

circle is given by

$$u(r,\phi) = (-1/2\pi)\int_{-\pi}^{\pi} \ell n\,[1-2r\,\cos(\phi-\theta)+r^2]f(\theta)d\theta+ C.$$

Proof. We calculate

$$u(r,\phi) = (1/2\pi)\int_{-\pi}^{\pi} f(\theta) [\sum_{n=1}^{\infty} r^n e^{in\phi-\theta)}/n +$$

$$+ \sum_{n=1}^{\infty} r^n e^{-in(\phi-\theta)}/n]d\theta+C,$$

Now

$$\sum_{n=1}^{\infty} r^n e^{in(\phi-\theta)}/n + \sum_{n=1}^{\infty} r^n e^{-in(\phi-\theta)}/n,$$

$$= \int_0^r \frac{1}{r}[\sum_{n=1}^{\infty} r^n e^{in(\phi-\theta)} + \sum_{n=1}^{\infty} r^n e^{-in(\phi-\theta)}]dr,$$

$$= \int_0^r \frac{1}{r}[\frac{re^{i(\phi-\theta)}}{1-re^{i(\phi-\theta)}} + \frac{re^{-i(\phi-\theta)}}{1-re^{-i(\phi-\theta)}}]dr,$$

$$= - \ln [(1-re^{i(\phi-\theta)})(1-re^{-i(\phi-\theta)})],$$

$$= - \ln [1-2r \cos(\phi-\theta)+r^2].$$

Substitution the completes the result.

Just as in the Dirichlet problem this can be strengthened substantially:

XV.5.2.   Theorem.   Let   f   be continuous on the unit circle and satisfy
$\int_{-\pi}^{\pi} f(\theta)d\theta = 0.$ Then

$$u(r,\phi) = (-1/2\pi)\int_{-\pi}^{\pi} \ln [1-2r \cos(\phi-\theta)+r^2]f(\theta)d\theta +C$$

satisfies $\nabla^2 u = 0$ when $r < 1$ and

$$\lim_{(r,\phi)\to(1,\phi_0)} \frac{\partial u}{\partial n}(r,\phi) = f(\phi_0).$$

Proof. We leave the verification that $\nabla^2 u = 0$ to the reader. We note that

$$\frac{\partial u}{\partial n} = \frac{\partial u}{\partial r} = (-1/2\pi)\int_{-\pi}^{\pi} \frac{-2\cos(\phi-\theta)+2r}{1-2r\cos(\phi-\theta)+r^2} f(\theta)d\theta.$$

Since $\int_{-\pi}^{\pi} f(\theta)d\theta = 0$, we add $\frac{1}{2\pi r}\int_{-\pi}^{\pi} f(\theta)d\theta$ to $\frac{\partial u}{\partial n}$ to find

$$\frac{\partial u}{\partial n} = (1/2\pi)\int_{-\pi}^{\pi} [\frac{1-r^2}{1-2r\cos(\phi-\theta)+r^2}]\frac{f(\theta)}{r} d\theta.$$

Using results from section XV.4, it is a trivial computation to show

$$\lim_{(r,\phi)\to(1,\phi_0)} \frac{\partial u(r,\phi)}{\partial n} = f(\phi_0).$$

The Neumann Problem in the Square $0 \le x \le \pi$, $0 \le y \le \pi$. We wish to solve

$$\frac{\partial^2 u}{\partial x^2} + \frac{\partial^2 y}{\partial y^2} = 0,$$

when $0 \le x \le \pi$, $0 \le y \le \pi$,

$$\frac{\partial u}{\partial x}(0,y) = 0,$$

$$\frac{\partial u}{\partial x}(\pi,y) = 0,$$

$$\frac{\partial u}{\partial y}(x,\pi) = 0,$$

$$\frac{\partial u}{\partial y}(x,0) = f(x),$$

where $f(0) = 0$, $f(\pi) = 0$, and $\int_0^\pi f(x)dx = 0$ in accordance with Theorem XV.2.6. Since an arbitrary constant can be obviously added to any solution, we should not expect a unique solution.

We employ separation of variables by assuming that $u(x,y) = X(x)Y(y)$. Then

$$X''Y + Y''X = 0,$$

or, dividing by XY,

$$\frac{X''}{Y} = -\frac{Y''}{Y} = -\lambda,$$

where λ is an arbitrary constant. Thus we have two linear differential equations,

$$X'' + \lambda X = 0,$$

$$Y'' - \lambda Y = 0,$$

to solve for X and Y.

Further, since $\frac{\partial u}{\partial x} = X'Y$, we find $X'(0) = 0$, $X'(\pi) = 0$. Similarly

$Y'(\pi) = 0$. We save the boundary condition $u(x,0) = f(x)$ until later.

Solving the equation in X, we find

$$X(x) = \begin{cases} A \sin \sqrt{\lambda}\ x + B \cos \sqrt{\lambda}\ x, & \text{when } \lambda \neq 0, \\[3mm] Ax + B, & \text{when } \lambda = 0. \end{cases}$$

Then

$$X'(x) = \begin{cases} A\sqrt{\lambda}\cos \sqrt{\lambda}\,x - B\sqrt{\lambda}\sin \sqrt{\lambda}\ x, & \text{when } \lambda \neq 0, \\[3mm] A, & \text{when } \lambda = 0. \end{cases}$$

Since $X'(0) = 0$, $A = 0$ in all cases. Since $X'(\pi) = 0$, we find $\sqrt{\lambda} = n$, where $n = 0, \pm 1, \dots$. Thus

$$X(x) = B \cos nx,$$

where $n = 0, \pm 1, \dots$.

Solving the equation in Y, we find

$$Y(y) = \begin{cases} C \sinh ny + D \cosh ny, & \text{when } n \neq 0, \\[3mm] Cy + D, & \text{when } n = 0. \end{cases}$$

Then

$$
Y'(y) = \begin{cases} Cn \cosh ny + Dn \sinh ny, & \text{when } n \neq 0, \\ \\ C, & \text{when } n = 0. \end{cases}
$$

Since $Y'(\pi) = 0$, when $n \neq 0$

$$C \cosh n\pi + D \sinh n\pi = 0,$$

or

$$C = K \sinh n\pi,$$

$$D = -K \cosh n\pi,$$

for some constant parameter K. When $n = 0$, $C = 0$. Thus

$$
Y(y) = \begin{cases} K \cosh n(\pi-y), & \text{when } n \neq 0, \\ \\ D, & \text{when } n = 0. \end{cases}
$$

Putting this all together we find

$$u(x,y) = A_n \cos nx \cosh n(\pi-y),$$

where $n = 0, 1, \ldots,$ and A_n is an arbitrary coefficient.

In order to satisfy the boundary condition

$$\frac{\partial u}{\partial y}(x,0) = f(x),$$

we attempt to write $u(x,y)$ as a sum of the solutions above:

$$u(x,y) = \sum_{n=0}^{\infty} A_n \cos nx \cosh n(\pi-y).$$

Then

$$\frac{\partial u}{\partial y}(x,y) = \sum_{n=1}^{\infty} -nA_n \cos nx \sinh n(\pi-y).$$

If $y = 0$,

$$f(x) = \sum_{n=1}^{\infty} [-nA_n \sinh n\pi]\cos nx.$$

The terms in brackets are the Fourier cosine coefficients on $[0,\pi]$ (See Examples IX.5.4). Thus

$$-nA_n \sinh n\pi = (2/\pi)\int_0^\pi \cos n\xi f(\xi)d\xi,$$

$n = 1, \ldots,$ (the coefficient of the constant term in the expansion is zero, since $\int_0^\pi f(x)dx = 0.$), and

$$u(x,y) = - \sum_{n=1}^{\infty} (2/\pi) \int_{0}^{\pi} \frac{\cos n\xi f(\xi)}{n \sinh n\pi} \, d\xi \, \cos nx \, \cosh n(\pi-y) + C.$$

Classically, if $y = 0$, the series expression for $\frac{\partial u}{\partial y}$ will not necessarily converge to $f(x)$ unless $f(x)$ is differentiable, or unless some sort of criterion different from pointwise convergence is used. Mere continuity is not sufficient.

The Nonhomogeneous Dirichlet Problem in the Square $0 \leq x \leq \pi$, $0 \leq y \leq 1$

We wish to solve

$$-(\frac{\partial^2 u}{\partial x^2} + \frac{\partial^2 u}{\partial y^2}) = F(x,y),$$

when $0 \leq x \leq \pi$, $0 \leq y \leq \pi$, where F is continuous, subject to the boundary conditions

$$u(0,y) = 0, \ u(\pi,y) = 0, \ u(x,0) = 0, \ u(x,\pi) = 0.$$

Since the boundary conditions show u vanishes at $x = 0$, $x = \pi$, this suggests we expand both u and F in a Fourier sine series (See Examples IX.5.4),

$$u(x,y) = \sum_{n=1}^{\infty} a_n(y)\sin nx,$$

$$F(x,y) = \sum_{n=1}^{\infty} b_n(y)\sin nx,$$

where

$$b_n(y) = \frac{2}{\pi} \int_0^\pi F(\xi,y)\sin n\xi d\xi.$$

The problem, of course, is to find the coefficients $\{a_n(y)\}_{n=1}^\infty$.

If the series are inserted in the differential equation, we find

$$a_n'' - n^2 a_n = -b_n,$$

$$a_n(0) = 0, \quad a_n(\pi) = 0.$$

A tedious computation shows

$$a_n = \int_0^y \frac{\sinh n\eta \sinh n(\pi-y)}{n \sinh n\pi} b_n(\eta)d\eta$$

$$+ \int_y^\pi \frac{\sinh ny \sinh n(\pi-\eta)}{n \sinh n\pi} b_n(\eta)d\eta.$$

Thus

$$u(x,y) = \int_0^\pi \int_0^\pi G(x,y,\xi,\eta)F(\xi,\eta)d\xi d\eta,$$

where

$$G(x,y,\xi,\eta) = \begin{cases} \sum_{n=1}^\infty \dfrac{2 \sinh n(\pi-y)\sinh n\eta \sin nx \sin n\xi}{\pi n \sin hn\pi}; & \eta < y \\ \sum_{n=1}^\infty \dfrac{2 \sin n(\pi-\eta)\sinh ny \sin nx \sin n\xi}{\pi n \sinh n\pi}; & \eta > y. \end{cases}$$

If we use the results following Theorem XV.3.10, we find

$$G(x,y,\xi,\eta) = \sum_{n=1}^{\infty} \sum_{m=1}^{\infty} \frac{4 \sin nx \sin n\xi \sin my \sin m\eta}{\pi^2 (m^2 + n^2)}$$

We leave as an exercise the problem of showing these expressions are equal.

The Dirichlet Problem in an Infinite Strip. The problem to follow does not quite fit the assumptions made previously. Only minor modifications are needed, however, to have the theory apply. One primary modification is that we now consider the functions involved as distributions of slow growth rather than ordinary distributions. The reason for this is that we wish to apply the Fourier transform.

We wish to solve

$$\frac{\partial^2 u}{\partial x^2} + \frac{\partial^2 u}{\partial y^2} = 0,$$

where $-\infty < x < \infty$, $0 \leq y \leq \pi$, together with the boundary conditions

$$u(x,0) = 0,$$

$$u(x,\pi) = f(x),$$

where $f(x)$ is a continuous distribution of slow growth. To do so we employ the Fourier transform in x, which yields

$$\frac{d^2u}{dy^2} - \omega^2 U = 0,$$

$$U(\omega,0) = 0,$$

$$U(\omega,\pi) := F(\omega),$$

where $U = F(u)$ and $F = F(f)$.

The solution to the differential equation with the first boundary condition is

$$U(\omega) = A(\omega) \sinh \omega y.$$

If $y = \pi$, we find

$$F(\omega) = A(\omega) \sinh \omega\pi,$$

and

$$U(\omega) = F(\omega) \sinh \omega y / \sinh \omega\pi.$$

Thus

$$u(x,y) = (1/\sqrt{2\pi}) \int_{-\infty}^{\infty} [F(\omega)\sinh \omega y/\sinh \omega\pi] e^{ix\omega} d\omega,$$

$$= (1/2\pi) \int_{-\infty}^{\infty} [\int_{-\infty}^{\infty} \{e^{i(x-v)\omega} \sinh \omega y/\sinh \omega\pi\} d\omega] f(v) dv.$$

If f is real valued, then so is u, and

$$u(x,y) = (1/2\pi)\int_{-\infty}^{\infty} [\int_{-\infty}^{\infty} \{\cos \omega(x-v)\sinh \omega y/\sinh \omega\pi\}d\omega]f(v)dv.$$

If f is truly well behaved (such as integrable and differentiable on $(-\infty,\infty)$), then the solution may be valid pointwise. If not, it still may be valid in $L^1(-\infty,\infty)$ or $L^2(-\infty,\infty)$. In any case it is always a solution when considered as a distribution.

The Dirichlet Problem in the Unit Sphere with One Axis of Symmetry. In this example we turn our attention to the interior of the sphere in three dimensions. In order to simplify the problem somewhat, we assume that the problem is symmetric with respect to the z axis, where x,y,z is a rectangular coordinate system. The problem in spherical coordinates is

$$\sin \theta \frac{\partial}{\partial r} (r^2 \frac{\partial u}{\partial r}) + \frac{\partial}{\partial \theta} (\sin \theta \frac{\partial u}{\partial \theta}) + \frac{1}{\sin \theta} \frac{\partial^2 u}{\partial \phi^2} = 0,$$

when r < 1, and

$$u(1,\phi,\theta) = f(\theta),$$

when r = 1. Since the problem is independent of the coordinate ϕ, $\frac{\partial^2 u}{\partial \phi^2} = 0$, and that term may be dropped from the differential equation. Letting $u(r,\phi,\theta) = R(r)\psi(\theta)$ in

$$\sin \theta \frac{\partial}{\partial r} (r^2 \frac{\partial u}{\partial r}) + \frac{\partial}{\partial \theta} (\sin \theta \frac{\partial u}{\partial \theta}) = 0,$$

we have

$$\frac{(r^2 R')'}{R} + (\frac{1}{\sin \theta}) \frac{(\sin \theta \psi')'}{\psi} = 0 .$$

Since each term is a function of a different variable, yet their sum is 0, they must both be constant. Letting these constants be μ and $-\mu$, we have

$$(r^2 R')' - \mu R = 0,$$

$$(\sin \theta \psi')' + \mu \sin \theta \psi = 0.$$

Further, since the solution must be bounded (by $\sup\limits_{0 \le \theta \le \pi} |f(\theta)|$), both R and ψ must be bounded.

If we let $z = \cos \theta$, the equation in ψ becomes

$$\frac{d}{dz}((1 - z^2)\frac{d\psi}{dz}) + \mu \psi = 0,$$

which we recognize as the Legendre equation. Since the solutions are bounded near $z = \pm 1$ ($\theta = 0$ or π) only when $\mu = n(n+1)$, where n is an integer, the solutions are $P_n(z)$, $n = 0,1,\ldots,$ the Legendre polynomials.

When $\mu = n(n+1)$, the solutions for R which are bounded, are easily seen to be r^n, $n = 0,\ldots$. Thus we attempt to write a solution in the form

$$u(r,\theta) = \sum_{n=0}^{\infty} A_n r^n P_n(\cos \theta).$$

Letting r = 1, we have

$$f(\theta) = \sum_{n=0}^{\infty} A_n P_n(\cos \theta).$$

Since the Legendre polynomials are orthogonal, and have norm $(2/(2n+1))^{1/2}$,

$$A_n = -\int_0^\pi P_n(\cos \theta) f(\theta) \sin \theta \, d\theta/(2/(2n+1)),$$

and the solution, therefore, is

$$u(r,\theta) = -\sum_{n=0}^{\infty} \frac{\int_0^\pi P_n(\cos \theta) f(\theta) \sin \theta \, d\theta}{2/(2n+1)} \, r^n P_n(\cos \theta).$$

If the problem is not symmetric with respect to an axis, the solution not only involves the Legendre polynomials, but also their derivatives.

XV. Exercises

1. Prove Theorem XV.2.7.

2. Verify in detail the remarks following Theorem XV.2.8 concerning the operator $[\nabla^2 - P(x)]u$.

3. Is the exterior Neumann problem well posed?

4. Extend Theorem XV.3.6 to dimensions n > 3.

5. Develop a Green's function theory for the Neumann and mixed boundary value problems similar to that of Section XV.3.

6. Show that in polar coordinates the two dimensional Laplace equation is

$$(1/r)\frac{\partial}{\partial r}(r\ \frac{\partial u}{\partial r}) + (1/r^2)\frac{\partial^2 u}{\partial \phi^2} = 0.$$

7. Show that in spherical coordinates the three dimensional Laplace equation is

$$\sin \theta\ \frac{\partial}{\partial r}\ (r^2\ \frac{\partial u}{\partial r}) + \frac{\partial}{\partial \theta}\ (\sin \theta\ \frac{\partial u}{\partial \theta}) + (1/\sin \theta)\frac{\partial^2 u}{\partial \phi^2} = 0.$$

8. Verify directly that Poisson's formula satisfies Laplace's equation.

9. Fill in the details in the proof of Theorem XV.5.2.

10. Show that the two expressions for the Green's function for the non-homogeneous Dirichlet problem in Section XV.5 are equivalent.

11. Show that the formula for $u(x,y)$, the solution to the Dirichlet problem in an infinite strip, is a generalized solution of Laplace's equation.

12. Solve the Neumann problem on the exterior of the unit circle.

13. Solve the Dirichlet problem for the interior of a cube. What is the Green's function ?

14. Solve the Dirichlet problem for the interior of a doubly infinite circular cylinder

    a. When the boundary data is symmetric about the axis of the cylinder.

    b. When the boundary data is not symmetric about the axis of the cylinder.

15. Show that on the unit circle the eigenvalues and eigenfunctions for the problem

$$-\nabla^2 u = \lambda u,$$

when $|r| < 1$, $u(1,\phi) = 0$, are $\lambda = \lambda_{nj}^2$,

$$u_{nj}(r,\phi) = e^{in\phi} J_{|n|}(\lambda_{nj}r)/(\sqrt{\pi} \ J_{|n|+1}(\lambda_{nj})\cdot|,$$

where $J_{|n|}(\lambda_{nj}) = 0$, n an integer, $j = 1,\ldots$. Using the result at

the end of the proof of Theorem XV.3.10, show that the Green's function

for the Dirichlet problem on the unit circle is

$$G(r,\phi,\rho,\theta) = \sum_{n=-\infty}^{\infty} \sum_{j=1}^{\infty} e^{in(\phi-\theta)} J_{|n|}(\lambda_{nj}r) J_{|n|}(\lambda_{nj}\rho)/\pi \ J_{|n|+1}(\lambda_{nj})^2.$$

References

1. P.W. Berg and J.L. McGregor, "Elementary Partial Differential Equations," Holden-Day, San Francisico, 1966.

2. R. Courant and D. Hilbert, "Methods of Mathematical Physics, vol. II," Interscience, New York, 1962.

3. R.V. Churchill, "Fourier Series and Boundary Value Problems," McGraw-Hill, New York, 1963.

4. D. Greenspan, "Introduction to Partial Differential Equations," McGraw-Hill, New York, 1961.

5. G. Hellwig, "Differentialoperatoren," Springer-Verlag, Berlin, 1964.

6. O.D. Kellogg, "Foundations of Potential Theory," Springer-Verlag, Berlin, 1929.

7. I.G. Petrovskii, "Partial Differential Equations," Scripta Technica, London, 1967.

8. M.H. Protter and H.F. Weinberger, "Maximum Principles in Differential
    Equations," Prentice-Hall, Englewood-Cliffs, N.J., 1967.

9. I. Stakgold, "Boundary Value Problems of Mathematical Physics, vol. II,"
    MacMillan, New York, 1968.

10. H.F. Weinberger, "A First Course in Partial Differential Equations,"
    Blaisdell, New York, 1965.

XVI.   THE HEAT EQUATION

The most frequently encountered equation of parabolic type is the heat or diffusion equation, which is the subject of this chapter. Our setting is again a region in E^n, but in addition a new variable t, which may be though of as time, also appears. Consequently the type of problem to be studied is not a boundary value problem, as it was with Laplace's equation, but an initia value problem.

Specifically we shall study the equation

$$\frac{\partial u}{\partial t} - \nabla^2 u = 0,$$

or in the nonhomogeneous case,

$$\frac{\partial u}{\partial t} - \nabla^2 u = F(x,t),$$

where F is locally integrable in both x and t. Note that if both u and F are independent of t, the equation is reduced to the Laplace or Poisson equation.

XVI.1. <u>Introduction, the Cauchy Problem</u>. The following is in the way of a re mainder to the reader.

XVI.1.1. <u>Definition</u>. <u>A problem is well posed if</u>

1. <u>There exists a solution.</u>

2. The solution is unique.

3. The solution depends continuously upon the boundary data and the nonhomogeneous term (if any).

The type of problem which arises most naturally when discussing the heat equation is called a Cauchy problem.

XVI.1.2. Definition. Let R be a bounded region in E^n with piecewise smooth boundary S. Let $f(x)$ be continuous on $R + S$, and let $g(x,t)$ be continuous on $S \times [0,\infty)$. The Cauchy problem for the heat equation is to determine the function u which satisfies.

1. u is defined and continuous on $(R + S) \times [0,\infty)$.

2. $\frac{\partial u}{\partial t} - \nabla^2 u = 0$. when x is in R, and $t > 0$.

3. $u(x,0) = f(x)$ for all x in $R + S$.

4. u satisfies a boundary condition of one of the following three types when x is on S:

    a. Dirichlet type:

       $u(x,t) = g(x,t)$, when x is on S, and $t > 0$.

    b. Neumann type:

       $\frac{\partial u(x,t)}{\partial n_x} = g(x,t)$, when x is on S, and $t > 0$.

c. **Mixed type:**

$$\frac{\partial u(x,t)}{\partial n_x} + \alpha u(x,t) = g(x,t),$$ where α is a real number, x is

on S, and $t > 0$.

There are also problems where one boundary condition holds on part of S
and another elsewhere. These may be included in the mixed category if
$\alpha = \alpha(x)$, is permitted to vary as x varies over S and α and g may
be simultaneously infinite. (So the term $\frac{\partial u}{\partial n_x}$ may be regarded as insignificant

There are also similarly defined problems over unbounded regions, over
the exterior regions, as well as their nonhomogeneous counterparts. The re-
acter is free to make his own definitions in these cases.

We now turn our attention toward a maximum principle.

XVI.1.3. Lemma. Let R be a bounded region in E^n with piecewise smooth
boundary S. Let $v = v(x,t)$ be continuous on $(R + S) \times [0,\infty)$ and suppose
that

$$\frac{\partial v}{\partial t} - \nabla^2 v < 0$$

when x is in R and $t > 0$. Then the maximum of v occurs when $t = 0$
or when x is in S.

Proof. If the maximum of v occurs when $t > 0$ and x is in R, then
$\frac{\partial v}{\partial t} = 0$ and $\frac{\partial^2 v}{\partial x_j^2} \leq 0$, $j = 1,\ldots,n$. Thus $\frac{\partial v}{\partial t} - \nabla^2 v \geq 0$, which is a
contradiction.

XVI.1.4. Theorem (The Maximum Principle). Let R be a bounded region in E^n with piecewise smooth boundary S. Let u be continuous on $(R + S) \times [0,\infty)$ and suppose that

$$\frac{\partial u}{\partial t} - \nabla^2 u = 0$$

when x is in R and $t > 0$. If there exist constants m and M such that

$$m \leqq u(x,t) \leqq M$$

for all x in $R + S$ when $t = 0$, and for all x in S when $t > 0$, then

$$m \leqq u(x,t) \leqq M$$

for all x in $R + S$ and $t \geqq 0$.

Proof. Since x is in $R + S$, $\|x\| \leqq r$ for some positive number r. Consider

$$v = u + \varepsilon\|x\|^2.$$

When x is in S, or when $t = 0$,

$$v < M + \varepsilon r^2.$$

When   x   is in   R   and   t > 0,

$$\frac{\partial v}{\partial t} - \nabla^2 v = -2n\,\varepsilon < 0.$$

Thus from Lemma XVI.1.3,

$$v \leq M + \varepsilon r^2,$$

for all   x   in   R + S   and   t ≥ 0.   Since   u ≤ v,

$$u \leq M + \varepsilon r^2.$$

Since   ε   is arbitrary,   u ≤ M.

   To show   m ≤ u,   consider   −u,   which is less than or equal to   −m.

   Notice that we have not excluded the possiblity of having   u = M   or
u = m   when   x   is in   R   and   t > 0.   This can occur when the initial value
for   u   as well as the boundary values for u   are all equal to the same constant.   It is possible to show, however, that unless this occurs, the interior
values of   u   (when   x   is in   R,   and   t > 0)   are strictly between   m   and
M.

XVI.2.   The Cauchy Problem with Dirichlet Boundary Data.   Let us consider the
problem

$$\frac{\partial u}{\partial t} - \nabla^2 u = 0,$$

when x is in R, and t > 0;

$$u(x,0) = f(x),$$

when x is in R + S, and t = 0;

$$u(x,t) = g(x,t),$$

when x is in S and t > 0. This is the Cauchy problem with Dirichlet boundary data.

XVI.2.1. Theorem. Let R be a bounded region in E^n with piecewise smooth boundary S. Let u satisfy the Cauchy problem with Dirichlet boundary data. Then u is unique.

Proof. If u_1 and u_2 are solutions, then $u = u_1 - u_2$ satisfies the Cauchy problem with 0 initial and boundary values. By the maximum principle, u = 0 everywhere.

XVI.2.2. Theorem. Let R be a bounded region in E^n with piecewise smooth boundary S. The solution to the Cauchy problem with Dirichlet boundary data is continuous with respect to the initial and boundary values.

Proof. Let u_1 satisfy

$$\frac{\partial u_1}{\partial t} - \nabla^2 u_1 = 0,$$

$$u_1(x,0) = f_1(x),$$

$$u(x,t) = g_1(x,t),$$

when x is in S. Let u_2 satisfy

$$\frac{\partial u_2}{\partial t} - \nabla^2 u_2 = 0,$$

$$u_2(x,0) = f_2(x),$$

$$u_2(x,t) = g_2(x,t)$$

when x is in S. Further suppose that for some $\varepsilon > 0$

$$\left| f_1(x) - f_2(x) \right| < \varepsilon,$$

and

$$\left| g_1(x,t) - g_2(x) \right| < \varepsilon$$

for all appropriate values of x and t. Then according to the maximum
principle $u = u_1 - u_2$ satisfies $|u| < \varepsilon$ also.

XVI.2.3. Theorem. Let R be a bounded region in E^n with piecewise smooth
boundary S. If the solution to the Cauchy problem for R with Dirichlet

boundary data exists for a set of initial and boundary values with a certain

range, then the homogeneous Cauchy problem is well posed.

    We invite the reader to formulate his own theorems concerning Neumann or

mixed boundary data. (A generalization of the second part of Theorem XV.1.1

is most useful.)

    As in the previous chapter we shall omit the proof of the existence of

solutions in a general setting. Rather we shall again be content with

exhibiting solutions under certain circumstances. Before considering various

examples, we shall first look at the Green's function for the Cauchy problem,

then use it to generate the solutions.

XVI.3. The Solution to the Nonhomogeneous Cauchy Problem.

XVI.3.1. Definition. Let R be a bounded region in E^n with piecewise
smooth boundary S, and let $t \geq 0$. The causal Green's function for the
Cauchy problem with Dirichlet boundary data is the generalized function
$G(x,t,\xi,\tau)$ which satisfies

    1. $\dfrac{\partial G}{\partial t} - \nabla_x^2 G = \delta(x-\xi)\delta(t-\tau)$, when x,ξ are in R, and $t,\tau > 0$.

    2. $G = 0$ if $t < \tau$, and x,ξ are in R.

    3. $G = 0$ when x is on S.

    There are similar Green's functions for the Cauchy problem with Neumann

or mixed boundary data. Only assumption 3 is modified. Further, there are also Green's functions for exterior and unbounded regions.

Before calculating some examples we wish to see how the Green's function

is used. To do so we need the following revision of Green's formula.

XVI.3.2. Theorem (Green's Formula for a Cylindrical Region). _Let R be_

a bounded region in E^n with piecewise smooth boundary S. Let t be in

$[T_1, T_2]$. _Then if u and v are sufficiently differentiable with respect to_

x and t,

$$\int_{T_1}^{T_2} \int_R [v\{\frac{\partial u}{\partial t} - \nabla^2 u\} - u\{-\frac{\partial v}{\partial t} - \nabla^2 v\}]dxdt =$$

$$- \int_R uv\big|_{t=T_1} dx + \int_R uv\big|_{t=T_2} dx + \int_{T_1}^{T_2} \int_S [u\frac{\partial v}{\partial n_x} - v\frac{\partial u}{\partial n_x}]dsdt.$$

We have been deliberately vague with regard to the assumptions concerning

the differentiability of u and v, since our solutions and boundary functions

are infinitely differentiable when considered as distributions.

As we inspect Green's formula we notice that it would be extremely con-

venient if the function v were to vanish when t increases beyond a cer-

tain point and if v could be calculated using the formal adjoint operator

$(-\frac{\partial}{\partial t} - \nabla^2)$. We show that the Green's function G is related to these

properties:

We observe that if t is replaced by $t - \tau$ in G, then

$$G(x,t,\xi,\tau) = G(x, t - \tau, \xi, 0).$$

Further, if t and τ are interchanged, and

$$G^*(x,t,\xi,\tau) = G(x,\tau,\xi,t),$$

then G^* satisfies

1. $- \dfrac{\partial G^*}{\partial t} - \nabla_x^2 G^* = \delta(x - \xi)\delta(t - \tau)$, when x,ξ are in R and $t,\tau > 0$.

2. $G^* = 0$, if $t > \tau$, and x,ξ are in R.

3. $G^* = 0$, if x is in S.

Thus G^* has the desired properties described in the previous paragraph.

If, in Green's formula, we let

$$u = G(x,t,\xi_1,\tau_1),$$

$$v = G^*(x,t,\xi_0,\tau_0),$$

then, when ξ_0,ξ_1 are in R, τ_0,τ_1 are in $[T_1,T_2]$,

$$\int_{T_1}^{T_2} \int_R [G^*(x,t,\xi_0,\tau_0)\delta(x - \xi_1)\delta(t - \tau_1)$$

$$- (x - \xi_0)\delta(t - \tau_0)G(x,t,\xi_1,\tau_1)]dxdt = 0.$$

Evaluating the integral above, the following result is derived:

XVI.3.3.   Theorem.   The causal Green's function for the Cauchy problem for a region R in E^n with piecewise smooth boundary S satisfies

1. $-\dfrac{\partial G(x,t,\xi,\tau)}{\partial \tau} - \nabla^2 G(x,t,\xi,\tau) = \delta(x - \xi)\delta(t - \tau)$, when x,$\xi$ are in R, and $t,\tau > 0$.

2. $G = 0$, if $\tau > t$, and x, ξ are in R.

3. $G = 0$, if ξ is in S.

Proof.   The application of Green's formula shows

$$G^*(\xi_1,\tau_1,\xi_0,\tau_0) = G(\xi_0,\tau_0,\xi_1,\tau_1).$$

The results follow from this equivalence and the properties of G^*.

Let us now consider the nonhomogeneous Cauchy problem with Dirichlet boundary data,

$$\frac{\partial u}{\partial t} - \nabla^2 u = F(x,t),$$

when x is in R and $t > 0$, and F is continuous in x and t,

$$u(x,0) = f(x),$$

when x is in $R + S$, and $t = 0$,

$$u(x,t) = g(x,t),$$

when x is an S, and $t > 0$.

XVI.3.4.   Theorem.   Let   R   be a bounded region in E^n with piecewise smooth

boundary   S,   and let $t \geq 0$.   If the causal Green's function for the Cauchy

problem with Dirichlet boundary data exists, then the solution to the non-

homogeneous Cauchy problem with Dirichlet boundary data is given by

$$u(x,t) = \int_0^t \int_R G(x,t,\xi,\tau)F(\xi,\tau)d\xi d\tau$$

$$+ \int_R G(x,t,\xi,0)f(\xi)d\xi$$

$$- \int_0^t \int_S \frac{\partial G}{\partial n_\xi}(x,t,\xi,\tau)g(\xi,\tau)ds_\xi d\tau.$$

Proof. We apply Theorem XVI.3.2, letting $u = u(x,t)$, $v = G(\xi,\tau,x,t)$,   and

letting $T_1 = 0$, $T_2 = t$.

The first integral generates the nonhomogeneous term, the second the

initial value, and the third the boundary term. Hence a knowledge of the

Green's function solves all three separate problems simultaneously. Further

by using the formula just derived as well as the maximum principle, it is

possible to show the nonhomogeneous Cauchy problem is also well posed. It is

for these reasons the Green's function is so important.

XVI.4.   Examples. We shall now give four examples of problems involving

bounded and unbounded regions. In addition, in the latter we employ a

Neumann boundary condition in part. We hope these examples will provide

sufficient insight so the reader may adapt the techniques to other problems.

The Green's Function for a Bounded Region R in E^n. We assume that
R is a bounded region in E^n with piecewise smooth boundary S. We note
that when F and g are zero, then

$$u(x,t) = \int_R G(x,t,\xi,0)f(\xi)d\xi$$

satisfies

$$\frac{\partial u}{\partial t} - \nabla^2 u = 0,$$

when x is in R, and t > 0,

$$u(x,0) = f(x),$$

when x is in R + S, and t = 0,

$$u(x,t) = 0,$$

when x is on S and t > 0. Thus, by solving this homogeneous problem,
we can find the Green's function.

We employ separation of variables, letting u = X(x)T(t). Then, dividing
by XT, we find

$$-\frac{T'}{T} = -\frac{\nabla^2 X}{X} = \lambda,$$

where λ is a constant. That is,

$$T' = -\lambda T,$$

$$-\nabla^2 X = \lambda X.$$

In addition, employing the boundary condition, we find that $X(x) = 0$, when x is on S. The problem in X is, therefore, a Dirichlet problem. There exist normalized eigenfunctions $\{u_j(x)\}_{j=1}^{\infty}$, with corresponding positive eigenvalues $\{\lambda_j\}_{j=1}^{\infty}$. For each λ_j, $j = 1,\ldots,$

$$T = c_j e^{-\lambda_j t}.$$

Thus, summing over these solutions, we find

$$u(x,t) = \sum_{j=1}^{\infty} c_j u_j(x) e^{-\lambda_j t}.$$

In order to satisfy the initial condition, we require

$$f(x) = \sum_{j=1}^{\infty} c_j u_j(x),$$

which implies

$$c_j = \int_R u_j(\xi) f(\xi) d\xi.$$

Inserting this in the expression for u, we have

$$u(x,t) = \sum_{j=1}^{\infty} [\int_R u_j(\xi)f(\xi)d\xi]u_j(x)e^{-\lambda_j t},$$

or,

$$u(x,t) = \int_R G(x,t,\xi,0)f(\xi)d\xi,$$

where

$$G(x,t,\xi,0) = \sum_{j=1}^{\infty} u_j(x)u_j(\xi)e^{-\lambda_j t}.$$

Thus the Green's function

$$G(x,t,\xi,\tau) = \sum_{j=1}^{\infty} u_j(x)u_j(\xi)e^{-\lambda_j(t-\tau)}.$$

The solution to the nonhomogeneous Cauchy problem with Dirichlet boundary data is

$$u(x,t) = \sum_{j=1}^{\infty} \left[\int_0^t \int_R u_j(\xi)e^{\lambda_j\tau}F(\xi,\tau)d\xi d\tau \right.$$

$$+ \int_R u_j(\xi)f(\xi)d\xi$$

$$\left. - \int_0^t \int_S \frac{\partial u_j(\xi)}{\partial n_\xi}e^{\lambda_j\tau}g(\xi,\tau)ds_\xi d\tau \right]u_j(x)e^{-\lambda_j t}.$$

The Green's Function for $0 \leq x < \infty$, $t \geq 0$ with Dirichlet Boundary Data. In order to find the Green's function when $0 \leq x < \infty$, $t > 0$, we solve the partial differential equation

$$\frac{\partial G}{\partial t} - \frac{\partial^2 G}{\partial x^2} = \delta(x - \xi)\delta(t - \tau)$$

when x, ξ are in $[0,\infty)$ and $t > \tau$, together with the boundary condition $G = 0$ when $x = 0$. The boundary condition suggests using the Fourier sine transform. Taking the transform of both sides of the differential equation, we have

$$\sqrt{\frac{2}{\pi}} \int_0^\infty \frac{\partial G}{\partial t} \sin xy dx - \sqrt{\frac{2}{\pi}} \int_0^\infty \frac{\partial^2 G}{\partial x^2} \sin xy dx =$$

$$\delta(t - \tau)\sqrt{\frac{2}{\pi}} \int_0^\infty \delta(x - \xi)\sin xy dx.$$

Letting $\Gamma(y,t,\xi,\tau) = \sqrt{\frac{2}{\pi}} \int_0^\infty G(x,t,\xi,\tau)\sin xy dx$, this yields

$$\frac{\partial \Gamma}{\partial t} + y^2\Gamma = \delta(t - \tau)\sqrt{\frac{2}{\pi}} \sin \xi y.$$

Thus

$$\Gamma(y,t,\xi,\tau) = \sqrt{\frac{2}{\pi}} H(t - \tau)e^{-y^2(t - \tau)} \sin \xi y,$$

when $t > \tau$ and $y \geq 0$. Taking the inverse transform,

$$G(x,t,\xi,\tau) = (2/\pi)\int_0^\infty H(t - \tau)e^{-y^2(t - \tau)} \sin \xi y \sin xy dy,$$

where x, ξ are in $[0, \infty)$ and $t > \tau$.

If $\sin \xi y \sin xy$ is replaced by $(1/2)[\cos(x - \xi)y - \cos(x + \xi)y]$, the integration above may be performed in a manner similar to that in Lemma XII.2.10 by writing the cosines in exponential form. The result is

$$G(x,t,\xi,\tau) = [H(t-\tau)/(4\pi[t-\tau])^{1/2}][e^{-(x-\xi)^2/4(t-\tau)} - e^{-(x+\xi)^2/4(t-\tau)}].$$

where x, ξ are in $[0, \infty)$. This may be inserted in the formula for the solution of the nonhomogeneous problem.

The Green's Function for $-\infty < x < \infty$, $t \geqq 0$. Here no boundary data is needed since the interval is doubly infinite. We recall from Theorem XIV.5.5 that

$$G(x,t,\xi,\tau) = H(t-\tau)e^{-(x-\xi)^2/e(t-\tau)}/(4\pi[t-\tau])^{1/2}.$$

Hence the solution to the nonhomogeneous Cauchy problem is

$$u(x,t) = \int_0^t \int_{-\infty}^{\infty} [e^{-(x-\xi)^2/4(t-\tau)}/(4\pi[t-\tau])^{1/2}]F(\xi,\tau)d\xi d\tau$$

$$+ \int_{-\infty}^{\infty} [e^{-(x-\xi)^2/4t}/(4\pi t)^{1/2}]f(\xi)d\xi.$$

The boundary term is missing since the interval in x is doubly infinite.

The Green's Function For $0 \leqq x_1 \leqq \pi$, $0 \leqq x_2 < \infty$, $-\infty < x_3 < \infty$ With Neumann Boundary Data When $x_1 = 0$, $x_1 = \pi$, Dirichlet Data When $x_2 = 0$.

The problem we wish to solve is

$$\frac{\partial G}{\partial t} - \frac{\partial^2 G}{\partial x_1^2} - \frac{\partial^2 G}{\partial x_2^2} - \frac{\partial^2 G}{\partial x_3^2} = \delta(x_1-\xi_1)\delta(x_2-\xi_2)\delta(x_3-\xi_3)\delta(t-\tau),$$

$$\frac{\partial G}{\partial x_1}\bigg|_{x_1=0} = 0, \qquad \frac{\partial G}{\partial x_1}\bigg|_{x_1=\pi} = 0,$$

$$G\bigg|_{x_2=0} = 0.$$

We take the Fourier transform of the differential equation with respect to x_3. If H is the Fourier transform of G with respect to x_3, then the resulting equation is

$$\frac{\partial H}{\partial t} - \frac{\partial^2 H}{\partial x_1^2} - \frac{\partial^2 H}{\partial x_2^2} + y_3^2 H = \delta(x_1-\xi_1)\delta(x_2-\xi_2)\frac{e^{-i\xi_3 y_3}}{\sqrt{2\pi}}\delta(t-\tau).$$

Next we take the Fourier sine transform with respect to x_2. If K is the Fourier sine transform of H with respect to x_2, then the resulting equation is

$$\frac{\partial K}{\partial t} - \frac{\partial^2 K}{\partial x_1^2} + y_2^2 K + y_3^2 K = \delta(x_1-\xi_1)\frac{\sin \xi_2 y_2}{\sqrt{\frac{\pi}{2}}}\frac{e^{-i\xi_3 y_3}}{\sqrt{2\pi}}\delta(t-\tau).$$

We now expand both sides in a Fourier cosine series with respect to x_1. If L_n is the nth coefficient of K, then the resulting equation is

$$\frac{\partial L_n}{\partial t} + (n^2 + y_2^2 + y_3^2)L_n = \frac{\cos n\xi_1}{\sqrt{\frac{\pi}{2}}}\frac{\sin \xi_2 y_2}{\sqrt{\frac{\pi}{2}}}\frac{e^{-i\xi_3 y_3}}{\sqrt{2\pi}}\delta(t-\tau).$$

Solving this <u>ordinary</u> differential equation, we find

$$L_n = H(t-\tau) \frac{\cos n\xi_1}{\sqrt{\frac{\pi}{2}}} \frac{\sin \xi_2 y_2}{\sqrt{\frac{\pi}{2}}} \frac{e^{-\xi_3 y_3}}{\sqrt{2\pi}} e^{-(n^2+y_2^2+y_3^2)(t-\tau)}$$

Thus

$$K = H(t-\tau) \frac{2}{\pi} [\frac{1}{2} + \sum_{n=1}^{\infty} \cos n\xi_1 \cos nx_1 e^{-n^2(t-\tau)}] .$$

$$\frac{\sin \xi_2 y_2}{\sqrt{\frac{\pi}{2}}} \frac{e^{-i\xi_3 y_3}}{\sqrt{2\pi}} e^{-(y_2^2+y_3^2)(t-\tau)},$$

$$H = H(t-\tau) (\frac{2}{\pi})(\frac{2}{\pi}) \int_0^{\infty} \sin \xi_2 y_2 \sin x_2 y_2 e^{-y_2^2(t-\tau)} dy_2 .$$

$$[\frac{1}{2} + \sum_{n=1}^{\infty} \cos n\xi_1 \cos nx_1 e^{-n^2(t-\tau)}] .$$

$$\frac{e^{-i\xi_3 y_3}}{\sqrt{2\pi}} e^{-y_3^2(t-\tau)},$$

and

$$G(x_1,x_2,x_2,t,\xi_1,\xi_2,\xi_3,\tau) = \frac{1}{2\pi} \int_{-\infty}^{\infty} e^{i(x_3-\xi_3)y_3} e^{-y_3^2(t-\tau)} dy_3 .$$

$$\frac{2}{\pi} \int_0^{\infty} \sin \xi_2 y_2 \sin x_2 y_2 e^{-y_2^2(t-\tau)} dy_2 .$$

$$\frac{2}{\pi} [\frac{1}{2} + \sum_{n=1}^{\infty} \cos n\xi_1 \cos nx_1 e^{-n^2(t-\tau)}].$$

Note that the three expressions involved are the Green's functions for the respective one-dimensional problems. This only occurs when considering the Green's function for the heat equation.

XVI. 5. <u>Homogeneous Problems</u>. In this last section we give some additional

examples concerning homogeneous problems. In so doing we wish to emphasize

that the techniques of the previous section may also be used to solve homogeneous

equations directly. As a bonus we will also be able to find the Green's

function merely by replacing t by t-τ in the expressions under the integral

signs.

   <u>The Cauchy Problem on 0 ≦ x ≦ π with Mixed Boundary Data</u>. The term

<u>mixed</u> in the heading is somewhat a misnomer, since the problem we wish to

solve is

$$\frac{\partial u}{\partial t} - \frac{\partial^2 u}{\partial x^2} \, ,$$

when $0 < x < \pi$, and $t > 0$,

$$u(x,0) = f(x),$$

when $0 \leqq x \leqq \pi$, and $t = 0$,

$$u\Big|_{x=0} = 0, \quad \frac{\partial u}{\partial x}\Big|_{x=\pi} = 0.$$

The boundary terms are not mixed in the strict sense as we defined mixed

boundary terms, but in a more general sense.

   To solve the problem, we again turn the crank by using separation of

variables. We let $u = X(x)T(t)$. Then

$$\frac{T'}{T} = \frac{X''}{X} = -\lambda,$$

where λ is a constant. The boundary data further implies that $X(0) = 0$, $X'(\pi) = 0$. We easily find that λ has the discrete values $(1/4)(2n-1)^2$, $n = 1,\ldots,$ and for each n,

$$X = \sin \frac{(2n-1)}{2} x,$$

$$T = e^{-(1/4)(2n-1)^2 t}.$$

Thus

$$u(x,t) = \sum_{n=1}^{\infty} b_n \sin \frac{(2n-1)}{2} x \; e^{-(1/4)(2n-1)^2 t}.$$

Letting $t = 0$,

$$f(x) = \sum_{n=1}^{\infty} b_n \sin \frac{(2n-1)}{2} x.$$

While this is not an ordinary Fourier series, it is a Sturm-Liouville expansion. Thus

$$b_n = (2/\pi)\int_0^{\pi} \sin \frac{(2n-1)}{2} \xi \; f(\xi) d\xi,$$

and

$$u(x,t) = \sum_{n=1}^{\infty} (2/\pi)[\int_0^{\pi} \sin \frac{(2n-1)}{2} \xi \; f(\xi) d\xi] \sin \frac{(2n-1)}{2} x \; e^{-(1/4)(2n-1)^2 t}$$

In view of Theorem XVI.3.4, we see that

$$G(x,t,\xi,\tau) = (2/\pi) \sum_{n=1}^{\infty} \sin \frac{(2n-1)}{2} \xi \sin \frac{(2n-1)}{2}x \; e^{-(1/4)(2n-1)^2(t-\tau)}$$

is the associated Green's function.

The Cauchy Problem on the Unit Circle With Dirichlet Boundary Data. We wish to solve in polar coordinates, the problem

$$\frac{\partial u}{\partial t} - ((1/r)\frac{\partial}{\partial r}(r \frac{\partial u}{\partial r}) + (1/r^2)\frac{\partial^2 u}{\partial \phi^2}) = 0,$$

when $r < 1$, $0 \le \phi \le 2\pi$,   and $t > 0$,

$$u(1,\phi,t) = 0,$$

for all $t > 0$,

$$u(r,\phi,0) = f(r,\phi),$$

where   f   is continuous when $r \le 1$, $0 \le \phi \le 2\pi$.   We let $u = T(t)R(r)\Phi(\phi)$ and use separation of variables. Separating   T   first, we find

$$\frac{T'}{T} = ((1/r) \frac{(rR')'}{R} + (1/r^2) \frac{\Phi''}{\Phi}) = -\lambda^2,$$

where λ^2 is a constant. Next separating Φ,

$$\frac{\Phi''}{\Phi} = -\left(\frac{r(rR')'}{R} + \lambda^2 r^2\right) = -\mu^2,$$

where μ^2 is a second constant. The separated equations are, therefore,

$$T' + \lambda^2 T = 0,$$

$$\Phi'' + \mu^2 \Phi = 0,$$

$$r(rR')' + (\lambda^2 r^2 - \mu^2)R = 0.$$

In addition we wish the solution to be bounded, periodic in ϕ with period 2π, and vanish when $r = 1$. Thus we require R to be bounded, Φ to have period 2π, and $R(1) = 0$.

These requirements force μ^2 to be of the form n^2, $n = 0,1,\ldots$, and

$$\Phi = (A \cos n\phi + B \sin n\phi),$$

$n = 0,1,\ldots$.

We recognize the equations in R as the Bessel equations. The only solutions which are bounded are the ordinary Bessel functions $J_n(\lambda r)$. Since u vanishes when $r = 1$, we require that $J_n(\lambda) = 0$. If λ_{nm}, $m = 1,\ldots$, are the zeros of $J_n(\lambda)$, then the parameter λ must be one of these. Thus, the solution u has the form

$$u(r,\phi,t) = \sum_{n=0}^{\infty} \sum_{m=1}^{\infty} (A_{mn} \cos n\phi + B_{mn} \sin n\phi) J_n(\lambda_{nm} r) e^{-\lambda_{nm}^2 t}.$$

Letting $t = 0$, we find

$$f(r,\phi) = \sum_{n=0}^{\infty} \sum_{m=1}^{\infty} (A_{mn} \cos n\phi + B_{mn} \sin n\phi) J_n(\lambda_{nm} r),$$

from which it follows that

$$A_{m0} = \frac{\displaystyle\int_0^{2\pi} \int_0^1 f(r,\phi) J_0(\lambda_{m0} r) r dr d\phi}{\pi J_1(\lambda_{0m})^2} \quad,$$

$$A_{mn} = \frac{\displaystyle\int_0^{2\pi} \int_0^1 f(r,\phi) J_n(\lambda_{mn} r) \cos n\phi \; r dr d\phi}{(\pi/2) J_{n+1}(\lambda_{mn})^2}$$

when $n > 0$, and

$$B_{mn} = \frac{\displaystyle\int_0^{2\pi} \int_0^1 f(r,\phi) J_n(\lambda_{mn} r) \sin n\phi \; r dr d\phi}{(\pi/2) J_{n+1}(\lambda_{mn})^2}$$

when $n > 0$. $B_{m0} = 0$. These, inserted in the expression for u, yield the solution in final form.

Note that the Bessel equations are singular at $r = 0$, just as in the last chapter the Legendre equations were singular at ± 1. In both cases, however, the spectrum is discrete.

XVI. Exercises

1. Is the Cauchy problem with Neumann or mixed boundary data well posed?
    Hint: Generalize the second part of Theorem XV.1.1 to fit the heat
    equation.

2. Adapt the proof of the maximum principle for the heat equation to fit
    Laplace's equation. Thus reprove the maximum principle for Laplace's
    equation.

3. What properties should the Green's function for the Cauchy problem with
    Neumann or mixed boundary data have? Develop a theory for such problems.

4. Let R be a bounded region in E^n with piecewise smooth boundary S.
    Find the Green's function for the Cauchy problem in R with Neumann
    or mixed boundary data.

5. Calculate the Green's function for the Cauchy problem on $0 \leqq x < \infty$,
    $t \geqq 0$ with Neumann boundary data.

6. Verify that the solution to the nonhomogeneous Cauchy problem on
    $-\infty < x < \infty$, $t \geqq 0$ is a generalized solution to the heat equation and
    satisfies the initial conditions.

References

1. P.W. Berg and J.L. McGregor, "Elementary Partial Differential Equations,"
    Holden-Day, San Francisco, 1966.

2. R. Courant and D. Hilbert, "Methods of Mathematical Physics, vol. II,"
    Interscience, New York, 1962.

3. R.V. Churchill, "Fourier Series and Boundary Value Problems," McGraw-Hill,
    New York, 1963.

4. D. Greenspan, "Introduction to Partial Differential Equations," McGraw-
    Hill, New York, 1961.

5. G. Hellwig, "Differentialoperatoren," Springer Verlag, Berlin, 1964.

6. I.G. Petrovskii, "Partial Differential Equations," Scripta Technica,
    London, 1967.

7. M.H. Protter and H.F. Weinberger, "Maximum Principles in Differential
    Equations," Prentice-Hall, Englewood Cliffs, N.J., 1967.

8. I. Stakgold, "Boundary Value Problems of Mathematical Physics, vol. II,"
    Macmillan, New York, 1968.

9. H.F. Weinberger, "A First Course in Partial Differential Equations,"
    Blaisdell, New York, 1965.

XVII.   THE WAVE EQUATION

The subject of this, the final chapter of this book, is perhaps the most
interesting of the classical partial differential equations of mathematical
physics.   The theory of the wave equation, the standard example of hyperbolic
equations, also builds upon the theory of Laplace's equation, but is somewhat
more complicated due to the absence of exponential decay in the time variable,
as was the case with the heat equation.   Our setting is again a region in E^n
with an additional variable t for time.   At various instances we shall restrict the number of dimensions for computational simplicity.

We shall consider the equation

$$\frac{\partial^2 u}{\partial t^2} - \nabla^2 u = 0,$$

or its nonhomogeneous counterpart

$$\frac{\partial^2 u}{\partial t^2} - \nabla^2 u = F(x,t),$$

where F is locally integrable in both x and t.   As with the heat equation,
if both u and F are independent of t, the equation is reduced to the
Laplace or Poisson equation.

XVII.1.   <u>Introduction,   the Cauchy Problem</u>.   The following is inserted for a
third time as a remainder to the reader.

XVII.1.1. Definition. A problem is well posed if

1. There exists a solution.

2. The solution is unique.

3. The solution depends continuously upon boundary data and the non-
    homogeneous term (if any).

As with the heat equation, the type of problem most frequently encountered
is an initial value or Cauchy problem.

XVII.1.2. Definition. Let R be a bounded region in E^n with piecewise
smooth boundary S. Let $f_1(x)$, $f_2(x)$ be continuous on R + S, and let
$g(x,t)$ be continuous on $S \times [0,\infty)$. The Cauchy problem for the wave equation
is to determine the function u which satisfies

1. u is defined and continuous on $(R + S) \times [0,\infty)$.

2. $\dfrac{\partial^2 u}{\partial t^2} - \nabla^2 u = 0$

    when x is in R, and t > 0.

3. $u(x,0) = f_1(x)$,

    $\dfrac{\partial u}{\partial t}(x,0) = f_2(x)$,

    for all x in R + S.

4. u satisfies a boundary condition of one of the following three types
    when x is on S:

a. Dirichlet type:

$$u(x,t) = g(x,t),$$

when x is on S, and $t > 0$.

b. Neumann type:

$$\frac{\partial u}{\partial n_x}(x,t) = g(x,t)$$

when x is on S, and $t > 0$

c. Mixed type:

$$\frac{\partial u}{\partial n_x}(x,t) + \alpha u(x,t) = g(x,t)$$

where α is a real number, x is on S, and $t > 0$.

There are also problems where one boundary condition holds on part of S and another elsewhere. These may be included in the mixed category if $\alpha = \alpha(x)$ is permitted to vary as x varies over S and α and g may be simultaneously infinite (so that the term $\frac{\partial u}{\partial n_x}$ may be regarded as insignificant).

There are also similarly defined problems over unbounded regions, over exterior regions, as well as their nonhomogeneous counterparts. The reader is free to make his own definitions in these cases.

Unlike Laplace's equation and the heat equation, the wave equation lacks a maximum principle. Therefore, in order to prove the uniqueness of solutions, a different procedure is necessary. We use what is called an energy function.

XVII.1.3. <u>Theorem.</u> <u>Let R be a bounded region in E^n with piecewise</u>
<u>smooth boundary S. Let u satisfy the Cauchy problem for the wave equation</u>
<u>with Dirichlet boundary data. Then u is unique.</u>

Proof. Let u_1 and u_2 be two solutions. Then $u = u_1 - u_2$ satisfies the
wave equation with zero initial and boundary values. We multiply

$$\frac{\partial^2 u}{\partial t^2} - \nabla^2 u = 0$$

by $\frac{\partial u}{\partial t}$. Noting that

$$0 = \frac{\partial u}{\partial t} \left[\frac{\partial^2 u}{\partial t^2} - \nabla^2 u \right]$$

$$= \frac{\partial}{\partial t} \left[(1/2) \left(\frac{\partial u}{\partial t} \right)^2 + (1/2) \| \text{grad}_x u \|^2 \right]$$

$$- \text{div} \left[\frac{\partial u}{\partial t} \, \text{grad}_x u \right],$$

we integrate from 0 to t and over R. Thus

$$0 = \int_R \left[(1/2) \left(\frac{\partial u}{\partial t} \right)^2 + (1/2) \| \text{grad}_x u \|^2 \right]_{t=0}^{t=t} dx$$

$$- \int_0^t \int_R \text{div} \left[\frac{\partial u}{\partial t} \, \text{grad}_x u \right] dx dt.$$

We apply the divergence theorem to the inner portion of the second integral
to find

$$0 = \int_R \left[(1/2)\left(\frac{\partial u}{\partial t}\right)^2 + (1/2)\|grad_x u\|^2\right]_{t=0}^{t=t} dx - \int_0^t \int_S \frac{\partial u}{\partial t} \frac{\partial u}{\partial n_x} \, ds\,dt.$$

On S, $u = 0$, so $\frac{\partial u}{\partial t} = 0$. Further, when $t = 0$,

$u = 0$ and $\frac{\partial u}{\partial t} = 0$, so both $\frac{\partial u}{\partial t}$ and $grad_x u$ are zero. Thus

$$\int_R \left[(1/2)\left(\frac{\partial u}{\partial t}\right)^2 + (1/2)\,grad_x u\,^2\right] dx = 0$$

when evaluated at the upper limit of the previous integral: t. Thus for arbitrary $t > 0$, $\frac{\partial u}{\partial t}$ and $grad_x u$ are zero. This implies that u is constant. Since $u(x,0) = 0$, $u = 0$ for all t.

The term energy function is used because

$$\int_R \left[(1/2)\left(\frac{\partial u}{\partial t}\right)^2 + (1/2)\|grad_x u\|^2\right] dx$$

can be thought of as representing a form of energy.

The region over which the integration was performed can be restricted somewhat. In "cylindrical" (x,t) space it can be represented pictorially by

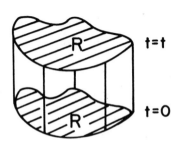

It is possible to show that at a point (x_0, t_0), u is affected only by a cone shpaed region determined by the characteristic surface passing through (x_0, t_0), the boundary S when $t \le t_0$, and the region R when $t = 0$.

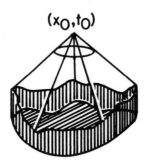

(x_0, t_0)

The equation of the cone shaped part is

$$t^2 - \|x\|^2 = t_0^2 - \|x_0\|^2.$$

The energy function technique can also be used to show uniqueness with other boundary data.

The final two theorems of this section, Stoke's rule and Duhamel's principle, show that when computing solutions, the work necessary may be substantially reduced. An initial position problem may be reduced to an initial velocity problem. Likewise, a nonhomogeneous problem is also equivalent to an intial velocity problem. Hence, it is sufficient to consider the homogeneous Cauchy problem with an initial velocity and zero initial position.

XVII.1.4. <u>Theorem (Stoke's Rule)</u>. <u>Let R be a bounded region in E^n with piecewise smooth boundary S. Let u be the solution to the Cauchy problem</u>

$$\frac{\partial^2 u}{\partial t^2} - \nabla^2 u = 0,$$

when x is in R, and t > 0,

$$u(x,0) = 0,$$

$$\frac{\partial u}{\partial t}(x,0) = f(x),$$

when x is in R + S, t = 0, and f is continuous, together with a

homogeneous boundary condition when x is in S and t > 0. Then

$$v = \frac{\partial u}{\partial t}$$

satisfies the problem

$$\frac{\partial^2 v}{\partial t^2} - \nabla^2 v = 0,$$

when x is in R, and t > 0,

$$v(x,0) = f(x),$$

$$\frac{\partial v}{\partial t}(x,0) = 0,$$

when x is in R + S, together with the same homogeneous boundary condition

on S.

Proof. If we differentiate the differential euqation with respect to t, we
find that v' satisfies the same equation. Clearly

$$v(x,0) = f(x),$$

since

$$\frac{\partial u}{\partial t}(x,0) = f(x).$$

Now

$$\frac{\partial v}{\partial t}(x,0) = \frac{\partial^2 u}{\partial t^2}(x,0)$$

$$= \nabla^2 u(x,0).$$

But when t = 0, u = 0 for all x in R + S. Thus $\nabla^2 u = 0$, and

$$\frac{\partial v}{\partial t}(x,0) = 0.$$

Note that in the course of the proof we have assumed that the partial
differential equation could be differentiated again. Classicially this would
not necessarily be possible. However, when u is considered as a distribution,
this assumption is quite reasonable.

XVII.1.5. Theorem (Duhamel's Principle). Let R be a bounded region in E^n
with piecewise smooth boundary S. Let u satisfy

$$\frac{\partial^2 u}{\partial t^2} - \nabla^2 u = 0,$$

when   x   is in   R,   and   t > 0,

$$u(x,0) = 0,$$

$$\frac{\partial u}{\partial t}(x,0) = f(x),$$

where   x   is in   R + S,   t = 0,   and   f   is continuous,   together with a
homogeneous boundary condition when   x   is in   S,   and   t > 0.   Let   u
be represented in the above problem as a linear transformation on   f.   That
is, let

$$u(x,t) = A[f(\xi)](x,t),$$

where   A   operates upon   f   as a function of ξ and generates a function of
(x,t).   Then, when   F(x,t)   is continuous,

$$w(x,t) = \int_0^t A[F(\xi,\tau)](x,t-\tau)d\tau$$

satisfies

$$\frac{\partial^2 w}{\partial t^2} - \nabla^2 w = F$$

when   x   is in   R,   and   t > 0,

$$w(x,0) = 0,$$

$$\frac{\partial w}{\partial t}(x,0) = 0,$$

when x is in R + S, together with the same homogeneous boundary condition

on S.

Proof. We compute

$$\frac{\partial w}{\partial t} = A[F(\xi,t)](x,0) + \int_0^t \frac{\partial}{\partial t}\{A[F(\xi,\tau)](x,t-\tau)\}d\tau.$$

$$\frac{\partial^2 w}{\partial t^2} = A[\frac{\partial F}{\partial t}(\xi,t)](x,0) + \frac{\partial}{\partial t}\{A[F(\xi,\tau)](x,t-\tau)\}\big|_{\tau=t}$$

$$+ \int_0^t \frac{\partial^2}{\partial t^2}\{A[F(\xi,\tau)](x,t-\tau)\}d\tau.$$

Further

$$\nabla^2 w = \int_0^t \nabla^2\{A[F(\xi,\tau)](x,t-\tau)\}d\tau.$$

It is evident from inspection that

$$w(x,0) = 0,$$

$$\frac{\partial w}{\partial t}(x,0) = 0.$$

Thus

$$\frac{\partial^2 w}{\partial t^2} - \nabla^2 w = \int_0^t [\frac{\partial^2}{\partial t^2} - \nabla^2]\{A[F(\xi,\tau)](x,t-\tau)\}d\tau$$

$$+ A[\frac{\partial F}{\partial t}(\xi,t)](x,0)$$

$$+ \frac{\partial}{\partial t}\{A[F(\xi,\tau)](x,t-\tau)\}\big|_{\tau=t}$$

$$= 0 + 0 + F(x,t).$$

Again we emphasize that Stoke's rule and Duhamel's principle show that it is sufficient to solve the homogeneous Cauchy problem with 0 initial position and fixed initial velocity. The uniqueness theorem (Theorem XVII.1.3) then shows the solutions generated by Stoke's rule and Duhamel's principle are unique.

Two steps remain to show the Cauchy problem is well posed:

1. Continuity with respect to the initial values the boundary data, and the nonhomogeneous term.

2. The existence of solutions.

We shall show by an explicit formula in certain special (but extremely important) cases that these criteria are met. In one, two or three dimensions, these formulas bear the names of d'Alembert, Poisson, and Kirchhoff (although it is due to Poisson). For the remaining situtions, we rely upon the existence of a Green's function, just as was done when considering the heat equation.

XVII.2. Solutions in 1, 2 And 3 Dimensions. As it turns out, it is more con-

venient to consider these examples in 1, 2, and 3 dimensions in reverse order.

Thus we begin with Kirchhoff's formula and conclude with those of d'Alembert.

XVII.2.1. Theorem (Kirchhoff's formula). Let $f_1(x_1,x_2,x_3)$ be in $C^3(E^3)$,
$f_2(x_1,x_2,x_3)$ be in $C^2(E^3)$, and let u satisfy

$$\frac{\partial^2 u}{\partial t^2} - \left(\frac{\partial^2 u}{\partial x_1^2} + \frac{\partial^2 u}{\partial x_2^2} + \frac{\partial^2 u}{\partial x_3^2}\right) = 0,$$

when $t > 0$,

$$u(x_1,x_2,x_3,0) = f_1(x_1,x_2,x_3),$$

$$\frac{\partial u}{\partial t}(x_1,x_2,x_3,0) = f_2(x_1,x_2,x_3).$$

Then

$$u(x_1,x_2,x_3,t) = (1/4\pi) \int_{S_t} [f_2(\xi_1,\xi_2,\xi_3)/t]ds$$

$$+ (1/4\pi) \frac{\partial}{\partial t} \int_{S_t} [f_1(\xi_1,\xi_2,\xi_3)/t]ds ,$$

where S_t is a sphere of radius t, centered at (x_1,x_2,x_3).

Proof. From Stoke's rule it is sufficient to consider the case where $f_1 = 0$.

Applying the three dimensional Fourier transform to the differential equation,

we have

$$\frac{d^2U}{dt^2} + (y_1^2 + y_2^2 + y_3^2)U = 0,$$

$$U(0) = 0,$$

$$U'(0) = F_2(y_1,y_2,y_3),$$

where U and F_2 are the Fourier transforms of u and f_2, respectively.
Thus

$$U = F_2(y_1,y_2,y_3)\sin(y_1^2+y_2^2+y_3^2)^{1/2}t/(y_1^2+y_2^2+y_3^2)^{1/2},$$

and

$$u(x_1,x_2,x_3,t) =$$

$$(1/\sqrt{2\pi})^3 \lim_{R\to\infty} \iiint_{y_1^2+y_2^2+y_3^2<R^2} F_2(y_1,y_2,y_3)\frac{\sin(y_1^2+y_2^2+y_3^2)^{1/2}t}{(y_1^2+y_2^2+y_3^2)^{1/2}} e^{i(x_1y_1+x_2y_2+x_3y_3)} dy_1 dy_2 dy$$

Substituting the integral for F_2,

$$u(x_1,x_2,x_3,t) =$$

$$(1/2\pi)^3 \lim_{R\to\infty} \iiint_{y_1^2+y_2^2+y_3^2<R^2} \int_{-\infty}^{\infty}\int_{-\infty}^{\infty}\int_{-\infty}^{\infty} f_2(\xi_1,\xi_2,\xi_3)e^{-i[y_1(\xi_1-x_1)+y_2(\xi_2-x_2)+y_3(\xi_3-x_3)]}$$

$$\cdot \frac{\sin(y_1^2+y_2^2+y_3^2)^{1/2}t}{(y_1^2+y_2^2+y_3^2)^{1/2}} d\xi_1 d\xi_2 d\xi_3 dy_1 dy_2 dy_3 .$$

We interchange the order of integration, then introduce spherical coordinates (ρ,ϕ,θ) in place of (y_1,y_2,y_3) with the direction $\theta = \pi$ the same as that of the vector $(\xi_1-x_1,\ \xi_2-x_2,\xi_3-x_3)$. The inner integral is

$$\int_0^R \int_0^\pi \int_0^{2\pi} e^{ir\rho\ \cos\ \theta}[\sin\ \rho t/\rho]\rho^2\ \sin\ \theta d\phi d\theta d\rho,$$

where $r = ((\xi_1-x_1)^2 + (\xi_2-x_2)^2 + (\xi_3-x_3)^2)^{1/2}$. This can be reduced. Integration with respect to ϕ and θ yields

$$2\pi \int_0^R [e^{-ir\rho\ \cos\ \theta}/ir\rho]_{\theta=0}^{\theta=\pi}\ \sin\ \rho t\rho d\rho$$

$$= (4\pi/r)\int_0^R\ \sin\ r\rho\ \sin\ \rho t d\rho.$$

We insert this in the expression for u, again reverse the order of integration, and introduce still another set of spherical coordinates

$$\xi_1-x_1 = r\ \sin\ \theta\ \cos\ \phi,$$

$$\xi_2-x_2 = r\ \sin\ \theta\ \sin\ \phi,$$

$$\xi_3-x_3 = r\ \cos\ \phi.$$

Then

$$u(x_1,x_2,x_3,t) =$$

$$(1/2\pi^2)\lim_{R\to\infty} \int_0^R\ \sin\ \rho t[\int_0^\infty\ \sin\ r\rho\{\int_0^\pi \int_0^{2\pi} f_2(\xi_1,\xi_2,\xi_3)r\ \sin\ \theta d\phi d\theta\}dr]d\rho.$$

We recognize this as the inversion formula for the Fourier sine transform

of the expression

$$\left\{ \int_0^\pi \int_0^{2\pi} f_2(\xi_1,\xi_2,\xi_3) r \sin \theta \, d\phi \, d\theta \right\}.$$

Thus

$$u(x_1,x_2,x_3,t) = (t/4\pi) \int_0^\pi \int_0^{2\pi} f_2(\xi_1,\xi_2,\xi_3) \sin \theta \, d\phi \, d\theta,$$

where, here, integration is over the sphere

$$(\xi_1-x_1)^2 + (\xi_2-x_2)^2 + (\xi_3-x_3)^2 = t^2,$$

which is S_t. Now, on S_t, $ds = t^2 \sin \theta \, d\phi \, d\theta$. So

$$u(x_1,x_2,x_3,t) = (1/4\pi) \int_{S_t} [f_2(\xi_1,\xi_2,\xi_3)/t] \, ds .$$

The general case then follows from Stoke's rule. We now note that our

formula is valid, even if f_1 and f_2 do not possess Fourier transform in

the classical sense. It is possible to show directly that u satisfies

the Cauchy problem over E^3.

Let us now assume that in the formula

$$u(x_1,x_2,x_3,t) = (1/4\pi) \int_{S_t} [f_2(\xi_1,\xi_2,\xi_3)/t] \, ds$$

the function f_2 is independent of the third variable. Letting

$$\rho = ((\xi_1 - x_1)^2 + (\xi_2 - x_2)^2)^{1/2},$$

if we project the sphere S_t onto a circle C_t in the (ξ_1, ξ_2) - plane, we find that an element of area in the circle C_t is given by

$$dA(\xi_1, \xi_2) = \rho d\rho d\phi,$$

where

$$\rho = t \sin \theta.$$

Thus

$$d\rho = t \cos \theta d\theta,$$

and

$$dA(\xi_1, \xi_2) = t^2 \sin \theta \cos \theta d\phi d\theta.$$

Since on S_t

$$ds = t^2 \sin \theta d\phi d\theta,$$

we have

$$dA(\xi_1,\xi_2) = \cos\theta ds .$$

Now,

$$\cos\theta = \sqrt{t^2-\rho^2}/t,$$

so, inserting all this in the expression for u,

$$u(x_1,x_2,t) = (1/2\pi)\int_{C_t} [f_2(\xi_1,\xi_2)/\sqrt{t^2-\rho^2}]dA(\xi_1,\xi_2)$$

$(1/2\pi,$ since both the top and bottom of S_t project onto C_t). The result is clearly independent of x_3 in every way. We have proved the following theorem, valid for two dimensions:

XVII.2.2. <u>Theorem (Poisson's Formula)</u>. <u>Let</u> $f_1(x_1,x_2)$ <u>be in</u> $C^3(E^2)$, $f_2(x_1,x_2)$ <u>be in</u> $C^2(E^2)$, <u>and let</u> u <u>satisfy</u>

$$\frac{\partial^2 u}{\partial t^2} - (\frac{\partial^2 u}{\partial x_1^2} + \frac{\partial^2 u}{\partial x_2^2}) = 0$$

<u>when</u> t > 0,

$$u(x_1,x_2,0) = f_1(x_1,x_2),$$

$$\frac{\partial u}{\partial t}(x_1,x_2,0) = f_2(x_1,x_2).$$

Then

$$u(x_1,x_2,t) = (1/2\pi)\int_{C_t} \frac{f_2(\xi_1,\xi_2)d\xi_1 d\xi_2}{(t^2-(\xi_1-x_1)^2-(\xi_2-x_2)^2)^{1/2}}$$

$$+ (1/2\pi)\frac{\partial}{\partial t}\int_{C_t} \frac{f_1(\xi_1,\xi_2)d\xi_1 d\xi_2}{(t^2-(\xi_1-x_1)^2-(\xi_2-x_2)^2)^{1/2}}$$

where C_t is a circle of radius t, centered at (x_1,x_2).

This formula can also be verified directly without resorting to Fourier transforms.

It is possible to reduce Poisson's formula to one dimension when f_1 and f_2 are functions only of x. We prefer, however, to derive d'Alembert's formula directly.

XVII.2.3. <u>Theorem (d'Alembert's Formula)</u>. <u>Let</u> $f_1(x)$ <u>be in</u> $c^2(E)$, $f_2(x)$ <u>be in</u> $C(E)$, <u>and let</u> u <u>satisfy</u>

$$\frac{\partial^2 u}{\partial t^2} - \frac{\partial^2 u}{\partial x^2} = 0$$

<u>when</u> $t > 0$,

$$u(x,0) = f_1(x),$$

$$\frac{\partial u}{\partial t}(x,0) = f_2(x),$$

Then

$$u(x,t) = (1/2)[f_1(x+t)+f_1(x-t)]+(1/2)\int_{x-t}^{x+t} f_2(\xi)d\xi.$$

Proof. This result can be derived in either of two ways, by reducing the equation to standard form and integrating directly, or by using the Fourier transform. We shall use the Fourier transform here. Next, when we discuss the problem over a bounded interval, we shall use the other technique.

We temporarily let $f_1 = 0$. Taking the Fourier transform of the differential equation and the initial conditions with respect to x_1, we have

$$\frac{d^2U}{dt^2} + y^2U = 0,$$

$$U(0) = 0,$$

$$U'(0) = F_2(y),$$

where U and F_2 are the Fourier transforms of u and f_2, respectively. Thus

$$U = [F_2(y)/y]\sin yt,$$

and

$$u(x,t) = \lim_{R\to\infty} (1/2\pi)\int_{-R}^{R} \int_{-\infty}^{\infty} f_2(\xi) \frac{\sin yt}{y} e^{iy(x-\xi)}d\xi dy.$$

If we replace sin yt/y by a complex integral in t, we find

$$u(x,t) = (1/2)\int_{-t}^{t} \lim_{R\to\infty} [(1/2\pi)\int_{-R}^{R} \int_{-\infty}^{\infty} f_2(\xi) e^{iy(\tau+x-\xi)} d\xi dy] d\tau.$$

We now note that as R approaches ∞, the term

$$(1/2\pi)\int_{-R}^{R} \int_{-\infty}^{\infty} f_2(\xi) e^{iy(\tau+x-\xi)} d\xi dy$$

converges to $f_2(x+\tau)$. Thus

$$u(x,t) = (1/2)\int_{-t}^{t} f_2(x+\tau) d\tau,$$

$$= (1/2)\int_{x-t}^{x+t} f_2(\xi) d\xi.$$

The general formula then follows from Stoke's rule.

Let us now consider the problem

$$\frac{\partial^2 u}{\partial t^2} - \frac{\partial^2 u}{\partial x^2} = 0$$

when $0 \leqq x \leqq 1$, and t > 0,

$$u(x,0) = f_1(x),$$

$$\frac{\partial u}{\partial t}(x,0) = f_2(x),$$

when $0 \leqq x \leqq 1$. With Dirichlet boundary data

$$u(0,t) = 0,$$

$$u(1,t) = 0,$$

when $t > 0$.

The Fourier transform is no longer applicable. In fact d'Alembert's formula does not have meaning for large t, since neither f_1 nor f_2 is defined outside of $[0,1]$. To have it make sense, we must extend the definitions of f_1 and f_2. to see how this should be done, we use the boundary conditions at 0 and 1.

We make a change of variables in to the characteristic coordinates

$$\xi = x+t,$$

$$\eta = x-t.$$

The differential equation then is equivalent to

$$\frac{\partial^2 u}{\partial \xi \partial \eta} = 0.$$

Integrating with respect to η, we find

$$\frac{\partial u}{\partial \xi} = \phi'(\xi),$$

where $\phi'(\xi)$ is an arbitrary function of ξ. Integrating with respect to ξ, we find

$$u = \phi(\xi) + \psi(\eta),$$

Reinserting the original coordinates, we have

$$u(x,t) = \phi(x+t) + \psi(x-t).$$

Now we employ the initial conditions. We note that

$$\frac{\partial u}{\partial t}(x,t) = \phi'(x+t) - \psi'(x-t).$$

Thus

$$f_1(x) = \phi(x) + \psi(x),$$

$$f_2(x) = \phi'(x) - \psi'(x).$$

Integrating the second from 0 to x, we have

$$f_1(x) = \phi(x) + \psi(x),$$

$$\int_0^x f_2(\xi)d\xi + C = \phi(x) - \psi(x),$$

where $C = \phi(0) - \psi(0)$. Solving for ϕ and ψ,

$$\phi(x) = (1/2)[f_1(x) + \int_0^x f_2(\xi)d\xi + C],$$

$$\psi(x) = (1/2)[f_1(x) - \int_0^x f_2(\xi)d\xi - C].$$

When these are inserted in the expression for u, the constant C drops out.

$$u(x,t) = (1/2)[f_1(x+t) + f_1(x-t)] + (1/2)\int_{x-t}^{x+t} f_2(\xi)d\xi.$$

This solution is valid as long as $x+t$ and $x-t$ are both in $[0,1]$. In order to extend the solution outside this range, we must use the boundary conditions. Let us write

$$u(x,t) = A(x+t) + B(x-t),$$

where

$$A(x+t) = (1/2)[f_1(x+t) + \int_0^{x+t} f_2(\xi)d\xi],$$

$$B(x-t) = (1/2)[f(x-t) + \int_{x-t}^0 f_2(\xi,d\xi].$$

Now since $u(0,t) = 0$ for all t, we have

$$A(t) = -B(-t).$$

$$A(1+t) = -B(1-t).$$

Now

1. The first determines B from −1 to 0 by requiring

$$B(t) = -A(-t).$$

2. In the second let s = 1+t. Then

$$A(s) = -B(2-s).$$

Since B is determined from −1 to 1, and −1 ≤ 2−s ≤ 1 implies 1 ≤ s ≤ 3, this determines A in the interval from 1 to 3. A is thus known from 0 to 3.

3. The first now determines B from −3 to 0. Hence B is determined from −3 to 1.

4. Since B is determined from −3 to 1, and −3 ≤ 2−s ≤ 1 implies 1 ≤ s ≤ 5, the second now determines A from 0 to 5.

This process may be continued indefinitely.

Similarly, reversing the roles of A and B, we may extend A to the left and B to the right. Thus determining the solution for all x and t.

If the expressions for A and B are inserted in the first of these boundary equations, we have

$$f_1(t) + \int_0^t f_2(\xi)d\xi = -f_1(-t) + \int_0^{-t} f_2(\xi)d\xi.$$

This implies that f_1 and f_2 should be extended as odd functions with respect to 0. In a similar should also be extended in an odd manner with respect to 1. With these extensions d'Alembert's formula is valid for all x in $[0,1]$ and $t > 0$, classically if the extensions are suitably differentiable, in the generalized sense if they are locally integrable.

In summary, we state :

XVII.2.4. <u>Theorem (d'Alembert's Formula for $[0,1]$)</u>. <u>Let $f_1(x)$ and $f_2(x)$</u>
<u>be defined on $[0,1]$ and be extended in an odd manner with respect to 0 and</u>
<u>1 over the real line E. Let the extension of $f_1(x)$ be in $C^2(E)$, and let</u>
<u>the extension of $f_2(x)$ be in $C^1(E)$. Then the solution to the Cauchy</u>
<u>problem</u>

$$\frac{\partial^2 u}{\partial t^2} - \frac{\partial^2 u}{\partial x^2} = 0$$

<u>when $0 \leq x \leq 1$, and $t > 0$,</u>

$$u(x,0) = f_1(x),$$

$$\frac{\partial u}{\partial t}(x,0) = f_2(x),$$

<u>when $0 \leq x \leq 1$, is given by</u>

$$u(x,t) = (1/2)[f_1(x+t) + f_1(x-t)] + (1/2)\int_{x-t}^{x+t} f_2(\xi)d\xi.$$

We see here the real necessity of distributions. If f_1 and f_2 are not differentiable in the classical sense, the classical solution will not exist. As generalized functions, however, they pose no problem.

If the boundary data is of Neumann type rather than Dirichlet type as just discussed, then f_1 and f_2 should be extended as even functions outside [0,1]. If the data is of Dirichlet type at $x = 0$ and Neumann type at $x = 1$, then f_1 and f_2 should be extended as odd functions with respect to 0 and as even functions with respect to 1.

This procedure also works in higher dimensions provided

1. The finite region is rectangular with sides parallel to the coordinate axes.

2. The boundary data is of Dirichlet or Neumann type.

Finally, we note that the solutions can be expressed quite elegantly by means of a Fourier sine series when the boundary data is of Dirichlet type, by means of a Fourier cosine series when the boundary data is of Neumann type, or by a Sturm-Liouville expansion when the boundary data is of mixed type.

We conclude this sequence of formulas with an extension of d'Alembert's formula to the nonhomogeneous problem on $(-\infty,\infty)$. There are similar extensions for finite intervals. We leave these to the reader.

XVII.2.5.   Theorem (D'Alembert's Formula for the Nonhomogeneous Equation).

Let $f_1(x)$ be in $C^2(E)$, $f_2(x)$ be in $C^1(E)$, and $F(x,t)$ be continuously

differentiable in x and continuous in t. Let u satisfy

$$\frac{\partial^2 u}{\partial t^2} - \frac{\partial^2 u}{\partial x^2} = F(x,t)$$

when $t > 0$,

$$u(x,0) = f_1(x),$$

$$\frac{\partial u}{\partial t}(x,0) = f_2(x).$$

Then

$$u(x,t) = (1/2)[f_1(x+t) + f_1(x-t)] + (1/2)\int_{x-t}^{x+t} f_2(\xi)d\xi$$

$$+ (1/2)\int_0^t \int_{x-(t-\eta)}^{x+(t-\eta)} F(\xi,\eta)d\xi d\eta.$$

Proof.   We transform into characteristic coordinates by letting

$$r = x+t,$$

$$s = x-t.$$

The differential equation then becomes

$$\frac{\partial^2 u}{\partial r \partial s}\left(\frac{r+s}{2},\frac{r-s}{2}\right) = -(1/4)F\left(\frac{r+s}{2},\frac{r-s}{2}\right).$$

Integrating with respect to r from s to r_0, this becomes

$$\frac{\partial u}{\partial s}\left(\frac{r_0+s}{2},\frac{r_0-s}{2}\right) = \frac{\partial u}{\partial s}(s,0) - (1/4)\int_s^{r_0} F\left(\frac{r+s}{2},\frac{r-s}{2}\right)dr.$$

$$= (1/2)\frac{\partial u}{\partial x}(s,0) - (1/2)\frac{\partial u}{\partial t}(s,0) - (1/4)\int_s^{r_0} F\left(\frac{r+s}{2},\left(\frac{r-s}{2}\right)\right)dr.$$

Integrating with respect to s from s_0 to r_0, we find

$$u(r_0,0) - u\left(\frac{r_0+s_0}{2},\frac{r_0-s_0}{2}\right) =$$

$$\int_{s_0}^{r_0} \left[(1/2)\frac{\partial u}{\partial x}(s,0) - (1/2)\frac{\partial u}{\partial t}(x,0)\right]ds$$

$$- (1/4)\int_{s_0}^{r_0}\int_s^{r_0} F\left(\frac{r+s}{2},\frac{r-s}{2}\right)drds.$$

The first integral on the right is

$$(1/2)[u(r_0,0) - u(s_0,0)] - (1/2)\int_{s_0}^{r_0} f_2(s)ds.$$

In the second we convert back to (x,t) coordinates by letting $\xi = \frac{r+s}{2}$, $\eta = \frac{r-s}{2}$. The Jacobian of this transformation is 2. Thus

$$\int_{s_0}^{r_0}\int_s^{r_0} F\left(\frac{r+s}{2},\frac{r-s}{2}\right)drds = 2\int_0^{\frac{r_0-s_0}{2}}\int_{s_0+\eta}^{r_0-\eta} F(\xi,\eta)d\xi d\eta$$

Letting $x_0 = \dfrac{r_0+s_0}{2}$, $t_0 = \dfrac{r_0-s_0}{2}$, and recalling that $u(x,0) = f_1(x)$, we find

$$u(x_0,t_0) = u(\frac{r_0+s_0}{2},\frac{r_0-s_0}{2}),$$

$$= u(r_0,0) - (1/2)[u(r_0,0) - u(s_0,0)]$$

$$+ (1/2)\int_{s_0}^{r_0} f_2(s)ds + (1/2)\int_0^{\frac{r_0-s_0}{2}} \int_{s_0+\eta}^{r_0-\eta} F(\xi,\eta)d\xi d\eta,$$

$$= (1/2)[f_1(x_0+t_0) + f_1(x_0-t_0)] + (1/2)\int_{x_0-t_0}^{x_0+t_0} f_2(s)ds$$

$$+ (1/2)\int_0^{t_0} \int_{x_0-(t_0-\eta)}^{x_0+(t_0-\eta)} F(\xi,\eta)d\xi d\eta.$$

We conclude this section with the following observation.

XVII.2.6. Theorem. In 1, 2, or 3 dimensions, where d'Alembert's, Poisson's, or Kirchhoff's formulas hold, the solution to the Cauchy problem is continuous with respect to the initial data and nonhomogeneous term. the Cauchy problem is well posed.

XVII.3. The Solution to the Nonhomogeneous Cauchy Problem. We now turn our attention to the solution of the Cauchy problem over a more general region R in E^n. For computational simplicity we restrict R by assuming that it is bounded and has piecewise smooth boundary S, just as was done earlier. What follows, however, can be easily adapted to other regions.

The key to the solution, as it was for the heat equation, is the Green's function.

XVII.3.1. Definition. Let R be a bounded region in E^n with piecewise smooth boundary S, and let $t \geqq 0$. The causal Green's function for the Cauchy problem with Dirichlet boundary data is the generalized function $G(x,t,\xi,\tau)$ which satisfies

1. $\dfrac{\partial^2 G}{\partial t^2} - \nabla_x^2 G = \delta(x-\xi)\delta(t-\tau)$

    when x, ξ are in R, and $t, \tau > 0$.

2. $G = 0$ if $t < \tau$, and x, ξ are in R.

3. $G = 0$ when x is on S.

As indicated, there are similar Green's functions for the Cauchy problem with Neumann or mixed boundary data. Only assumption 3 is modified. Further, there are also Green's functions for exterior and unbounded regions.

Before calculating these Green's functions for various regions, we wish to show how it is used. As with the heat equation, we need the following modification of Green's formula.

XVII.3.2. Theorem (Green's Formula for a Cylindrical Region). Let R be a bounded region in E^n with piecewise smooth boundary S. Let t be in $[T_1, T_2]$. Then if u and v are sufficiently differentiable with respect to x and t,

$$\int_{T_1}^{T_2} \int_R [v\{\frac{\partial^2 u}{\partial t^2} - \nabla^2 u\} - u\{\frac{\partial^2 v}{\partial t^2} - \nabla^2 v\}]dxdt =$$

$$- \int_R [v \frac{\partial u}{\partial t} - u \frac{\partial v}{\partial t}]\Big|_{t=T_1} dx + \int_R [v \frac{\partial u}{\partial t} - u \frac{\partial v}{\partial t}]_{t=T_2} dx$$

$$+ \int_{T_1}^{T_2} \int_S [u \frac{\partial v}{\partial n} - v \frac{\partial u}{\partial n}]dsdt.$$

We have again been deliberately vague with regard to the assumptions concerning the differentiability of u and v, since our solutions and boundary functions are infinitely differentiable when considered as distributions.

We continue to follow the path previously taken when discussing the heat equation. We observe that if t is replaced by t-τ in G, then

$$G(x,t,\xi,\tau) = G(x,t-\tau,\xi,0).$$

Further, if t and τ are interchanged, and

$$G^*(x,t,\xi,\tau) = G(x,\tau,\xi,t),$$

then G^* satisfies

1. $\dfrac{\partial^2 G^*}{\partial t^2} - \nabla_x^2 G^* = \delta(x-\xi)\delta(t-\tau)$

   when x,ξ are in R, and $t,\tau > 0$.

2. $G^* = 0$ if $t > \tau$, and x, ξ are in R.

3. $G^* = 0$ if x is in S.

Now, in Green's formula let

$$u = G(x,t,\xi_1,\tau_1),$$

$$v = G^*(x,t,\xi_0,\tau_0).$$

Then, when ξ_0,ξ_1 are in R, and τ_0,τ_1 are in $[T_1,T_2]$,

$$\int_{T_1}^{T_2} \int_R [G^*(x,t,\xi_0,\tau_0)\delta(x-\xi_1)\delta(t-\tau_1)$$

$$- \delta(x-\xi_0)\delta(t-\tau_0)G(x,t,\xi_1,\tau_1)]dxdt = 0.$$

Evaluating the integral above, the following result is derived:

XVII.3.3. Theorem. The causal Green's function for the Cauchy problem for

a region R in E^n with piecewise smooth boundary S satisfies

1. $\dfrac{\partial^2 G(x,t,\xi,\tau)}{\partial\tau^2} - \nabla^2 G(x,t,\xi,\tau) = \delta(x-\xi)\delta(t-\tau)$

    when x, ξ are in R and $t,\tau > 0$.

2. $G = 0$, if $\tau > t$, and x,ξ are in R.

3. $G = 0$, if ξ is in S.

Proof. The application of Green's formula shows

$$G^*(\xi_1,\tau_1,\xi_0,\tau_0) = G(\xi_0,\tau_0,\xi_1,\tau_1).$$

The results follow from this equivalence and the properties of G^*.

Let us now consider the nonhomogeneous Cauchy problem with Dirichlet boundary data,

$$\frac{\partial^2 u}{\partial t^2} - \nabla^2 u = F(x,t),$$

when x is in R and $t > 0$, and F is locally integrable in x and t,

$$u(x,0) = f_1(x),$$

$$\frac{\partial u}{\partial t}(x,0) = f_2(x),$$

when x is in $R + S$ and $t = 0$,

$$u(x,t) = g(x,t),$$

when x is on S, and $t > 0$.

XVII.3.4. Theorem. Let R be a bounded region in E^n with piecewise smooth boundary S, and let $t > 0$. If the causal Green's function for the Cauchy problem with Dirichlet boundary data exists, then the solution to the Cauchy problem with Dirichlet boundary data is given by

$$u(x,t) = \int_0^t \int_R G(x,t,\xi,\tau)F(\xi,\tau)d\xi d\tau$$

$$+ \int_R G(x,t,\xi,0)f_2(\xi)d\xi$$

$$+ \int_R \frac{\partial}{\partial t} G(x,t,\xi,0)f_1(\xi)d\xi$$

$$- \int_0^t \int_S \frac{\partial G}{\partial n_\xi} (x,t,\xi,\tau)g(\xi,\tau)ds_\xi d\tau.$$

Proof. We apply Theorem XVII.3.2, letting $u = u(x,t)$, $v = G(\xi,\tau,x,t)$, and letting $T_1 = 0$, $T_2 = t$. We note that

$$\frac{\partial G}{\partial \tau}(\xi,\tau,x,t) = - \frac{\partial G}{\partial t}(\xi,\tau,x,t),$$

then interchange the roles of (x,t) and (ξ,τ).

The first integral generates the nonhomogeneous term, the second the initial velocity, the third the initial position, and the fourth the boundary term. Hence a knowledge of the Green's function solves all these separate problems simultaneously.

Viewed with Stoke's rule in mind, we see that when the second integral is differentiated, it yields the third. With Duhamel's principle in mind, we see that the second will also yield the first.

XVII.3.5. Theorem. If the Green's function for the Cauchy problem with Dirichlet boundary data exists for a region R in E^n with piecewise smooth boundary S, then the solution to the Cauchy problem is continuous with respect to the initial and boundary data and the nonhomogeneous term. The Cauchy problem is well posed.

Proof. This follows immediately from Theorems XVII.1.3 and XVII.3.4.

XVII.4. Examples. Rather than present a general existence theorem, we shall present several examples in which we shall calculate the Green's function. Then by using either the formula of Theorem XVII.3.4, or similar formulas for Neumann or mixed boundary data, the general problem may be solved.

The Green's Function for a Bounded Region in E^n. We assume that R is a bounded region in E^n with piecewise smooth boundary S. We note that when F, f_1, and g are zero then

$$u(x,t) = \int_R G(x,t,\xi,0)f_2(\xi)d\xi$$

satisfies

$$\frac{\partial^2 u}{\partial t^2} - \nabla^2 u = 0$$

when x is in R, and t > 0,

$$u(x,0) = 0,$$

$$\frac{\partial u}{\partial t}(x,0) = f_1(x),$$

when   x   is in   R + S,   and   t = 0,

$$u(x,t) = 0,$$

when   x   is in   S,   and   t > 0.   Thus, by solving this homogeneous problem, we can find the Green's function.

We employ separation of variables, letting   u = X(x)T(t).   Dividing by   XT,   the differential equation yields

$$-\frac{T''}{T} = -\frac{\nabla^2 X}{X} = \lambda,$$

where λ is a constant.   Thus

$$T'' + \lambda T = 0$$

and

$$-\nabla^2 X = \lambda X.$$

Further the boundary condition   u = 0   on   S   implies that

$$X(x) = 0$$

for all x on S.

The problem in X is a Dirichlet problem. There exist normalized eigenfunctions $\{u_j(x)\}_{j=1}^{\infty}$, with corresponding positive eigenvalues $\{\lambda_j\}_{j=1}^{\infty}$. For each λ_j, j = 1,...,

$$T = A_j \cos \sqrt{\lambda_j}\, t + B_j \sin \sqrt{\lambda_j}\, t.$$

Thus, summing over j,

$$u(x,t) = \sum_{j=1}^{\infty} [A_j \cos \sqrt{\lambda_j}\, t + B_j \sin \sqrt{\lambda_j}\, t] u_j(x),$$

and

$$\frac{\partial u}{\partial t}(x,t) = \sum_{j=1}^{\infty} \sqrt{\lambda_j}\, [-A_j \sin \sqrt{\lambda_j}\, t + B_j \cos \sqrt{\lambda_j}\, t] u_j(x).$$

Letting t = 0, the initial conditions imply

$$0 = \sum_{j=1}^{\infty} [A_j \cdot 1 + B_j \cdot 0] u_j(x),$$

and

$$f_2(x) = \sum_{j=1}^{\infty} \sqrt{\lambda_j}\, [A_j \cdot 0 + B_j \cdot 1] u_j(x).$$

Thus $A_j = 0$, $j = 1,\ldots,$ and

$$B_j = (1/\sqrt{\lambda_j})\int_R u_j(\xi)f_2(\xi)d\xi.$$

$j = 1,\ldots$. So we find

$$u(x,t) = \sum_{j=1}^{\infty} [(1/\sqrt{\lambda_j})\int_R u_j(\xi)f_2(\xi)d\xi \; \sin \sqrt{\lambda_j} \; t]u_j(x).$$

or

$$u(x,t) = \int_R G(x,t,\xi,0)f_2(\xi)d\xi,$$

where

$$G(x,t,\xi,0) = \sum_{j=1}^{\infty} u_j(x)u_j(\xi)\sin \sqrt{\lambda_j} \; t/\sqrt{\lambda_j}.$$

Thus the Green's function

$$G(x,t,\xi,\tau) = \sum_{j=1}^{\infty} u_j(x)u_j(\xi)\sin \sqrt{\lambda_j}(t-\tau)/\sqrt{\lambda_j}.$$

The solution to the nonhomogeneous Cauchy problem with Dirichlet boundary
data is

$$u(x,t) = \sum_{j=1}^{\infty} \left[\int_0^t \int_R u_j(\xi) \sin \sqrt{\lambda_j}(t-\tau) F(\xi,\tau) d\xi d\tau \right.$$

$$+ \int_R u_j(\xi) f_2(\xi) d\xi \sin \sqrt{\lambda_j} \, t$$

$$+ \int_R u_j(\xi) f_1(\xi) d\xi \cos \sqrt{\lambda_j} \, t$$

$$\left. - \int_0^t \int_S \frac{\partial u_j(\xi)}{\partial n_\xi} \sin \sqrt{\lambda_j} (t-\tau) g(\xi,\tau) ds_\xi d\tau \right] u_j(x) / \sqrt{\lambda_j}.$$

The Green's Function for $0 \leq x \leq \pi$ with Dirichlet Boundary Data.   Let us apply the results of the example just completed to the case where $R = [0,\pi]$.   The homogeneous problem solved is now

$$\frac{\partial^2 u}{\partial t^2} - \frac{\partial^2 u}{\partial x^2} = 0,$$

when   x   is in $[0,\pi]$,   and $t > 0$,

$$u(x,0) = 0,$$

$$\frac{\partial u}{\partial t}(x,0) = f_2(x),$$

when   x   is in $[0,\pi]$,   and $t = 0$,

$$u(0,t) = 0, \quad u(\pi,t) = 0,$$

for all $t > 0$.

Letting u = X(x)T(t),   we find

$$T'' + \lambda T = 0,$$

$$X'' + \lambda X = 0,$$

$$X(0) = 0, \quad X(\pi) = 0.$$

Thus

$$X(x) = \sqrt{2/\pi} \, \sin nx,$$

n = 1,..., and

$$T(t) = A_n \cos nt + B_n \sin nt.$$

So

$$u(x,t) = \sum_{n=1}^{\infty} [A_n \cos nt + B_n \sin nt]\sqrt{2/\pi} \, \sin nx,$$

and

$$\frac{\partial u}{\partial t}(x,t) = \sum_{n=1}^{\infty} (-n)[A_n \sin nt - B_n \cos nt]\sqrt{2/\pi} \, \sin nx,$$

when $t \geq 0$. Letting $t = 0$, we find

$$0 = \sum_{n=1}^{\infty} A_n \sqrt{2/\pi} \sin nx,$$

$$f_2(x) = \sum_{n=1}^{\infty} n B_n \sqrt{2/\pi} \sin nx,$$

from which it follows that $A_n = 0$, $n = 1,\ldots,$ and

$$B_n = n\sqrt{2/\pi} \int_0^{\pi} \sin n\xi \, f_2(\xi)d\xi,$$

$n = 1,\ldots$.

$$u(x,t) = \sum_{n=1}^{\infty} (2/n\pi) \int_0^{\pi} \sin n\xi \, f_2(\xi)d\xi \, \sin nt \sin nx,$$

$$= \int_0^{\pi} G(x,t,\xi,0)f_2(\xi)d\xi,$$

where

$$G(x,t,\xi,0) = \sum_{n=1}^{\infty} (2/n\pi)\sin n\xi \sin nt \sin nx.$$

Thus the Green's function is

$$G(x,t,\xi,\tau) = \sum_{n=1}^{\infty} (2/n\pi)\sin n\xi \sin n(t-\tau)\sin nx.$$

Particularly interesting is the problem with initial position given by

$$
f_1(x) = \begin{cases} x, & \text{when } 0 \leq x \leq \pi/2, \\[2em] \pi - x, & \text{when } \pi/2 \leq x \leq \pi, \end{cases}
$$

and 0 initial velocity.   In this case

$$
u(x,t) = \sum_{k=0}^{\infty} \frac{2(-1)^k}{(2k+1)^2} \cos (2k+1)t \sin (2k+1)x.
$$

This series converges uniformly for all x   and   t.   However, if it is twice differentiated, either with respect to x   or   t,   the result is

$$
- \sum_{k=0}^{\infty} 2(-1)^k \cos(2k+1)t \sin (2k+1)x.
$$

This fails to converge in the classical sense, although it does converge as a distribution.   The series $u(x,t)$ is therefore a weak solution of the Cauchy problem.   It is problems just such as this which **make distributions necessary.**

The Green's Function for $0 \leq x < \infty$, $t \geq 0$ with Neumann Boundary Data

At $x = 0$.   The Green's function   G   satisfies

$$
\frac{\partial^2 G}{\partial t^2} - \frac{\partial^2 G}{\partial x^2} = \delta(x-\xi)\delta(t-\tau)
$$

when $0 < x, \xi < \infty$ and $t,\tau > 0$,

$$\frac{\partial G}{\partial x}(0,t,\xi,\tau) = 0$$

for all $t > \tau$, and $G = 0$, when $\tau > t$. If we take the Fourier cosine transform with respect to x on both sides of the differential equation, we have

$$\sqrt{2/\pi} \int_0^\infty \frac{\partial^2 G}{\partial t^2} \cos xy dx - \sqrt{2/\pi} \int_0^\infty \frac{\partial^2 G}{\partial t^2} \cos xy dx =$$

$$\delta(t-\tau) \sqrt{2/\pi} \int_0^\infty \delta(x-\xi) \cos xy dx.$$

If

$$\Gamma(y,t,\xi,\tau) = \sqrt{2/\pi} \int_0^\infty G(x,t,\xi,\tau) \cos xy dx,$$

then, this implies

$$\Gamma'' + y^2 \Gamma = \sqrt{2/\pi} \, \delta(t-\tau) \cos \xi y.$$

Thus

$$\Gamma(y,t,\xi,\tau) = A \cos yt + B \sin yt$$

$$+ \sqrt{2/\pi} \cos \xi y \int_0^t \delta(s-\tau) \frac{\sin y(t-s)}{y} ds.$$

If $\tau > t$, $G = 0$. Thus $\Gamma = 0$ also. This in turn implies that $A = 0$
and $B = 0$. Thus when $t > \tau$,

$$\Gamma(y,t,\xi,\tau) = \sqrt{2/\pi} \ \cos \ \xi y \ \sin \ y(t-\tau)/y.$$

To find G, we write

$$\Gamma(y,t,\xi,\tau) = \sqrt{2/\pi}[\sin \ y(t-\tau+\xi)+\sin \ y(t-\tau-\xi)]/2y,$$

$$= \sqrt{2/\pi}[(1/2)\int_{0}^{t-\tau+\xi} 1 \ \cos \ xydx$$

$$+ \ (1/2)\int_{0}^{t-\tau-\xi} 1 \ \cos \ xydx],$$

$$= \sqrt{2/\pi} \ \int_{0}^{\infty} [(1/2)H(t-\tau-\{x-\xi\}) + (1/2)H(t-\tau-\{x+\xi\})]\cos \ xydx,$$

where H is the Heaviside function

$$H(x) = \begin{cases} 1, & \text{when} \ \ x \geqq 0, \\ \\ 0, & \text{when} \ \ x < 0. \end{cases}$$

Thus

$$G(x,t,\xi,\tau) = (1/2)H(t-\tau-\{x-\xi\}) + (1/2)H(t-\tau-\{x+\xi\}).$$

The Green's Function for $-\infty < x < \infty$, $t \geqq 0$. In this problem the

Green's function satisfies

$$\frac{\partial^2 G}{\partial t^2} - \frac{\partial^2 G}{\partial x^2} = \delta(x-\xi)\delta(t-\tau),$$

when $t,\tau > 0$, $G = 0$ when $\tau > t$. We take the ordinary Fourier transform on both sides of the differential equation. If

$$\Gamma(y,t,\xi,\tau) = (1/\sqrt{2\pi})\int_{-\infty}^{\infty} e^{-ixy}G(x,t,\xi,\tau)dx,$$

then this yields

$$\Gamma'' + y^2\Gamma = (1/\sqrt{2\pi})e^{-i\xi y}\delta(t-\tau).$$

Thus

$$\Gamma(y,t,\xi,\tau) = A \cos yt + B \sin yt + (1/\sqrt{2\pi})e^{-i\xi y}\int_0^t \delta(s-\tau)\frac{\sin y(t-s)}{y}ds.$$

When $\tau > t$, $G = 0$, and $\Gamma = 0$. Hence $A = 0$, and $B = 0$. This further impliε that when $t > \tau$,

$$\Gamma(y,t,\xi,\tau) = (1/\sqrt{2\pi})e^{-i\xi y}\sin y(t-\tau)/y.$$

To find G, we write

$$\Gamma(y,t,\xi,\tau) = (1/\sqrt{2\pi})\int_{-(t-\tau)+\xi}^{(t-\tau)+\xi} (1/2)e^{-ixy}dx,$$

$$= (1/\sqrt{2\pi})\int_{-\infty}^{\infty} (1/2)H(t-\tau-|x-\xi|)e^{-ixy}dx.$$

Thus

$$G(x,t,\xi,\tau) = (1/2)H(t-\tau-|x-\xi|).$$

According to Theorem XVII.3.4, when modified to fit the doubly infinite interval, the solution to the homogeneous wave equation with initial position $f_1(x)$ and initial velocity $f_2(x)$ is

$$u(x,t) = \int_{-\infty}^{\infty} G(x,t,\xi,0)f_2(\xi)d\xi + \int_{-\infty}^{\infty} \frac{\partial}{\partial t} G(x,t,\xi,0)f_1(\xi)d\xi.$$

Since

$$\frac{\partial G}{\partial t} = (1/2)\delta(t-\tau-|x-\xi|),$$

$$u(x,t) = (1/2)\int_{-\infty}^{\infty} H(t-|x-\xi|)f_2(\xi)d\xi + (1/2)\int_{-\infty}^{\infty} \delta(t-|x-\xi|)f_1(\xi)d\xi,$$

$$= (1/2)[f_1(x+t) + f_1(x-t)] + (1/2)\int_{x-t}^{x+t} f_2(\xi)d\xi,$$

and we have rederived d'Alembert's formula.

The Green's Function for $0 \le x_1 \le \pi$, $0 \le x_2 < \infty$, $-\infty < x_3 < \infty$, with Neumann Boundary Data $x_1 = 0$, $x_1 = \pi$, Dirichlet Boundary Data at $x_2 = 0$.

In this problem the Green's function satisfies

$$\frac{\partial^2 G}{\partial t^2} - \left(\frac{\partial^2 G}{\partial x_1^2} + \frac{\partial^2 G}{\partial x_2^2} + \frac{\partial^2 G}{\partial x_3^2}\right) = \delta(x_1 - \xi_1)\delta(x_2 - \xi_2)\delta(x_3 - \xi_3)\delta(t - \tau),$$

when $0 < x_1 < \pi$, $0 < x_2 < \infty$, $-\infty < x_3 < \infty$, and

$$\frac{\partial G}{\partial x_1}\bigg|_{x_1 = 0} = 0, \quad \frac{\partial G}{\partial x_1}\bigg|_{x_1 = \pi} = 0, \quad G\bigg|_{x_2 = 0}.$$

We take the Fourier transform in x_3, the Fourier sine transform in x_2, then expand the result in a Fourier cosine series. If $\{L_n\}_{n=0}^{\infty}$ are the resulting coefficients, then

$$L_n'' + (n^2 + y_2^2 + y_3^2)L_n = \frac{\cos n\xi_1}{\sqrt{\frac{\pi}{2}}} \frac{\sin \xi_2 y_2}{\sqrt{\frac{\pi}{2}}} \frac{e^{-i\xi_3 y_3}}{\sqrt{2\pi}} \delta(t - \tau).$$

Thus

$$L_n = A \cos (n^2 + y_2^2 + y_3^2)^{1/2} t + B \sin (n^2 + y_2^2 + y_3^2)^{1/2} t$$

$$+ \frac{\cos n\xi_1}{\sqrt{\frac{\pi}{2}}} \frac{\sin \xi_2 y_2}{\sqrt{\frac{\pi}{2}}} \frac{e^{-i\xi_3 y_3}}{\sqrt{2\pi}} \frac{\sin(n^2 + y_2^2 + y_3^2)^{1/2}(t - \tau)}{(n^2 + y_2^2 + y_3^2)^{1/2}}$$

when $t > \tau$. Now since G is zero when $t < \tau$, so is L_n. Therefore, letting $t < \tau$, we see that $A = 0$ and $B = 0$, and that for $t < \tau$

$$L_n = \frac{\cos n\xi_1}{\sqrt{\frac{\pi}{2}}} \frac{\sin \xi_2 y_2}{\sqrt{\frac{\pi}{2}}} \frac{e^{-i\xi_3 y_3}}{\sqrt{2\pi}} \frac{\sin(n^2 + y_2^2 + y_3^2)^{1/2}(t - \tau)}{(n^2 + y_2^2 + y_3^2)^{1/2}}.$$

Thus, inverting we find

$G(x_1,x_2,x_3,t,\xi_1,\xi_2,\xi_3,\zeta) =$

$$\left[(1/\pi)(2/\pi)(1/2\pi) \int_0^\infty \int_{-\infty}^\infty \sin \xi_2 y_2 \, \sin x_2 y_2 \cdot \right.$$

$$\left. e^{i(x_3-\xi_3)y_3} \frac{\sin(y_2^2+y_3^2)^{1/2}(t-\tau)}{(y_2^2+y_3^2)^{1/2}} \, dy_2 dy_3 \right]$$

$$+ \left[\sum_{n=1}^\infty (2/\pi)(2/\pi)(1/2\pi) \int_0^\infty \int_{-\infty}^\infty \cos n\xi_1 \, \cos nx_1 \, \sin \xi_2 y_2 \, \sin x_2 y_2 \cdot \right.$$

$$\left. e^{i(x_3-\xi_3)y_3} \frac{\sin(n^2+y_2^2+y_3^2)^{1/2}(t-\tau)}{(n^2+y_2^2+y_3^2)^{1/2}} \, dy_2 dy_3 \right].$$

The result here should be compared with the third example in Section XVI.4. Note that the Green's function here is not a product, as it was there.

The Green's Function for the Unit Circle with Dirichlet Boundary Data.
Here we choose to solve the homogeneous problem

$$\frac{\partial^2 u}{\partial t^2} - \left[(1/r)\frac{\partial}{\partial r}(r \frac{\partial u}{\partial r}) + (1/r^2)\frac{\partial^2 u}{\partial \phi^2} \right] = 0,$$

when $r < 1$, and $t > 0$,

$$u(1,\phi,t) = 0,$$

when $t > 0$, and

$$u(r,\phi,0) = 0,$$

$$\frac{\partial u}{\partial t}(r,\phi,0) = f(r,\phi),$$

when $t = 0$, where f is continuous in (r, ϕ). To employ separation of variables, we let

$$u = T(t)R(r)\Phi(\phi).$$

Then, separating T, we find

$$\frac{T''}{T} = [\frac{(1/r)(rR')'}{R} + \frac{(1/r^2)\Phi''}{\Phi}] = -\lambda^2.$$

Thus

$$T'' + \lambda^2 T = 0.$$

Further, since $u = 0$ when $t = 0$, we require

$$T(0) = 0.$$

Next we multiply by r^2 and separate Φ. We have

$$\frac{r(rR')'}{R} + \lambda^2 r^2 = -\frac{\Phi''}{\Phi} = \mu^2.$$

Further since u is smooth when $r < 1$, we require

$$\Phi(0) = \Phi(2\pi),$$

$$\Phi'(0) = \Phi'(2\pi).$$

The resulting equation in R is

$$r^2R'' + rR' + (\lambda^2 r^2 - \mu^2)R = 0.$$

This problem is singular at $r = 0$. We therefore require that R be bounded as r approaches 0. Since $u(1,\phi,t) = 0$, we also require $R(1) = 0$.

The solutions are

$$\Phi(\phi) = e^{in\phi},$$

where n is any integer $(\mu = n)$.

$$R(r) = J_n(\lambda r),$$

where λ is a zero of the Bessel function $J_n(\lambda)$. Since for each n, there exist an infinite number of these $\{\lambda_{nm}\}_{m=1}^{\infty}$, λ has a doubly infinite range

$$\lambda = \lambda_{nm},$$

n an integer, $m = 1,\ldots$. Finally

$$T = \sin \lambda_{nm} t.$$

Thus the solution u has the form

$$u(r,\phi,t) = \sum_{n=-\infty}^{\infty} \sum_{m=1}^{\infty} A_{nm} J_n(\lambda_{nm} r) e^{in\phi} \sin \lambda_{nm} t.$$

Differentiating,

$$\frac{\partial u}{\partial t}(r,\phi,t) = \sum_{n=-\infty}^{\infty} \sum_{m=1}^{\infty} \lambda_{nm} A_{nm} J_n(\lambda_{nm} r) e^{in\phi} \cos \lambda_{nm} t.$$

Letting $t = 0$, we find

$$f(r,\phi) = \sum_{n=-\infty}^{\infty} \sum_{m=1}^{\infty} \lambda_{nm} A_{nm} J_n(\lambda_{nm} r) e^{in\phi},$$

from which it follows that

$$A_{nm} = \frac{\int_0^{2\pi} \int_0^1 f(\rho,\theta) e^{-in\theta} J_n(\lambda_{nm}\rho) \rho \, d\rho \, d\theta}{\pi \lambda_{nm} J_{|n|+1}(\lambda_{nm})^2}.$$

Inserting this in the series for u, we find

$$u(r,\phi,t) = \int_0^{2\pi} \int_0^1 G(r,\phi,t,\rho,\theta,0) f(\rho,\theta) \rho \, d\rho \, d\theta,$$

where

$$G(r,\phi,t,\rho,\theta,\tau) = \sum_{n=-\infty}^{\infty} \sum_{m=1}^{\infty} \frac{J_n(\lambda_{nm} r) J_n(\lambda_{nm}\rho) e^{in(\phi-\theta)} \sin \lambda_{nm}(t-\tau)}{\pi \lambda_{nm} J_{|n|+1}(\lambda_{nm})^2}.$$

XVII.   Exercises

1. Show that the solution to the Cauchy problem with mixed boundary data
    is unique when $\alpha \geqq 0$.

2. Verify directly that Kirchhoff's formula satisfies the Cauchy problem in E^3. Verify that it yields a generalized solution when f_1, f_2 are not differentiable in the classical sense.

3. Verify directly that Poisson's formula satisfies the Cauchy problem in E^2. Verify that it yields a generalized solution when f_1, f_2 are not differentiable in the classical sense.

4. Using Fourier transforms, derive Poisson's formula directly, as was done for Kirchhoff's formula.

5. Verify directly that d'Alembert's formula satisfies the Cauchy problem in $(-\infty,\infty)$. Verify that it yields a generalized solution when f_1 and f_2 are not differentiable in the classical sense.

6. Show that

$$u(x,t) = \phi(x+t) + \psi(x-t),$$

where ϕ, ψ are locally integrable, is a generalized solution of

$$\frac{\partial^2 u}{\partial t^2} - \frac{\partial^2 u}{\partial x^2} = 0.$$

7. Solve the Cauchy problems with Dirichlet, Neumann or mixed boundary data on $[0,1]$ by separation of variables. Assume that the initial value functions are sufficiently well behaved.

8. Extend d'Alembert's formula for $[0,1]$ when the boundary data is of Neumann type.

9. Show that the nonhomogeneous d'Alembert formulas follow directly from the homogeneous formulas and Duhamel's principle.

10. Use Duhamel's principle to solve the nonhomogeneous Cauchy problems in E^2 and E^3.

11. Prove in detail a theorem showing in some sense that the solution to the nonhomogeneous Cauchy problem in E, E^2, or E^3 is continuous with respect to the nonhomogeneous term.

12. What properties should the Green's function for the Cauchy problem with Neumann or mixed boundary data have? Develop a theory for such problems.

13. Let R be a bounded region in E^n with piecewise smooth boundary S. Find the Green's function for the Cauchy problem with Neumann or mixed boundary data.

14. Find the solution to the nonhomogeneous Cauchy problem with Dirichlet boundary data on $[0,\pi]$.

15. Show how the Green's function on $[0,\infty)$ with Neumann boundary data at 0 generates a solution to the Cauchy problem with Neumann boundary data.

16. Adapt the procedure used in the example of Section XVII.4 concerning the interval $(-\infty,\infty)$ to E^2 and E^3 to rederive Poisson's and Kirchhoff's formulas.

References

1. P.W. Berg and J.L. McGregor, "Elementary Partial Differential Equation," Holden-Day, San Francisco, 1966.

2. R. Courant and D. Hilbert, "Methods of Mathematical Physics, vol. II," Interscience, New York, 1962.

3. R.V. Churchill, "Fourier Series and Boundary Value Problems," McGraw-Hill, New York, 1963.

4. D. Greenspan, "Introduction to Partial Differential Equations," McGraw-Hill, New York, 1961.

5. G. Hellwig, "Differentialoperatoren," Springer Verlag, Berlin, 1964.

6. I.G. Petrovskii, "Partial Differential Equations," Scripta Technica, London, 1967.

7. M.H. Protter and H.F. Weinberger, "Maximum Principles in Differential Equations," Prentice-Hall, Englewood Cliffs, N.J., 1967.

8. I. Stakgold, "Boundary Value Problems of Mathematical Physics, vol. II," Macmillan, New York, 1968.

9. H.F. Weinberger, "A First Course in Partial Differential Equations," Blaisdell, New York, 1965.

APPENDIX I

The Spectral Resolution of an Unbounded Self-Adjoint Operator.

Chapter VIII contains the derivation of spectral resolution or integral representation of bounded self-adjoint, normal, and unitary operators. The self-adjoint expansion was later modified in the special case the operator was compact. This subsequently led to representations of the regular and singular Sturm-Liouville operators. However, since the Sturm-Liouville operators are differential operators, they are not bounded, and do not fit into the category originally assumed.

The major reason the spectral resolution could be extended to these unbounded operators is that the spectral resolution for a bounded self-adjoint operator

$$A = \int_{m-}^{M} \lambda E(d\lambda)$$

can be extended in <u>general</u> to be valid when the operator, although still self-adjoint, is no longer bounded. The formula looks almost the same:

$$A = \int_{-\infty}^{\infty} \lambda E(d\lambda),$$

but its interpretation is a bit different. It is this formula which we have already seen in the special cases of the regular and singular Sturm-Liouville problem.

1. <u>Unbounded Linear Operators</u>. The linear operators of Chapter VIII were defined on all elements x in H. This is no longer the case with unbounded operators. In fact, if the domain of an operator A is all of H, it is bounded. For this reason if we wish to study an unbounded operator, we must pay strict attention to its domain.

In addition, the concept of adjoint operator is not quite so elementary. We must redefine what we mean by it.

1.1. <u>Definition. Let A be a linear operator with domain D, which is dense in a Hilbert space H. Let y be an element in H for which there corresponds an element y^* in H such that</u>

$$(Ax,y) = (x,y^*)$$

<u>for all x in D. We then define the adjoint operator A^* by letting</u>

$$A^*y = y^*.$$

1.2. <u>Theorem</u>. A^* <u>is uniquely defined, closed, and linear</u>.

Proof. If $A^*y = y^*$, and $A^*y = y_1^*$, then

$$(x,y^*-y_1^*) = 0$$

for all x in D. Since D is dense, we can find a sequence $\{x_n\}_{n=1}^{\infty}$ in

D such that $\lim_{n \to \infty} x_n = y^* - y_1^*$. Choosing this sequence shows that $\|y^* - y_1^*\| = 0$, and $y^* = y_1^*$.

To show that A^* is closed, we must show that if y_n approaches y within its domain D^*, and $A^* y_n$ has a limit z, then y is also in D^*, and $A^* y = z$. This is easy, since for all x in D

$$(Ax, y) = \lim_{n \to \infty} (Ax, y_n),$$

$$= \lim_{n \to \infty} (x, A^* y_n),$$

$$= (x, z).$$

Thus y is in D^*, and $A^* y = z$.

The linearity also follows from the linearity of A by using the inner product definition.

In general it is impossible to say how large D^* is. Only the zero element θ in H is guaranteed to be in D^*. When A is closed, however, then D^* is dense in H. To show this, we need to introduce the graph of the operator.

2.   The Graph of an Operator.

2.1.   Definition.   Let $H = H \times H$, the set of all pairs $\{x, y\}$ of elements in H. If we define

$$c\{x,y\} = \{cx,cy\}$$

where   c   is a complex (real) number,

$$\{x_1,y_1\} + \{x_2,y_2\} = \{x_1+x_2,y_1+y_2\},$$

$$(\{x_1,y_1\},\{x_2,y_2\}) = (x_1,x_2) + (y_1,y_2),$$

then H is a Hilbert space.

2.2. Definition. The set of all elements $\{x,Ax\}$, where x is in D, is the graph of A, and is denoted by G_A.

In addition we shall need the following so-called mixing operators:

2.3. Definition. Let the operators U and V on H be defined by

$$U\{x,y\} = \{y,x\},$$

$$V\{x,y\} = \{y,-x\}.$$

2.4. Theorem.

$$UV = - VU,$$

$$U^2 = I = - V^2.$$

We leave the proof to the reader.

We note that the equation defining A^*,

$$(Ax,y) = (x,y^*)$$

can be written as

$$(V\{x, Ax\}, \{y,y^*\}) = 0,$$

2.5. Theorem. Let G_{A^*} denote the graph of A^*, and let $[X]$ denote the closed subspace generated by a set X. Then

$$G_{A^*} = H - V[G_A].$$

Proof. The equation above shows that G_{A^*} is orthogonal to $V(G_A)$ and thus to $[V(G_A)]$. We have only to note that

$$[V(G_A)] = V[G_A].$$

Then

$$H = G_{A^*} + V[G_A],$$

and the result is immediate.

2.6. Theorem. If A is a linear operator on H, then the existence of
A^{-1}, A^*, $(A^{-1})^*$ implies the existence of $(A^*)^{-1}$ and, further,

$$(A^*)^{-1} = (A^{-1})^*.$$

Proof. We first note that $G_{A^{-1}} = U G_A$. Then

$$G_{(A^{-1})^*} = H - V[G_{A^{-1}}] = H - VU[G_A],$$

$$= U(UH - V[G_A]) = U(H - V[G_A]),$$

$$= U G_{A^*} = G_{(A^*)^{-1}},$$

which shows $(A^*)^{-1}$ exists and equals $(A^{-1})^*$.

2.7. Theorem. Let A be a closed linear operator with domain D dense in
H. Then the domain of A^*, D^*, is also dense in H. Thus A^{**} exists.
Further $A^{**} = A$.

Proof. Suppose D^* is not dense in H. Then there exists an element x in
H orthogonal to D^*. Then $\{\theta, x\}$ is orthogonal to all elements of the form
$\{A^* y, -y\}$ where y is in D^*, so $\{\theta, x\}$ is orthogonal to VG_{A^*}. Since the
orthogonal complement of G_{A^*} is $V[G_A]$, the orthogonal complement of VG_{A^*}
is

$$V^2[G_A] = - I[G_A] = [G_A].$$

But since A is closed, $[G_A] = G_A$. Thus $\{\theta, x\}$ is in G_A, $A\theta = x$. This shows that $x = \theta$, which is a contradiction.

Upon a reapplication to A^*, we find A^{**} exists. Since $G_{A^{**}}$ is the orthogonal complement of VG_{A^*}, we find $G_{A^{**}} = G_A$ and $A^{**} = A$.

3. Symmetric and Self-Adjoint Operators.

3.1. Definition. A linear operator A is symmetric if it has domain D dense in H and satisfies $A \subseteqq A^*$ ($D \subset D^*$ and $Ax = A^*x$ for all x in D).

It is easy to show that $A \subseteqq A^*$ implies $A^{**} \subseteq A^*$. Thus

$$A^{**} \subseteqq A^* = (A^*)^{**} = (A^{**})^*.$$

So every symmetric operator has a closed symmetric extension A^{**}.

3.2. Definition. If a linear operator A equals its adjoint A^*, then A is self-adjoint.

3.3. Theorem. If A is self-adjoint and A^{-1} exists, then A^{-1} is also self-adjoint.

Proof. We know

$$(A^{-1})^* = (A^*)^{-1} = A^{-1}.$$

The only problem arises if the domain of A^{-1} is not dense in H. If that is the case, there exists an element z such that for all y in the domain of A^{-1}

$$(y,z) = 0.$$

But the domain of A^{-1} consists of elements of the form y = Ax, where x is in D. Thus

$$(Ax,z) = 0$$

for all x in D. This implies that z is in D^* (= D) and Az = 0. Since A^{-1} exists, z = 0.

Before actually deriving the spectral resolution for an unbounded self-adjoint operator we need to discuss two other operators for use at that time.

3.4. <u>Theorem.</u> <u>Let A be a closed linear operator with domain D dense in H. Then the operators</u>

$$B = (I + A^*A)^{-1},$$

$$C = A(I + A^*A)^{-1},$$

<u>are defined for all x in H, $\|B\| \leq 1$, $\|C\| \leq 1$. B is symmetric and positive.</u>

Proof. We again turn to the graph of A. Since G_A and VG_{A^*} are orthogonal

complements, any element x in H can be uniquely decomposed by

$$\{x, \theta\} = \{y, Ay\} + \{A^*z, -z\}.$$

This implies

$$x = y + A^*z,$$

$$\theta = Ay - z,$$

can be uniquely solved for y and z. If y = Bx and z = Cx, then

$$I = B + A^*C, \quad 0 = AB - C.$$

These yield

$$I = [I + A^*A]B, \quad C = AB.$$

Further

$$\|x\|^2 = \|\{x, 0\}\|^2 = \|\{y, Ay\}\|^2 + \|\{A^*z, -z\}\|^2,$$

$$= \|y\|^2 + \|Ay\|^2 + \|A^*z\|^2 + \|z\|^2.$$

Thus

$$\|x\|^2 \geq \|y\|^2 + \|z\|^2 = \|Bx\|^2 + \|Cx\|^2,$$

and

$$\|B\| \leq 1, \quad \|C\| \leq 1.$$

If x is in the domain of $I + A^*A$, then

$$([I + A^*A]x,x) = \|x\|^2 + \|Ax\|^2 \geq \|x\|^2.$$

Thus if $(I + A^*A)x = \theta$, $x = \theta$, and $(I + A^*A)^{-1}$ exists. Since $(I + A^*A)B = I$,

$$B = (I + A^*A)^{-1}.$$

Finally,

$$(Bx,y) = (Bx,(I + A^*A)By),$$

$$= ((I + A^*A)Bx,By),$$

$$= (x,By),$$

so B is symmetric, and

$$(Bx,x) = (Bx,(I + A^*A)Bx),$$

$$= (BX,Bx) + (ABx,ABx),$$

$$\geq 0,$$

so B is positive.

4. The Spectral Resolution of an Unbounded Self-Adjoint Operator.

4.1. Theorem. Let A be a self-adjoint operator on H. Then

$$R_{\pm i} = (A \pm iI)^{-1}$$

exists, is everywhere defined, and is bounded.

Proof. We note formally that

$$C \pm iB = (A \pm iI)(I + A^2)^{-1},$$

$$= (A \mp iI)^{-1}.$$

If x is in D, then

$$\|(A \mp iI)x\|^2 = \|Ax\|^2 \mp i(x,Ax) \pm i(Ax,x) + \|x\|^2,$$

$$= \|Ax\|^2 + \|x\|^2.$$

So

$$\|(A \mp iI)x\|^2 \geq \|x\|^2,$$

which shows that $(A \mp iI)x = \theta$ implies $x = \theta$. Thus $R_{\pm i}$ exists. If $y = (A \mp iI)x$, then

$$\|y\|^2 \geq \|R_{\mp i}y\|^2,$$

so $R_{\pm i}$ is bounded by 1.

To show that R_{+i} is every where defined, we recall that Theorem 3.4 shows this is true for both B and C, and hence for $R_{\pm i}$.

4.2. <u>Definition</u>. <u>Let A be a self-adjoint operator on H. The operator</u>

$$T = (A - iI)(A + iI)^{-1}$$

<u>is the Cayley transform of A.</u>

4.3. <u>Theorem</u>. <u>The Cayley transform of a self-adjoint operator A on H is a unitary operator.</u>

Proof. We note from the proof of Theorem 4.1 that

$$\| (A + iI)x \| = \| (A - iI)x \|$$

for all x in D. If $y = (A + iI)x$, then

$$\|y\| = \| (A - iI)(A + iI)^{-1}y \| = \|Ty\|.$$

T is, therefore, isometric. It is defined for all elements of the form

$$y = (A + iI)x$$

by

$$Ty = (A - iI)x,$$

where x is in D. According to Theorem 4.1, as x varies through D,
y and Ty vary through H. The result then follows from Theorem XI.3.1.

4.4. <u>Theorem</u>. <u>Let T be the Cayley transform of the self-adjoint operator</u>
<u>A. Then</u>

$$A = i(I + T)(I - T)^{-1}.$$

Proof. Let y be in H. Then there exists x in D such that

$$y = (A + iI)x,$$

$$Ty = (A - iI)x.$$

Adding and subtracting,

$$(I + T)y = 2Ax,$$

$$(I - T)y = 2ix.$$

If $(I - T)y = \theta$, then $x = \theta$, and $y = \theta$. Thus $(I - T)^{-1}$ exists and

$$Ax = i(I + T)(I - T)^{-1}x.$$

4.5. **Theorem.** Let

$$T = \int_0^{2\pi} e^{i\theta}E(d\theta),$$

$$E(0) = 0, \quad E(2\pi) = I,$$

be the spectral decomposition of T. Then $E(\theta)$ is continuous at $\theta = 0$ and $\theta = 2\pi$.

Proof.   Continuity at 0 follows from Theorem VIII.9.2.   If

$$\Delta E = E(2\pi) - E(2\pi-) \neq 0,$$

then there exist x, y in H, y $\neq$ θ, such that

$$y = \Delta Ex.$$

Since E is a family of projections,

$$\Delta Ey = \Delta E^2 x = \Delta Ex = y.$$

So 1 is an eigenvalue of T with eigenfunction y.   But then $(I - T)^{-1}$ would not exist.

4.6. Theorem. Let the interval $(0,2\pi)$ be decomposed by $\{\theta_m\}_{m=-\infty}^{\infty}$ determined by

$$- \cot \theta_m/2 = m.$$

Then the projections

$$P_m = E(\theta_m) - E(\theta_{m-1})$$

are pairwise orthogonal, commute with T and A, and satisfy

$$\sum_{m=-\infty}^{\infty} P_m = I.$$

Proof. From Theorem VIII.9.2 $E(\theta)$ commutes with T and hence with A.
Further

$$\sum_{m=-\infty}^{\infty} P_m = \lim_{\theta \to 2\pi} E(\theta) - \lim_{\theta \to 0} E(\theta) = I - 0 = I.$$

Orthogonality also follows from Theorem VIII.9.2.

4.7.   Theorem.   Let $H_m = P_m H.$ If x is in H_m, then

$$Ax = \int_{m-1}^{m+} \lambda E_A(d\lambda)x,$$

where $\lambda = -\cot \theta/2,$ and

$$E_A(\lambda) = E(-2 \cot^{-1} \lambda) = E(\theta).$$

Proof. If x is in H_m,

$$Ax = i(I + T)(I - T)^{-1} P_m x,$$

$$= \int_0^{2\pi} [i(1 + e^{i\theta})/(1 - e^{i\theta})]K_{(\theta_{m-1}, \theta_m]}(\theta)E(d\theta)x,$$

where

$$K_{(\theta_{m-1},\theta_m]}(\theta) = \begin{cases} 1, & \text{when } \theta \in (\theta_{m-1},\theta_m], \\ \\ 0, & \text{when } \theta \notin (\theta_{m-1},\theta_m]. \end{cases}$$

Thus

$$Ax = \int_{\theta_{m-1}}^{\theta_{m+}} [i(1 + e^{i\theta})/(1 - e^{i\theta})]E(d\theta)x,$$

$$= \int_{\theta_{m-1}}^{\theta_{m+}} - \cot \theta/2 \; E(d\theta)x.$$

If we let $\lambda = - \cot \theta/2$ and

$$E_A(\lambda) = E(- 2 \cot^{-1} \lambda),$$

then

$$Ax = \int_{m-1}^{m+} \lambda E_A(d\lambda)x.$$

4.8. Theorem. Let H be a Hilbert space. Let A be an unbounded self-adjoint operator with domain D in H. Then there exists a collection of projections $E_A(\lambda)$, $-\infty < \lambda < \infty$, which are strong limits of the Cayley transform of A, satisfying

1. $E_A(\lambda) \le E_A(\mu)$ when $\lambda \le \mu$,

2. $\lim\limits_{\lambda \to -\infty} E_A(\lambda) = 0$,

3. $\lim\limits_{\lambda \to +\infty} E_A(\lambda) = I$,

4. $E_A(\lambda)$ is continuous from above,

such that

$$Ax = \int_{-\infty}^{\infty} \lambda E_A(d\lambda)x,$$

for all x in D.

Proof. If x is in D, then

$$Ax = A \sum_{m=-\infty}^{\infty} P_m x,$$

$$= \sum_{m=-\infty}^{\infty} AP_m x$$

$$= \sum_{m=-\infty}^{\infty} \int_{m-1}^{m+} \lambda E_A(d\lambda)x,$$

$$= \int_{-\infty}^{\infty} \lambda E_A(d\lambda)x.$$

Note that the equality between A and its integral representation is in the strong sense. In the bounded case this equality was uniform.

Appendix I, Exercises

1. Show that the regular Sturm–Liouville operator is self-adjoint and that
    its spectral resolution is the expansion previously derived.

2. Show that the singular Sturm–Liouville operator is self-adjoint and that
    its spectral resolution is the expansion previously derived.

References

1. N.I. Akhiezer and I.M. Glazman, "Theory of Linear Operators in Hilbert
    Space, vols. I and II," Frederick Ungar, New York, 1966.

2. N. Dunford and J.T. Schwartz, "Linear Operators, vols. I and II,"
    Interscience, New York, 1958 and 1964.

3. F. Riesz and B. Sz.-Nagy, "Functional Analysis," Frederick Ungar, New
    York, 1955.

4. M.H. Stone, "Linear Transformations in Hilbert Space," American
    Mathematical Society, New York, 1966.

APPENDIX II

The Derivation of the Heat, Wave and Laplace Equations.

1. **The Heat Equation.** Let us consider a rod of length L, which conducts heat uniformly throughout. Let x denote the distance along the rod from one end, and let A(x) denote the cross sectional area at each point x on the rod.

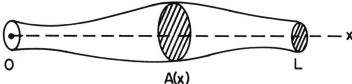

Let u(x,t) denote the temperature at time t in the cross section A(x) with coordinate x, let ρ denote the density of the substance from which the rod is made, and let c denote the specific heat of the substance (the energy necessary to raise one gram one degree centigrade).

1.1. **Lemma.** Let a and b be two points in the rod. The energy contained in the portion from a to b (the amount needed to raise its temperature from 0 to u(x,t)) is

$$Q(t) = \int_a^b u(x,t)c\rho(x)A(x)dx.$$

Proof. This is obviously just a summation over the interval [a,b].

1.2. Lemma. The rate of change of $Q(t)$ with respect to time is given by

$$\frac{dQ}{dt} = \text{flux} + \text{source},$$

where flux denotes heat (energy) gained or lost across the boundaries of the portion, and source denotes heat obtained from within the rod itself.

This is commonly called the conservation of energy. We assume it.

1.3. Lemma. Let $F(x,t)$ denote the heat per unit area passing in a positive direction through the cross section with coordinate x. Then

$$\text{flux} = A(a)F(a,t) - A(b)F(b,t).$$

Proof. This is trivial once we assume that no heat is exhanged on a direction other than that indicated by x.

To proceed further we assume a law which can be verified experimentally:

$$F(x,t) = -k(x) \frac{\partial u(x,t)}{\partial x},$$

where $k(x) > 0$ is called the heat conductivity.

1.4. Lemma.

$$\text{flux} = \int_a^b \frac{\partial}{\partial x} \left(A(x)k(x) \frac{\partial u(x,t)}{\partial x} \right) dx.$$

Proof.   According to the formula just assumed,

$$\text{flux} = - A(a)k(a) \frac{\partial u}{\partial x}(a,t) + A(b)k(b) \frac{\partial u}{\partial x}(b,t),$$

$$= \int_a^b \frac{\partial}{\partial x}(A(x)k(x) \frac{\partial u}{\partial x}(x,t))dx.$$

1.5.   Lemma.   Let   q(x,t,u)   denote the rate of heat per unit volume produced within the rod per unit time.   Then over the interval   [a,b],

$$\text{source} = \int_a^b q(x,t,u)A(x)dx.$$

1.6.   Theorem.   The flow of heat throughout a uniform rod satisfies the
equation

$$c\rho(x)A(x) \frac{\partial u}{\partial t} - \frac{\partial}{\partial x}(A(x)k(x) \frac{\partial u}{\partial x}) - A(x)q(x,t,u) = 0,$$

provided each component on the left is continuous for all   x.

Proof.   From the previous Lemmas

$$\int_a^b [c\rho A \frac{\partial u}{\partial t} - \frac{\partial}{\partial x}(Ak \frac{\partial u}{\partial x}) - Aq]dx = 0.$$

Since the interval [a,b] is arbitrary, the result follows by differentiating
with respect to   b.

1.7. Corollary. If the cross sectional area A is constant, then

$$c\rho(x) \frac{\partial u}{\partial t} - \frac{\partial}{\partial x} (k(x) \frac{\partial u}{\partial x}) = q(x,t,u).$$

1.8. Corollary. If q is independent of u, then

$$c\rho(x) \frac{\partial u}{\partial t} - \frac{\partial}{\partial x} (k(x) \frac{\partial u}{\partial x}) = q(x,t).$$

1.9. Corollary. If ρ and k are constants and q = 0, then

$$\frac{\partial u}{\partial t} - a^2 \frac{\partial^2 u}{\partial x^2} = 0,$$

where $a^2 = k/c\rho.$

The reader is invited to state some further variations.

2. Boundary Conditions. It is rather easy to give physical examples to
illustrate Dirichlet, Neumann and mixed boundary data for the one dimensional
heat equation. The first two impose no problem at all. If x = a is the
boundary point in question, then the Dirichlet condition

$$u(a,t) = g(t)$$

tells us that the end is kept at temperature g(t).

$$u(a,t) = 0$$

would correspond to the end at a being kept in contact with a block of ice.

The Neumann condition

$$\frac{\partial u}{\partial x}(0,t) = g(t)$$

illustrates the possibility that the end is insulated, and that heat escapes
or enters as determined by a temperature   gradient $g(t)$. If $g(t) = 0$,
then the end at a is completely insulated.

To properly interpret a mixed boundary condition is a bit more difficult.
First let us suppose that two rods are joined end to end.

2.1. Definition. Two rods, joined end to end, are in perfect thermal contact if both the temperature and the flux are continuous at the boundary.

2.2. Theorem. If two rods satisfy the heat equation continuously, and are
in perfect thermal contact at $x = a$, then

$$u_1(a,t) = u_2(a,t),$$

and

$$k_1(a) \frac{\partial u_1}{\partial x}(a,t) = k_2(a) \frac{\partial u_2}{\partial x}(a,t),$$

where the subscripts 1 and 2 denote the appropriate functions in the first
and second rods, respectively.

Proof.   The first is immediate.   From Lemma 1.4,

$$\left| A(a+\varepsilon)k_2(a+\varepsilon)\,\frac{\partial u_2}{\partial x}\,(a+\varepsilon,t) = A(a-\varepsilon)k_1(a-\varepsilon)\,\frac{\partial u_1}{\partial x}\,(a-\varepsilon,t)\right.$$

$$= \left| \int_{a-\varepsilon}^{a+\varepsilon} \frac{\partial}{\partial x}\,(A(x)k(x)\,\frac{\partial u}{\partial x}\,(x,t))\,dx\right|,$$

$$\le \left| \int_{a-\varepsilon}^{a+\varepsilon} M\,dx = 2\,\varepsilon M.\right.$$

We then let ε approach zero and divide by the common value of A in each term.

Now let a rod with a thin oxide film be in contact with an object with temperature determined by $g(t)$.   The statement

$$u(a,t) = g(t)$$

is not accurate because of the film.   Therefore, let the film be of thickness ε and extend to the left of the end at $x = a$.   If we assume that the rod is in perfect thermal contact with the film, then

$$u_f(a,t) = u(a,t),$$

and

$$k_f\,\frac{\partial u_f}{\partial x}\,(a,t) = k\,\frac{\partial u}{\partial x}\,(a,t),$$

where the subscript f denotes the film.   Further

$$u_f(a-\varepsilon,t) = g(t).$$

Now since ε is assumed to be small, we may approximate $\dfrac{\partial u_f}{\partial x}$ above by the

expression $[u_f(-\varepsilon,t) - u_f(a,t)]/[a-\varepsilon-a]$,   which equals

$$(g(t) - u(a,t))/(-\varepsilon).$$

Thus, upon substitution in the second equation at   a,

$$u(a,t) - (\varepsilon k/k_f) \frac{\partial u}{\partial x} (a,t) = g(t).$$

Clearly, if $\varepsilon = 0$,   this becomes a Dirichlet condition.

3. <u>The Wave Equation</u>. We wish to derive the equation which governs the
motion of a transverse plane vibration of a perfectly flexible uniform string.
To do so let the independent variable x vary in a direction of equilibrium
of the string and let u be orthogonal to x in the plane of the vibration.

We let $T(x,t)$, $\theta(x,t)$, and $\rho(x)$ denote tension along the string, the
angle of inclination of the graph of the string in the x-u plane, and the
density of the string per unit length when measured in an equilibrium position.

In addition we let $q(x,t,u,\frac{\partial u}{\partial x},\frac{\partial u}{\partial t})$ denote the external transverse force
per unit mass acting on the string.

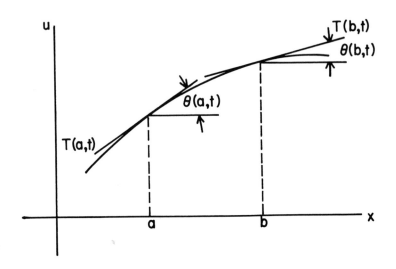

We now make one vital assumption: <u>At no point does the motion of the</u> <u>string   have   a horizontal component</u>.

Note we do not assume that the displacement   u   is small.

3.1.   <u>Theorem</u>. <u>Let   q   and   ρ   be continuous.   Then the transverse vibration</u> <u>of a perfectly flexible uniform string satisfies</u>

$$\frac{\partial^2 u}{\partial t^2} = \frac{T_0(t)}{\rho(x)} \frac{\partial^2 u}{\partial x^2} + q,$$

<u>where T_0 is the horizontal component of the tension</u>.

Proof. If we sum the forces acting upon the segment of the string above above the interval (a,b) we have

$$T(b,t)\cos \theta(b,t) - T(a,t)\cos \theta(a,t) = 0$$

in the horizontal direction.

$$T(b,t)\sin \theta(b,t) - T(a,t)\sin \theta(a,t) + \int_a^b q(x,t,u,\frac{\partial u}{\partial x},\frac{\partial u}{\partial t})\rho(x)\,dx$$

$$= \int_a^b \frac{\partial^2 u}{\partial t^2}(x,t)\rho(x)\,dx.$$

in the transverse direction. Applying the mean value theorem to the two integrals,

$$T(b,t)\sin \theta(b,t) - T(a,t)\sin \theta(a,t)$$

$$+ q(x_0,t,u(x_0,t),\frac{\partial u}{\partial x}(x_0,t),\frac{\partial u}{\partial t}(x_0,t))\rho(x_0)(b-a)$$

$$= \frac{\partial^2 u}{\partial t^2}(x_1,t)\rho(x_1)(b-a),$$

where x_0 and x_1 are between a and b. Now let

$$T(b,t)\cos \theta(b,t) = T(a,t)\cos \theta(a,t) = T_0(t).$$

Then eliminating $T(a,t)$ and $T(b,t)$,

$$T_0(t)[\tan \theta(b,t) - \tan \theta(a,t)] +$$

$$q(x_0,t,\ldots)\rho(x_0)(b-a) = \frac{\partial^2 u}{\partial t^2}(x_1,t)\rho(x_1)(b-a).$$

Now $\tan \theta = \frac{\partial u}{\partial x}$. Further since a and b are arbitrary, we may divide by

(b-a) and then let b approach a. x_0 and x_1 also approach a, and

$$T_0(t)\lim_{b\to a} \frac{\frac{\partial u}{\partial x}(b,t) - \frac{\partial u}{\partial x}(a,t)}{b-a} + q(a,\ldots)\rho(a) = \frac{\partial^2 u}{\partial t^2}(a,t)\rho(a),$$

or, dividing by ρ also and replacing a by x,

$$\frac{\partial^2 u}{\partial t^2} = \frac{T_0(t)}{\rho(x)}\frac{\partial^2 u}{\partial x^2} + q(x,t,u,\frac{\partial u}{\partial x},\frac{\partial u}{\partial t}).$$

We would like to emphasize that $T_0(t)$ is independent of x, and

that $\rho(x)$ describes the string in an equilibrium position. This seems to

be most natural. The equation is valid even if the tangent $\frac{\partial u}{\partial x}$ is large,

or if the displacement is large. This is not usually true of other derivations.

The technique presented here is due to Professor Orrin Frink.

3.2. <u>Corollary.</u> <u>If no external forces are present, then the plane vibration</u>

<u>of a perfectly flexible uniform string satisfies</u>

$$\frac{\partial^2 u}{\partial t^2} = \frac{T_0}{\rho}\frac{\partial^2 u}{\partial x^2}.$$

4. <u>Boundary Conditions.</u> If for a vibrating string the point x = a is a

boundary point, then

$$u(a,t) = g(t)$$

obviously describes a prescribed motion at x = a. On the other hand if the
end is free, but the transverse component of the tension is prescribed by
f(t),   then

$$T \sin \theta = T \frac{\frac{\partial u}{\partial x}}{(1 + (\frac{\partial u}{\partial x})^2)^{1/2}} = f(t).$$

If this is solved for $\frac{\partial u}{\partial x}$,

$$\frac{\partial u}{\partial x} (a,t) = g(t),$$

where

$$g(t) = [f(t)/T]/[1 - (f(t)/T)^2]^{1/2}$$

A mixed condition is found if a very short string is joined smoothly to a
larger one. If the smaller has a prescribed end, then, as was so with the
heat at the end of a rod, a mixed boundary condition results.

5. Laplace's Equation. The simplest circumstances which are described by
Laplace's equation are heat or vibrational problems under which a steady state
condition, independent of time, has been reached. In two space dimensions
either has the form

$$\frac{\partial^2 u}{\partial x^2} + \frac{\partial^2 u}{\partial y^2} = 0$$

since the time derivative vanishes.   Higher dimensions are similar.   Boundary

conditions are also similar.

Appendix II, Exercises

1. Derive the heat equation in 2 and 3 dimensions.

2. Show what form boundary conditions take for heat transfer problems in 2

    and 3 dimensions.

3. Derive the wave equation in 2 and 3 dimensions.

4. Show what form boundary conditions take for vibrational problems in 2 and

    3 dimensions.

References

1. P.W. Berg and J.L. McGregor, "Elementary Partial Differential Equations,"
    Holden-Day, San Francisco, 1966.

INDEX